"十三五"江苏省高等学校重点教材

飞行器设计与工程力学品牌专业系列教材

# Mechanics of Materials
# 材料力学
## （双语版）

王开福　编著

科学出版社

北京

## 内 容 简 介

This book is a bilingual textbook for mechanics of materials, written independently in English and Chinese, respectively. The main contents of the book include mechanics of materials fundamentals, axial tension and compression, torsion, bending internal forces, bending stresses, bending deformation, stress analysis and strength theories, combined loadings, stability of column, unsymmetrical bending, energy methods, impact loading, statically indeterminate structures, etc.

The book can be used as a bilingual textbook for mechanics of materials for engineering students majoring in aeronautical, mechanical, civil engineering, etc.

本书是材料力学双语教材，分别由英文和中文独立编写。本书主要内容包括材料力学基础、轴向拉伸与压缩、扭转、弯曲内力、弯曲应力、弯曲变形、应力分析与强度理论、组合载荷、压杆稳定、非对称弯曲、能量方法、冲击载荷和静不定结构等。

本书可作为航空航天工程、机械工程和土木工程等工科专业本科生的材料力学双语教材。

---

图书在版编目(CIP)数据

材料力学 ＝Mechanics of Materials：汉英对照／王开福编著. —北京：科学出版社，2018.9

"十三五"江苏省高等学校重点教材·飞行器设计与工程力学品牌专业系列教材

ISBN 978-7-03-058936-1

Ⅰ.①材… Ⅱ.①王… Ⅲ.①材料力学-高等学校-教材-汉、英 Ⅳ.①TB301

中国版本图书馆 CIP 数据核字(2018)第 220929 号

责任编辑：余 江 朱晓颖／责任校对：郭瑞芝
责任印制：张 伟／封面设计：迷底书装

---

科学出版社 出版

北京东黄城根北街 16 号
邮政编码：100717
http://www.sciencep.com

北京九州迅驰传媒文化有限公司印刷
科学出版社发行 各地新华书店经销

*

2018 年 9 月第 一 版　开本：787×1092 1/16
2024 年 3 月第五次印刷　印张：18 3/4
字数：480 000

**定价：79.00 元**
(如有印装质量问题，我社负责调换)

# Preface

"Mechanics of materials" is a required subject for engineering students majoring in aeronautical, mechanical and civil engineering, and it is usually taught during the sophomore year. This book is intended to provide these students with the theory and application of mechanics of materials, and mainly includes the analysis and design of strength, rigidity, and stability of a bar in axial tension or compression, a shaft in torsion, and a beam in bending.

The book is organized into 13 chapters and 3 appendixes. Chapter 1 introduces the fundamental concepts often used in mechanics of materials, such as internal forces, stresses, strains, etc., and the tensile and compressive properties of ductile and brittle materials subjected to axial loads. In Chapter 2 the normal and shearing stresses in a bar subjected to axial loads are analyzed, and the deformation of an axially loaded bar is also analyzed in this chapter. The normal and shearing stresses produced in a shaft in torsion, together with the angle of twist of the shaft, are discussed in Chapter 3. The shearing force and bending moment, normal and shearing stresses, and deflection and slope of a beam in pure or transverse-force bending are analyzed, respectively, in Chapters 4, 5 and 6. Chapter 7 mainly covers stress analysis, such as stress transformation, principal stress, maximum shearing stress, etc., and strength theories, including the maximum normal stress, normal strain, shearing stress and distortion energy criteria. Chapter 8 analyzes the stress of a member subjected to combined loadings, such as a bar in eccentric tension or compression, a beam in bending and tension/compression, a shaft in torsion and bending, etc. The stability of a column subjected to a centric compressive load is discussed in Chapter 9, including analysis of the critical loading and critical stress for a long or an intermediate column. The unsymmetrical pure bending, along with the unsymmetrical transverse-force bending, is covered in Chapter 10. The main energy methods including the principle of work and energy, reciprocal theorem, Castigliano's theorem, principle of virtual work, unit load method, etc., are presented in Chapter 11. Chapter 12 analyzes the vertical and horizontal impact of a member. The force method used for solving a statically indeterminate structure is presented in Chapter 13.

Kaifu WANG
Nanjing, June 2018

# 前　言

"材料力学"是航空航天工程、机械工程和土木工程等工科专业本科生的必修课，通常在大二学年讲授。本书主要讲述材料力学的理论及其应用，内容包括拉压杆、扭转轴、弯曲梁的强度、刚度和稳定分析与设计等。

本书共 13 章。第 1 章介绍材料力学常用基本概念，如内力、应力和应变等，以及轴向载荷作用下塑性和脆性材料的拉压性能。第 2 章分析轴向拉压杆正应力和剪应力，同时分析轴向拉压变形。第 3 章讨论扭转轴正应力、剪应力以及扭转角。第 4～6 章分别讨论纯弯曲梁和横力弯曲梁的剪力与弯矩、正应力与剪应力、挠度与转角。第 7 章主要讨论应力分析，如应力变换、主应力和最大剪应力等，同时涉及强度理论，包括最大正应力准则、最大线应变准则、最大剪应力准则和最大畸变能准则。第 8 章分析组合载荷作用构件的应力，如偏心拉压杆、拉压弯曲梁和弯曲扭转轴等。第 9 章讨论压杆稳定，包括分析细长压杆和中长压杆的临界载荷和临界应力。第 10 章涉及非对称纯弯曲以及非对称横力弯曲。第 11 章介绍主要的能量方法，包括功能原理、互等定理、卡氏定理、虚功原理和单位载荷法等。第 12 章分析构件的垂直和水平冲击。第 13 章介绍力法，用于求解静不定结构。

王开福

2018 年 6 月于南京

# Contents

**Chapter 1 Mechanics of Materials Fundamentals** ········· 1
    1.1 External Forces ········· 1
    1.2 Internal Forces ········· 2
    1.3 Stresses ········· 3
    1.4 Strains ········· 7
    1.5 Hooke's Law ········· 9
    1.6 Tensile Properties of Low-Carbon Steel ········· 9
    1.7 Stress-Strain Curve of Ductile Materials Without Distinct Yield Point ········· 11
    1.8 Ductile and Brittle Materials ········· 12
    1.9 Properties of Materials in Compression ········· 12
    Problems ········· 13

**Chapter 2 Axial Tension and Compression** ········· 15
    2.1 Axial Force ········· 15
    2.2 Normal Stress on Cross Section ········· 15
    2.3 Normal and Shearing Stresses on Oblique Section ········· 16
    2.4 Normal Strain ········· 18
    2.5 Tensile and Compressive Deformation ········· 19
    2.6 Statically Indeterminate Bar in Tension and Compression ········· 21
    2.7 Design of Tensile and Compressive Bar ········· 22
    Problems ········· 22

**Chapter 3 Torsion** ········· 25
    3.1 Torsional Moment ········· 25
    3.2 Hooke's Law in Shear ········· 25
    3.3 Shearing Stress on Cross Section ········· 26
    3.4 Normal and Shearing Stresses on Oblique Section ········· 29
    3.5 Angle of Twist ········· 29
    3.6 Statically Indeterminate Shaft ········· 31
    3.7 Design of Torsional Shaft ········· 31
    Problems ········· 32

**Chapter 4 Bending Internal Forces** ········· 35
    4.1 Shearing-Force and Bending-Moment Diagrams ········· 35

4.2　Relations Between Distributed Load, Shearing Force, and Bending Moment ············ 38
　　4.3　Relations Between Concentrated Load, Shearing Force, and Bending Moment ·········· 40
　　Problems ·············································································································· 42

## Chapter 5　Bending Stresses ·········································································· 44
　　5.1　Normal Stresses on Cross Section in Pure Bending ················································ 44
　　5.2　Normal and Shearing Stresses on Cross Section in Transverse-Force Bending ·········· 48
　　5.3　Design of Bending Beam ························································································ 52
　　Problems ·············································································································· 53

## Chapter 6　Bending Deformation ································································ 56
　　6.1　Method of Integration ··························································································· 57
　　6.2　Method of Superposition ······················································································ 58
　　6.3　Statically Indeterminate Beam ·············································································· 59
　　Problems ·············································································································· 60

## Chapter 7　Stress Analysis and Strength Theories ······································ 62
　　7.1　Stress Transformation ·························································································· 62
　　7.2　Principal Stresses ·································································································· 64
　　7.3　Maximum Shearing Stress ····················································································· 66
　　7.4　Pressure Vessels ···································································································· 67
　　7.5　Generalized Hooke's Law ······················································································ 69
　　7.6　Strength Theories ·································································································· 71
　　Problems ·············································································································· 75

## Chapter 8　Combined Loadings ··································································· 77
　　8.1　Bar in Eccentric Tension or Compression ····························································· 77
　　8.2　I-Beam in Transverse-Force Bending ···································································· 79
　　8.3　Beam in Bending and Tension/Compression ························································· 82
　　8.4　Shaft in Torsion and Bending ··············································································· 83
　　Problems ·············································································································· 86

## Chapter 9　Stability of Column ··································································· 88
　　9.1　Critical Load of Long Column with Pin Supports ················································· 88
　　9.2　Critical Load of Long Column with Other Supports ············································ 89
　　9.3　Critical Stress of Long Column ············································································ 90
　　9.4　Critical Stress of Intermediate Column ································································· 91
　　9.5　Design of Column ································································································· 92
　　Problems ·············································································································· 94

## Chapter 10　Unsymmetrical Bending ························································· 96
　　10.1　Unsymmetrical Pure Bending ·············································································· 96

  10.2 Unsymmetrical Transverse-Force Bending ·················································· 99
  Problems ······························································································································· 104

## Chapter 11 Energy Methods ··················································································· 106
  11.1 External Work ······················································································································ 106
  11.2 Stain-Energy Density ········································································································· 107
  11.3 Strain Energy ······················································································································· 108
  11.4 Principle of Work and Energy ·························································································· 111
  11.5 Reciprocal Theorem ·········································································································· 112
  11.6 Castigliano's Theorem ······································································································ 113
  11.7 Principle of Virtual Work ································································································· 116
  11.8 Unit Load Method ············································································································· 118
  11.9 Applications of Energy Methods ······················································································ 120
  Problems ······························································································································· 121

## Chapter 12 Impact Loading ···················································································· 125
  12.1 Vertical Impact ··················································································································· 125
  12.2 Horizontal Impact ·············································································································· 127
  Problems ······························································································································· 129

## Chapter 13 Statically Indeterminate Structures ··················································· 131
  13.1 Static Indeterminacy ········································································································· 131
  13.2 Force Method for Analysis of Statically Indeterminate Structures ························· 132
  13.3 Force Method for Analysis of Symmetrical Statically-Indeterminate Structures ····· 137
  Problems ······························································································································· 139

## Appendix I Properties of Area ············································································· 143
  I.1 First Moment (Static Moment) ······························································································ 143
  I.2 Moment of Inertia and Polar Moment of Inertia ································································ 144
  I.3 Radius of Gyration and Polar Radius of Gyration ···························································· 144
  I.4 Product of Inertia ··················································································································· 145
  I.5 Parallel-Axis Theorem ·········································································································· 145
  I.6 Properties of Commonly-Used Areas ················································································· 146

## Appendix II Shape Steels ························································································ 147
  II.1 I-Steel ···································································································································· 147
  II.2 Channel Steel ······················································································································ 148
  II.3 Equal Angle Steel ··············································································································· 149
  II.4 Unequal Angle Steel ·········································································································· 151

## Appendix III Deflection Curves ············································································· 154

## References ······································································································································ 155

# 目 录

**第1章 材料力学基础** ·············· 156
  1.1 外力 ·············· 156
  1.2 内力 ·············· 157
  1.3 应力 ·············· 158
  1.4 应变 ·············· 160
  1.5 胡克定律 ·············· 162
  1.6 低碳钢拉伸性能 ·············· 162
  1.7 无明显屈服点塑性材料的应力应变曲线 ·············· 164
  1.8 塑性材料和脆性材料 ·············· 165
  1.9 材料压缩性能 ·············· 165
  习题 ·············· 165

**第2章 轴向拉伸与压缩** ·············· 167
  2.1 轴力 ·············· 167
  2.2 横截面正应力 ·············· 167
  2.3 斜截面正应力和剪应力 ·············· 168
  2.4 线应变 ·············· 170
  2.5 拉压变形 ·············· 171
  2.6 静不定拉压杆 ·············· 172
  2.7 拉压杆设计 ·············· 173
  习题 ·············· 173

**第3章 扭转** ·············· 176
  3.1 扭矩 ·············· 176
  3.2 剪切胡克定律 ·············· 176
  3.3 横截面剪应力 ·············· 177
  3.4 斜截面正应力和剪应力 ·············· 179
  3.5 扭转角 ·············· 180
  3.6 静不定轴 ·············· 180
  3.7 扭转轴设计 ·············· 181
  习题 ·············· 182

**第4章 弯曲内力** ·············· 184
  4.1 剪力图和弯矩图 ·············· 184

4.2 分布载荷、剪力和弯矩之间的关系 ………………………………………………… 186
4.3 集中载荷、剪力和弯矩之间的关系 ………………………………………………… 188
习题 …………………………………………………………………………………………… 189

## 第 5 章 弯曲应力 ……………………………………………………………………………… 191
5.1 纯弯曲横截面正应力 ……………………………………………………………………… 191
5.2 横力弯曲横截面正应力和剪应力 ………………………………………………………… 194
5.3 弯曲梁设计 ………………………………………………………………………………… 197
习题 …………………………………………………………………………………………… 198

## 第 6 章 弯曲变形 ……………………………………………………………………………… 201
6.1 积分法 ……………………………………………………………………………………… 202
6.2 叠加法 ……………………………………………………………………………………… 203
6.3 静不定梁 …………………………………………………………………………………… 203
习题 …………………………………………………………………………………………… 204

## 第 7 章 应力分析与强度理论 ………………………………………………………………… 206
7.1 应力变换 …………………………………………………………………………………… 206
7.2 主应力 ……………………………………………………………………………………… 208
7.3 最大剪应力 ………………………………………………………………………………… 209
7.4 压力容器 …………………………………………………………………………………… 211
7.5 广义胡克定律 ……………………………………………………………………………… 212
7.6 强度理论 …………………………………………………………………………………… 214
习题 …………………………………………………………………………………………… 217

## 第 8 章 组合载荷 ……………………………………………………………………………… 218
8.1 偏心拉压杆 ………………………………………………………………………………… 218
8.2 横力弯曲工字梁 …………………………………………………………………………… 219
8.3 拉压弯曲梁 ………………………………………………………………………………… 221
8.4 弯曲扭转轴 ………………………………………………………………………………… 223
习题 …………………………………………………………………………………………… 225

## 第 9 章 压杆稳定 ……………………………………………………………………………… 227
9.1 两端铰支细长压杆临界载荷 ……………………………………………………………… 227
9.2 其他支撑细长压杆临界载荷 ……………………………………………………………… 228
9.3 细长压杆临界应力 ………………………………………………………………………… 228
9.4 中长压杆临界应力 ………………………………………………………………………… 229
9.5 压杆设计 …………………………………………………………………………………… 230
习题 …………………………………………………………………………………………… 232

## 第 10 章 非对称弯曲 ………………………………………………………………………… 233
10.1 非对称纯弯曲 …………………………………………………………………………… 233

10.2 非对称横力弯曲 ································································· 236
习题 ·············································································· 239

# 第 11 章 能量方法 ································································ 241
11.1 外功 ······································································· 241
11.2 应变能密度 ······························································· 242
11.3 应变能 ···································································· 243
11.4 功能原理 ································································· 245
11.5 互等定理 ································································· 246
11.6 卡氏定理 ································································· 247
11.7 虚功原理 ································································· 250
11.8 单位载荷法 ······························································· 251
11.9 能量方法应用 ···························································· 252
习题 ·············································································· 254

# 第 12 章 冲击载荷 ································································ 257
12.1 垂直冲击 ································································· 257
12.2 水平冲击 ································································· 259
习题 ·············································································· 261

# 第 13 章 静不定结构 ···························································· 263
13.1 静不定 ···································································· 263
13.2 力法分析静不定结构 ··················································· 263
13.3 力法分析对称静不定结构 ············································· 268
习题 ·············································································· 270

# 附录 I 截面性质 ································································ 273
I.1 静矩 ········································································· 273
I.2 惯性矩与极惯性矩 ······················································· 274
I.3 惯性半径与极惯性半径 ················································· 274
I.4 惯性积 ······································································ 275
I.5 平行移轴定理 ····························································· 275
I.6 常用截面几何性质 ······················································· 276

# 附录 II 型钢 ······································································ 277
II.1 工字钢 ···································································· 277
II.2 槽钢 ········································································ 278
II.3 等边角钢 ································································· 279
II.4 不等边角钢 ······························································· 281

# 附录 III 挠度曲线 ······························································ 284

# 参考文献 ··········································································· 285

# Chapter 1  Mechanics of Materials Fundamentals

In theoretical mechanics, bodies are assumed to be perfectly rigid. The deformations of bodies are important, however, as far as the resistance of the structures and machines to failure is concerned. Therefore, the bodies in mechanics of materials will no longer be assumed to be perfectly rigid as considered in theoretical mechanics.

Mechanics of materials studies the ability of structures and machines to resist failure, and mainly involves the following tasks: ① strength, i.e., the ability of members to support a specified load without experiencing excessive stresses; ② rigidity, i.e., the ability of members to support a specified load without undergoing unacceptable deformations; ③ stability, i.e., the ability of members to support a specified axial compressive load without causing a sudden lateral deflection.

Any material dealt with in mechanics of materials is assumed to be: ① continuous, i.e., the material consists of a continuous distribution of matter without voids; ② homogeneous, i.e., the material possesses the same mechanical properties at all points in the matter; ③ isotropic, i.e., the material has the same mechanical properties in all directions at any one point of the matter.

The strength and rigidity of a material depend on its abilities to support a specified load without experiencing both excessive stresses and unacceptable deformations. These abilities are inherent in the material itself and must be determined by experimental methods. One of the most important tests to determine the mechanical properties of a material is the tensile or compressive test. This test is often used to determine the stress-strain relation of the material used.

## 1.1  External Forces

Any external force applied to a body can be classified as either a surface force or a body force.

**1. Surface Force**

An external force that is applied to the surface of a body is called a surface force.

If the surface force is distributed over a finite area of the body, it is said to be a distributed load on a surface, Fig. 1.1(a). If the surface force is applied along a narrow area, this force is defined as a distributed load along a line, Fig. 1.1(b). If the area subjected to a surface force is very small, compared with the surface area of the body, then this surface force can be regarded as a concentrated load, Fig. 1.1(c).

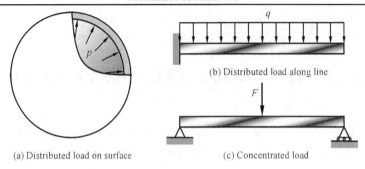

(a) Distributed load on surface  (b) Distributed load along line  (c) Concentrated load

Fig. 1.1

**2. Body Force**

An external force that is applied to every point within a body is called a body force. A gravitational force is an excellent example of the body force since it acts upon each of the particles forming the body.

## 1.2 Internal Forces

When various external loads are applied to a member, the corresponding distributed internal forces will be developed at any point within the member. The distributed internal forces on any section within the member can be determined by using the method of sections.

We imagine to use a plane $\Pi$, Fig. 1.2(a), to section the member where the distributed internal forces need to be determined. For determination of the distributed internal forces on the cut plane, the portion of the member to the right of the cut plane is removed, and it is replaced by the distributed internal forces acting on the left portion, Fig. 1.2(b).

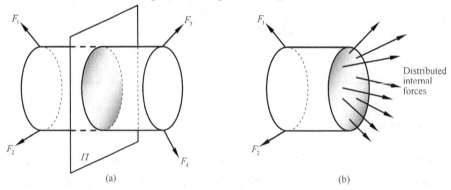

Fig. 1.2

For equilibrium of the remaining portion of the member, the distributed internal forces can be determined by using the equations of static equilibrium. Although the exact distribution of internal forces may be unknown, we can use the equations of static equilibrium to relate the applied external loads to the resultant force **R** and resultant couple $M_O$ about point $O$ on the cut plane, which are caused by the distributed internal forces, Fig. 1.3(a).

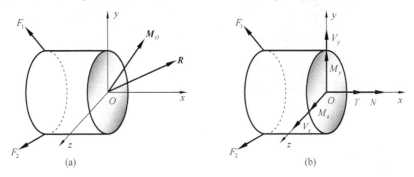

Fig. 1.3

Generally speaking, the resultant force **R** and resultant couple $M_O$ have arbitrary directions, neither perpendicular nor parallel to the cut plane. However, we can resolve the resultant force and couple into six components, respectively along the $x$, $y$, and $z$ axes, Fig. 1.3(b).

(1) Axial force. The normal component, along the $x$ direction, of the resultant force is called the axial force (normal force), $N$. It is developed when the external loads tend to pull or push the two segments of the member.

(2) Shearing force. The tangential components, respectively along the $y$ and $z$ directions, of the resultant force are regarded as the shearing forces, denoted by $V_y$ and $V_z$, which are developed when the external loads tend to cause the two segments of the member to slide over one another.

(3) Torsional moment. The normal component, rotating about the $x$ axis, of the resultant couple is called the torsional moment (twisting moment, or torque), $T$, and developed when the external loads tend to twist one segment of the member with respect to the other.

(4) Bending moment. The tangential components of the resultant couple tend to bend the member about the $y$ and $z$ axes, respectively. These two components, $M_y$ and $M_z$, rotating about the $y$ and $z$ axes respectively, are called the bending moments.

## 1.3 Stresses

The distributed internal forces are developed at any point within the member subjected to external loads. To define the stress at a given point $P$ of the section, Fig. 1.4(a), we consider a small area $\Delta A$ containing $P$ and assume that the resultant force is $\Delta F$ on the area $\Delta A$. In general, the force $\Delta F$ has a unique direction at a given point on the section and can be resolved into three components $\Delta N$, $\Delta V_y$ and $\Delta V_z$ respectively along the $x$, $y$, and $z$ axes, Fig. 1.4(b). $\Delta N$ is the normal component perpendicular to the area $\Delta A$, $\Delta V_y$ and $\Delta V_z$ are the two tangential components within the area $\Delta A$.

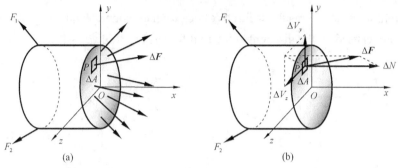

Fig. 1.4

### 1. Normal Stress

The intensity of the normal force, the normal force per unit area, acting normal to the area is defined as the normal stress, denoted by $\sigma$. The normal stress at the given point $P$ on the section of the member, Fig. 1.5, can be expressed as

$$\sigma_x = \lim_{\Delta A \to 0} \frac{\Delta N}{\Delta A} \tag{1.1}$$

where $\sigma_x$ is perpendicular to the section and the subscript represents the outward normal of the section and the direction of the normal stress. A positive sign is usually used to indicate a tensile stress and a negative sign to indicate a compressive stress. From SI units, with $\Delta N$ expressed in N and $\Delta A$ in m$^2$, the normal stress $\sigma_x$ is expressed in Pascal (Pa).

### 2. Shearing Stress

The intensity of the tangential force, the tangential force per unit area, acting tangent to the area is called the shearing stress, denoted by $\tau$. The two shearing stress components at the given point $P$ on the section of the member, Fig. 1.6, can be written, respectively, as

$$\tau_{xy} = \lim_{\Delta A \to 0} \frac{\Delta V_y}{\Delta A}, \quad \tau_{xz} = \lim_{\Delta A \to 0} \frac{\Delta V_z}{\Delta A} \tag{1.2}$$

where $\tau_{xy}$ and $\tau_{xz}$ lie in the section. Two subscripts are used for the shearing stress components: the first represents the direction of the outward normal line of the section; and the second indicates the direction of the shearing stress. In SI units, the shearing stresses $\tau_{xy}$ and $\tau_{xz}$ are also measured in Pa.

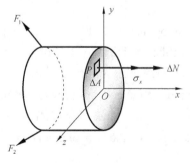

Fig. 1.5

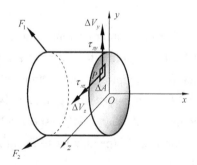

Fig. 1.6

In order to show how shearing stresses develop, we will consider a member subjected to two transverse forces of magnitude $F$, Fig. 1.7(a).

Sectioning the member at $CC'$ between the points of application of the two forces, and considering the equilibrium of the left portion, Fig. 1.7(b), we conclude that distributed internal forces must exist in the cross section. The resultant of these distributed internal forces is called direct shearing force, denoted by $V$. Dividing the direct shearing force $V$ in the cross section by the area $A$ of the cross section, we obtain the direct shearing stress in the section. Denoting the direct shearing stress by the letter $\tau$, we have

$$\tau = \frac{V}{A} \quad (1.3)$$

We should note that the value obtained is an average value of the shearing stress over the entire section.

Direct shearing stresses are commonly found in bolts, pins, and rivets used to connect various structural members and machine components. Consider the two plates, which are connected by a bolt, Fig. 1.8(a). If the plates are subjected to two tension forces of magnitude $F$, a direct shearing stress will develop in the section of bolt corresponding to the contacting surface of the plates. Drawing the diagram of the bolt located below the contacting surface, Fig. 1.8(b), we conclude that the direct shearing stress $\tau$ in the section is equal to $V/A$.

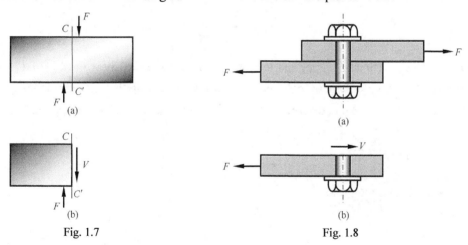

Fig. 1.7      Fig. 1.8

**Example 1.1** A load $F$ is applied to a steel rod supported as shown in Fig. 1.9(a) by a plate into which a 15 mm diameter hole has been drilled. Knowing that the shearing stress must not exceed 120 MPa in the steel, determine the largest load $F_{max}$ which may be applied to the rod.

**Solution** The shearing plane is a cylindrical surface, Fig. 1.9(b), and its shearing area is equal to $A = \pi dt$. Since the maximum shearing stress $\tau_{max} = 120$ MPa, then the largest load $F_{max}$ can, from Eq. (1.3), be obtained by

$$F_{max} = \tau_{max} A = 56.5 \text{ kN}$$

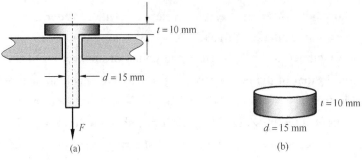

Fig. 1.9

### 3. Bearing Stress

Bolts, pins, and rivets create stresses on the bearing surface of the members they connect. Consider two plates connected by a bolt, Fig. 1.10(a). The bolt exerts on the upper plate a force $F_{bs}$, Fig. 1.10(b), equal and opposite to the force $F'_{bs}$ exerted by the upper plate on the bolt, Fig. 1.10(c). The force $F_{bs}$ exerted by the bolt represents the resultant of distributed forces on the inside surface of a half-cylinder.

Since the distribution of forces on the contacting surface of the members is quite complicated, the average value of the stress, obtained by dividing the resultant $F_{bs}$ by the projected area $A_{bs}$ of the bolt on the plate section, is regarded as the bearing stress $\sigma_{bs}$. Since this projected area $A_{bs}$ is equal to $td$, where $t$ is the plate thickness and $d$ is the diameter of the bolt, we have

$$\sigma_{bs} = \frac{F_{bs}}{A_{bs}} = \frac{F_{bs}}{td} \tag{1.4}$$

**Example 1.2** A load $F$ is applied to a steel rod supported as shown in Fig. 1.11(a) by a plate into which a 15 mm diameter hole has been drilled. Knowing that the bearing stress of the steel must not exceed 150 MPa, determine the largest load $F_{max}$ which may be applied to the rod.

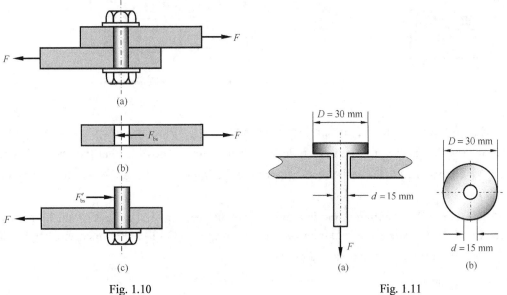

Fig. 1.10　　　　　　　　　　Fig. 1.11

**Solution** The bearing plane is an annular surface of inner diameter $d = 15$ mm and outer diameter $D = 30$ mm, Fig. 1.11(b), and thus the bearing area is equal to $A_{bs} = \frac{1}{4}\pi(D^2 - d^2)$. Since the maximum bearing stress $\sigma_{bs} = 150$ MPa, then the largest load $F_{max}$ can, from Eq. (1.4), be obtained by

$$F_{max} = \sigma_{bs} A_{bs} = 79.5 \text{ kN}$$

## 1.4  Strains

When external loads are applied to a member, they will tend to change the size and shape of the member. These changes are called deformations of the member.

**1. Normal Strain**

The elongation or contraction of a line segment per unit of length is referred to as the normal strain denoted by $\varepsilon$. Consider a line segment $\Delta x$ passing through the given point $P$ within the member, Fig. 1.12(a). After the member is deformed under the action of external loads, the line segment $\Delta x$ has length of $\Delta x'$, Fig. 1.12(b). The deformation of the line segment $\Delta x$ is equal to $\Delta \delta = \Delta x' - \Delta x$.

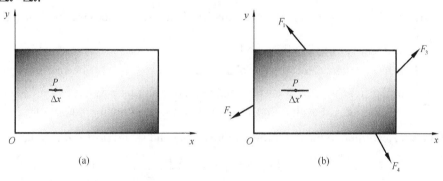

Fig. 1.12

As $\Delta x$ approaches zero, the quotient of $\Delta \delta$ and $\Delta x$ will approach a finite limit. This limit is called the normal strain at the given point $P$ along the direction of the line segment, which can be expressed as

$$\varepsilon_x = \lim_{\Delta x \to 0} \frac{\Delta \delta}{\Delta x} \tag{1.5}$$

where the normal strain $\varepsilon_x$ is a dimensionless quantity. The normal strain is said to be positive if the line segment elongates and negative if the line segment contracts.

**2. Shearing Strain**

The change of angle that occurs between two perpendicular line segments is referred to as the shearing strain, denoted by $\gamma$. Consider the two perpendicular line segments $\Delta x$ and $\Delta y$

intersecting at the given point P, Fig. 1.13(a). After the member is deformed under the action of external loads, the included angle between the two line segments is changed from $\frac{1}{2}\pi$ into $\theta'$, Fig. 1.13(b).

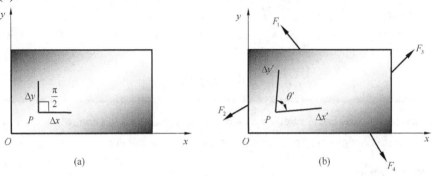

Fig. 1.13

The shearing strain at the given point P in the plane containing the two line segments can be written as

$$\gamma_{xy} = \frac{1}{2}\pi - \lim_{\substack{\Delta x \to 0 \\ \Delta y \to 0}} \theta' \qquad (1.6)$$

The shearing strain $\gamma_{xy}$ is positive if $\theta'$ is smaller than $\frac{1}{2}\pi$; otherwise, it is negative. The shearing strain $\gamma_{xy}$ is measured in radians (rad).

**Example 1.3**  Consider a rectangular plate $OACB$, Fig. 1.14(a). The size and shape of the plate is shown in Fig. 1.14(b) after it is subjected to a uniform plane deformation. Determine the normal strains $\varepsilon_x$ and $\varepsilon_y$, as well as the shearing strain $\gamma_{xy}$, at the point $O$ of the plate.

**Solution**  According to the definition of the normal and shearing strains, we have

$$\varepsilon_x = \frac{OA' - OA}{OA} = 2.84 \times 10^{-4}, \quad \varepsilon_y = \frac{OB' - OB}{OB} = 3.64 \times 10^{-4}$$

$$\gamma_{xy} = \frac{1}{2}\pi - \angle A'OB' \approx \frac{\Delta_y}{OA} + \frac{\Delta_x}{OB} = 3.51 \times 10^{-3} \text{ rad}$$

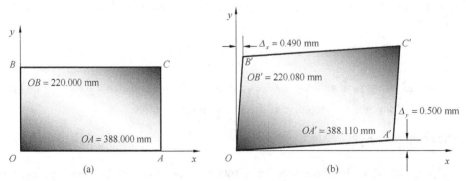

Fig. 1.14

## 1.5 Hooke's Law

Most engineering structures are designed to undergo relatively small deformations, involving only the straight-line portion of the corresponding stress-strain curve. For the straight-line portion of the stress-strain curve, the stress $\sigma$ is proportional to the strain $\varepsilon$, and we have

$$\sigma = E\varepsilon \tag{1.7}$$

This relation is known as Hooke's law. The coefficient $E$ is called the modulus of elasticity of the material, or also Young's modulus. Since the strain $\varepsilon$ is a dimensionless quantity, the modulus of elasticity $E$ is expressed in the same units as the stress $\sigma$.

## 1.6 Tensile Properties of Low-Carbon Steel

**1. Standard Specimen**

A standard specimen having certain size and shape must be used in the tensile test of low-carbon steel for purpose of the comparability of experimental results.

One type of standard tensile specimen of low-carbon steel commonly used is shown in Fig. 1.15(a). The cross-sectional area $A_0$ of the central portion of the specimen has been accurately determined and two gauge marks have been inscribed on the portion at a distance $l_0$ from each other. The distance $l_0$ is known as the gauge length of the specimen.

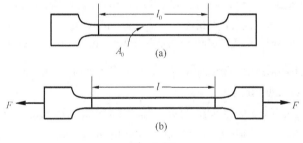

Fig. 1.15

**2. Tensile Diagram**

The specimen is then placed in a universal testing machine, which is used to apply a centric load $F$. As the load $F$ increases, the distance $l$ between the two gauge marks also increases, Fig. 1.15(b). The elongation $\delta = l - l_0$ is recorded for each value of $F$.

Plotting the magnitude $F$ of the load against the deformation $\delta$, we obtain a load-deformation diagram, Fig. 1.16. This diagram is called the tensile diagram of low-carbon steel.

### 3. Stress-Strain Curve

From each pair of readings $F$ and $\delta$, the stress $\sigma$ is computed by dividing $F$ by the original cross-sectional area $A_0$ of the specimen, and the strain $\varepsilon$ by dividing the elongation $\delta$ by the original distance $l_0$ between the two gauge marks.

Plotting $\varepsilon$ as an abscissa and $\sigma$ as an ordinate, we obtain a curve that is characteristic of the mechanical properties of low-carbon steel and that is independent of the dimensions of the particular specimen used. This curve is called a stress-strain curve of low-carbon steel, Fig. 1.17.

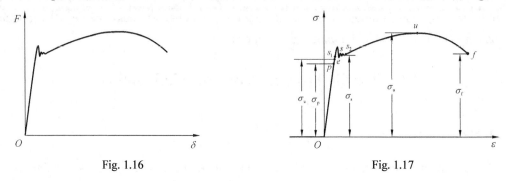

Fig. 1.16        Fig. 1.17

From the stress-strain curve of low-carbon steel shown in Fig. 1.17, we can identify four different ranges in which the material behaves.

(1) Elastic range ($Oe$). The region $Oe$ of the stress-strain curve is called the elastic range, Fig. 1.17. The material behaves elastically in this region. The largest stress corresponding to the elastic range is called the elastic limit, denoted by $\sigma_e$.

The straight-line region $Op$ within the elastic range $Oe$ is called the linearly elastic range. In this range, the material is linearly elastic, i.e., the stress is proportional to the strain. The maximum stress limit to this linear relationship is called the proportional limit, $\sigma_p$. If the stress slightly exceeds the proportional limit, the material still respond elastically. This continues until the stress reaches the elastic limit.

In the elastic range, if the load is removed, the specimen will return back to its original size and shape. Normally for low-carbon steel, the elastic limit is very close to the proportional limit and therefore rather difficult to determine.

(2) Yielding range ($s_1s_2$). A slight increase in stress above the elastic limit will result in a failure of the material and cause it to deform permanently, Fig. 1.17. This phenomenon is called yielding, which is indicated by the region $s_1s_2$ of the stress-strain curve. The stress at which yielding is initiated is called the yield strength, yield stress, or yield point, denoted by $\sigma_s$, and the deformation that occurs during the yielding range is called plastic deformation. This deformation is caused by slippage of the material along oblique surfaces and is due, therefore, primarily to shearing stresses. Once the yield point is reached, the specimen will continue to elongate without any increase in load.

For low-carbon steels and those that are hot rolled, the yield point is often distinguished by two values. The upper yield point occurs first, followed by a sudden decrease in load-carrying capacity to a lower yield point.

(3) Strain hardening range ($s_2u$). When yielding has ended, a further load can be applied to the specimen, resulting in a curve that rises continuously but becomes flatter until it reaches a maximum stress referred to as the ultimate strength or ultimate stress, $\sigma_u$. The rise in the curve in this manner is called strain hardening, and it is identified in Fig. 1.17 as the region $s_2u$.

(4) Necking range ($uf$). At the ultimate strength, the cross-sectional area begins to decrease in a local region of the specimen, Fig. 1.17. This phenomenon, which is called necking, is caused by slip planes formed within the material, and the actual strains produced are caused by shearing stress. As a result, a constriction or neck gradually tends to form in this region as the specimen elongates further. Since the cross-sectional area within this region is continuously decreasing, the smaller area can only carry an ever-decreasing load. Hence the stress-strain curve tends to curve downward until the specimen breaks at the fracture stress $\sigma_f$.

We note that rupture occurs along a cone-shaped surface which forms an angle of approximately $45°$ with the original surface of the specimen.

### 4. Percent Elongation and Percent Reduction in Area

A standard measure of the ductility of a material is its percent elongation $\delta_1$, which is defined as

$$\delta_1 = \frac{l_1 - l_0}{l_0} \times 100\% \tag{1.8}$$

where $l_0$ and $l_1$ denote, respectively, the initial gauge length of the tensile test specimen and its final gauge length at rupture.

Another measure of ductility which is sometimes used is the percent reduction in area $\psi_1$, defined as

$$\psi_1 = \frac{A_0 - A_1}{A_0} \times 100\% \tag{1.9}$$

where $A_0$ and $A_1$ denote, respectively, the initial cross-sectional area of the specimen and its minimum cross-sectional area at rupture.

## 1.7 Stress-Strain Curve of Ductile Materials Without Distinct Yield Point

In the case of many ductile materials, the onset of yield is not characterized by a fluctuating or horizontal portion of the stress-strain curve. Instead, the stress keeps increasing until the ultimate strength is reached. Necking then begins, leading eventually to rupture. For such

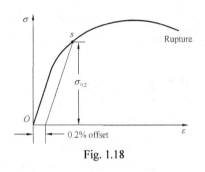

Fig. 1.18

materials, the yield strength can be defined by the offset method. A point $s$ on the stress-strain curve can be obtained by drawing through the point of the horizontal axis of abscissa $\varepsilon = 0.2\%$, an oblique straight-line parallel to the initial straight-line portion of the stress-strain curve, Fig. 1.18. The stress $\sigma_{0.2}$ corresponding to the point $s$ obtained in this fashion is defined as the offset yield strength of the material at 0.2% offset.

## 1.8 Ductile and Brittle Materials

Stress-strain curves of various materials vary widely. It is possible, however, to distinguish some common characteristics among the stress-strain curves of various groups of materials and to divide materials into two broad categories on the basis of these characteristics, namely, the ductile materials and brittle materials.

Ductile materials, which comprise structural steel, as well as many alloys of other metals, are characterized by the ability to yield at normal temperatures, Fig. 1.17.

Brittle materials, which comprise cast iron, glass, and stone, are characterized by the fact that rupture occurs without any noticeable prior change in the rate of elongation, Fig. 1.19. We note the absence of any necking of the specimen in the case of a brittle material, and observe that rupture occurs along a surface perpendicular to the load.

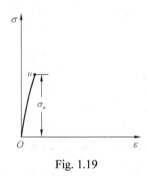

Fig. 1.19

## 1.9 Properties of Materials in Compression

When a specimen made of a ductile material is loaded in compression, the stress-stain curve obtained is essentially the same through its initial straight-line portion and through the beginning of the portion corresponding to yield and stain-hardening. Particularly noteworthy is the fact that for low-carbon steel, the yield strength is the same in both tension and compression. For large value of the strain, the tension and compression stress-strain curve diverge, and it should be noted that necking cannot occur in compression, Fig. 1.20(a).

For most brittle materials, say a cast iron, we find that the ultimate strength in compression is much larger than the ultimate strength in tension, Fig. 1.20(b).

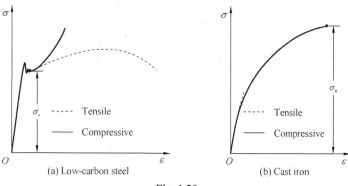

(a) Low-carbon steel　　　　(b) Cast iron

Fig. 1.20

## Problems

1.1 Two wooden planks, each $t = 20$ mm thick and $b = 80$ mm wide, are joined by the glued mortise joint, Fig. 1.21. Knowing that the joint will fail when the average shearing stress in the glue reaches $\tau = 1.2$ MPa, determine the largest axial load $F_{max}$ if the length of the cuts is equal to $a = 30$ mm.

1.2 A load $F$ is uniformly applied to a concrete foundation by a square plate, Fig. 1.22. Knowing that the bearing stress on the concrete foundation must not exceed 12 MPa, determine the side $a$ of the plate which will provide the most economical and safe design.

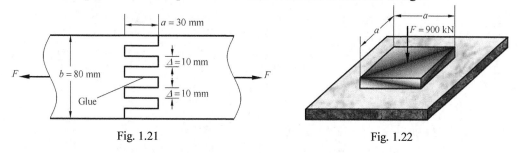

Fig. 1.21　　　　Fig. 1.22

1.3 The stress-strain diagram for an aluminum alloy that is used for making aircraft parts is shown in Fig. 1.23. If a specimen of this material is stressed to 550 MPa, determine the permanent strain that remains in the specimen when the load is released.

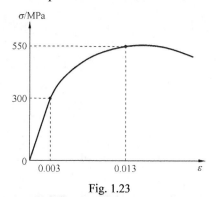

Fig. 1.23

1.4 A circular cross section rod shown in Fig. 1.24(a) is subjected to an axial load $F$. If a portion of the stress-strain diagram is shown in Fig. 1.24(b), determine the elongation of the rod when the load is applied. Knowing that $a = 50$ mm, $d = 4$ mm, and $F = 1131$ N.

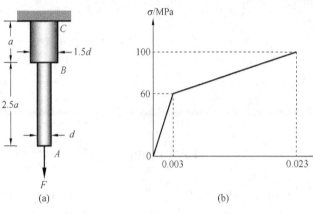

Fig. 1.24

# Chapter 2  Axial Tension and Compression

Consider a prismatic bar loaded at its ends by a pair of equal and opposite external forces coinciding with the longitudinal centroidal axis of the bar. If the external forces are directed away from the bar, the bar is said to be in tension, Fig. 2.1(a); if the external forces are directed toward the bar, the bar is in compression, Fig. 2.1(b).

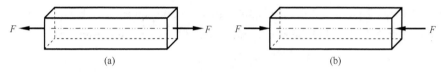

Fig. 2.1

## 2.1  Axial Force

Under the action of a pair of external forces, Fig. 2.1(a), distributed internal forces are set up within the bar. The resultant of these distributed internal forces is called the axial or normal force of the bar denoted by $N$, Fig. 2.2.

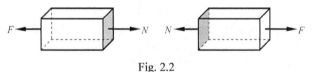

Fig. 2.2

The axial force can be determined by imagining a plane to section the bar anywhere along its length and oriented perpendicular to the longitudinal axis of the bar. For determination of the axial force on the cut plane, the right (or left) portion of the bar of the cut plane is removed and replaced by the axial force acting on the left (right) portion. By this method of introducing a cut plane, the original axial force now becomes an external force with the remaining portion of the bar. For equilibrium of the remaining portion of the bar, the axial force can be determined by using the equation of static equilibrium along the bar.

## 2.2  Normal Stress on Cross Section

It is necessary to make some assumption regarding the distribution of the axial force on the cross section of a bar subjected to tension or compression, Fig. 2.3(a). Since the external forces act through the centroid of the cross section, it is commonly assumed that the axial force caused by the applied external forces is uniformly distributed across the cross section of the

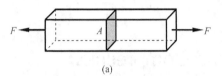

(a)

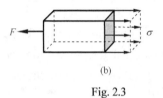

(b)

Fig. 2.3

bar, Fig. 2.3(b). Therefore, the normal stress on the cross section of the bar can be expressed as

$$\sigma = \frac{N}{A} \quad (2.1)$$

where $A$ is the cross-sectional area, and $N$ is the axial force on the cross section. A positive sign will be used to indicate a tensile stress (bar in tension) and a negative sign to indicate a compressive stress (bar in compression).

**Example 2.1** Each of the two vertical links $BD$ and $CE$ has a 16 mm×26 mm uniform rectangular cross section and each of the three pins has a 6 mm diameter, Fig. 2.4(a). Determine the maximum value of the average normal stress in the links $BD$ and $CE$.

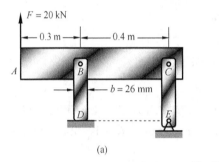
(a)

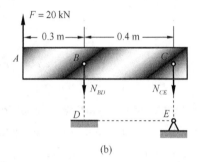
(b)

Fig. 2.4

**Solution** Considering the equilibrium of horizontal beam, Fig. 2.4(b), we have

$$N_{BD} = F\frac{l_{AC}}{l_{BC}} = 35 \text{ kN}, \quad N_{CE} = F - N_{BD} = -15 \text{ kN}$$

Since the link $BD$ is in tension, its minimum cross-sectional area is equal to

$$A_{BD} = t(b-d) = 16(26-6) = 320 \text{(mm}^2\text{)}$$

Using Eq. (2.1), we obtain the average normal stress in the link $BD$, which is equal to

$$\sigma_{BD} = \frac{N_{BD}}{A_{BD}} = 109.4 \text{ MPa}$$

Similarly, the average normal stress in the link $CE$ is

$$\sigma_{CE} = \frac{N_{CE}}{A_{CE}} = \frac{-15 \times 10^3}{16 \times 26 \times 10^{-6}} = -36.1 \text{(MPa)}$$

## 2.3 Normal and Shearing Stresses on Oblique Section

Consider a two-force bar of Fig. 2.5(a), which is subjected to a pair of tensile forces of magnitude $F$. If we section the bar through a plane forming an angle $\theta$ with the cross section and draw the free-body diagram of the left portion of the bar, we find from the equilibrium conditions of the free body that the axial force $N$ acting on the oblique section must be equal to $F$ in tension. Resolving $N$ into components $N_n$ and $N_t$, respectively normal and tangent to the oblique section, Fig. 2.5(b), we have

$$N_n = N\cos\theta \\ N_t = N\sin\theta \quad (2.2)$$

where the force $N_n$ represents the resultant of normal forces distributed over the section, and the force $N_t$ the resultant of tangential distributed forces. The corresponding normal and shearing stresses on the oblique section are obtained by dividing, respectively, $N_n$ and $N_t$ by the area $A_\theta$ of the oblique section:

$$\sigma_\theta = \frac{N_n}{A_\theta}, \quad \tau_\theta = \frac{N_t}{A_\theta} \quad (2.3)$$

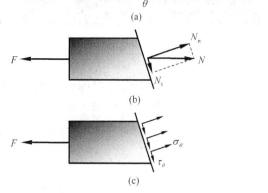

Fig. 2.5

Observing from Fig. 2.5(a) that $A_\theta = \dfrac{A}{\cos\theta}$, where $A$ denotes the cross-sectional area of the bar, we obtain

$$\sigma_\theta = \frac{N}{A}\cos^2\theta = \sigma\cos^2\theta, \quad \tau_\theta = \frac{N}{2A}\sin(2\theta) = \frac{1}{2}\sigma\sin(2\theta) \quad (2.4)$$

where $\sigma$ is the normal stress on the cross section.

We note from Eq. (2.4) that $\sigma_\theta = \sigma_{max} = \sigma$ and $\tau_\theta = 0$ when $\theta = 0$, and that $\sigma_\theta = \tau_\theta = \tau_{max} = \dfrac{1}{2}\sigma$ as $\theta = 45°$. Eq. (2.4) also shows that $\sigma_\theta = \tau_\theta = 0$ for $\theta = 90°$.

Ductile materials are weaker in shear than in tension. Therefore, when subjected to tension, a specimen made of a ductile material will fail along a plane forming a 45° angle with the cross section of the specimen, Fig. 2.6(a). Brittle materials are, however, weaker in tension than in shear. Thus when subjected to tension, a specimen made of a brittle material will fail along the cross section of the specimen, Fig. 2.6(b).

Ductile materials are weaker in shear than in compression. Therefore, when subjected to compression, a specimen made of a ductile material will fail along a plane forming a 45° angle with the cross section of the specimen, Fig. 2.7(a). Brittle materials are also weaker in shear than in compression. Thus when subjected to compression, a specimen made of a brittle material will break along a plane forming a 45°~55° angle with the cross section of the specimen, Fig. 2.7(b).

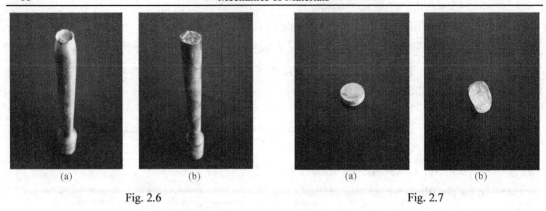

(a)      (b)          (a)      (b)

Fig. 2.6          Fig. 2.7

**Example 2.2** Two wooden members having a 90 mm×120 mm uniform rectangular cross section are joined by the glued splice shown in Fig. 2.8. Knowing that $F=100$ kN, determine the normal and shearing stresses in the glued splice.

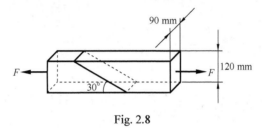

Fig. 2.8

**Solution** Using Eq. (2.4), we have

$$\sigma_{60°} = \sigma \cos^2 60° = \frac{100 \times 10^3}{90 \times 120 \times 10^{-6}} \times \cos^2 60° = 2.31 \text{(MPa)}$$

$$\tau_{60°} = \frac{\sigma}{2}\sin(2 \times 60°) = \frac{100 \times 10^3}{2 \times 90 \times 120 \times 10^{-6}} \times \sin 120° = 4.01 \text{(MPa)}$$

## 2.4 Normal Strain

**1. Longitudinal Normal Strain**

Consider a prismatic bar, of length $l$ and cross-sectional area $A$, which is suspended from the upper end, Fig. 2.9(a). If we apply an external load $F$ to the lower end, the bar will elongate, Fig. 2.9(b).

The longitudinal normal strain of the bar under axial loading can be expressed as

$$\varepsilon = \frac{\delta}{l} \qquad (2.5)$$

where $\delta$ is the longitudinal deformation of the bar, and $l$ is the length of the bar.

In the case of a bar of variable cross-sectional area, the normal stress on the cross section varies along the bar, and it is necessary to define the longitudinal normal strain at a given point by considering a small element of undeformed length $\Delta x$, Fig. 2.10(a). Denoting by $\Delta\delta$ the deformation of the element under the given loading, Fig. 2.10(b), the longitudinal normal strain at the given point located on the x-section can be written as

$$\varepsilon = \lim_{\Delta x \to 0} \frac{\Delta\delta}{\Delta x} = \frac{d\delta}{dx} \tag{2.6}$$

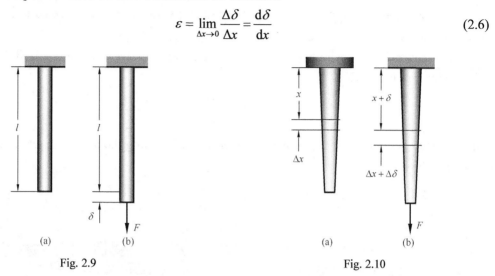

Fig. 2.9    Fig. 2.10

### 2. Lateral Normal Strain

When a bar is axially loaded, the normal stress on the cross section and the longitudinal normal strain satisfy Hooke's law, as long as the proportional limit of the material is not exceeded. And all materials considered will be assumed to be both homogeneous and isotropic, i.e., their mechanical properties will be assumed independent of both position and direction. Therefore, for the loading shown in Fig. 2.11 we have $\varepsilon_y = \varepsilon_z$, where $\varepsilon_y$ and $\varepsilon_z$ are referred to as the lateral strains respectively along the $y$ and $z$ axes.

An important constant for the given material is its Poisson's ratio, denoted by the letter $\mu$. It is defined as

$$\mu = -\frac{\varepsilon_y}{\varepsilon_x} = -\frac{\varepsilon_z}{\varepsilon_x} \tag{2.7}$$

Therefore, we have the lateral normal strains:

$$\varepsilon_y = \varepsilon_z = -\mu\varepsilon_x \tag{2.8}$$

Fig. 2.11

## 2.5  Tensile and Compressive Deformation

Consider a prismatic bar of length $l$ and cross-sectional area $A$ subjected to a centric axial load $F$, Fig. 2.12. If the normal stress on the cross section does not exceed the proportional limit of the material, we can apply Hooke's law, i.e., $\sigma = E\varepsilon$. Using $\sigma = N/A$ and $\varepsilon = \delta/l$, we have

$$\delta = \frac{Nl}{EA} \tag{2.9}$$

where $EA$ is the axial (tensile or compressive) rigidity of the bar. Eq. (2.9) can be used only if the bar is homogeneous and isotropic, has a uniform cross section, and is subjected to the centric axial loads at its ends.

In the case of a bar having variable cross section, Fig. 2.13, we can express the deformation of an element of length d$x$ as

$$d\delta = \frac{Ndx}{EA} \tag{2.10}$$

Thus the total deformation $\delta$ of the bar is obtained by integrating the above expression over the length $l$ of the bar:

$$\delta = \int \frac{Ndx}{EA} \tag{2.11}$$

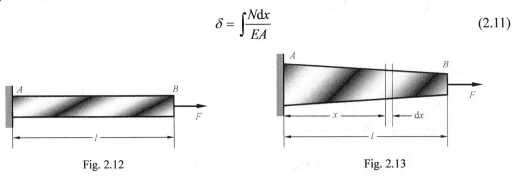

Fig. 2.12            Fig. 2.13

**Example 2.3** The 4-mm-diameter cable $BC$ is made of a steel with $E$=200 GPa. Knowing that the maximum stress in the cable must not exceed 190 MPa and that the elongation of the cable must not exceed 6 mm, find the maximum load $F_{max}$ that can be applied as shown in Fig. 2.14(a).

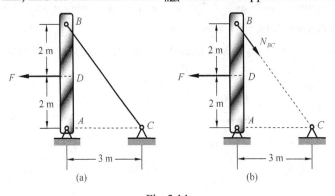

Fig. 2.14

**Solution** Using $AB$ as a free body and considering its equilibrium, Fig. 2.14(b), then we have

$$\sum M_A = 0: \ Fl_{AD} - \left(N_{BC} \frac{l_{AC}}{l_{BC}}\right) l_{AB} = 0$$

i.e.,

$$F = 1.2 N_{BC}$$

Determining the maximum load based on the stress in the cable, we have

$$F_{\max}^{\sigma} = 1.2 N_{BC} = 1.2 \sigma_{\max} A_{BC} = 1.2 \sigma_{\max} \left(\frac{1}{4}\pi d_{BC}^2\right) = 2.87 \text{ kN}$$

Determining the maximum load based on the deformation of the cable, we have

$$F_{\max}^{\delta} = 1.2 N_{BC} = 1.2 \frac{\delta E A_{BC}}{l_{BC}} = 1.2 \frac{\delta E \left(\frac{1}{4}\pi d_{BC}^2\right)}{l_{BC}} = 3.62 \text{ kN}$$

The smaller value of the two maximum loads should be chosen, that is, the maximum load $F_{\max}$ that can be applied to the structure is equal to $\min[F_{\max}^{\sigma}, F_{\max}^{\delta}] = 2.87 \text{ kN}$.

## 2.6 Statically Indeterminate Bar in Tension and Compression

In the problems considered in the preceding sections, we could use the equations of static equilibrium to determine the internal forces produced in the various portions of a bar under given loads. There are some problems encountered in engineering, however, in which the internal forces cannot be determined from the equations of static equilibrium alone. For these problems, the equations of static equilibrium must be complemented with additional equations based on the deformation of the bar. Because the equations of static equilibrium are not sufficient to determine either the reactions or the internal forces, problems of this type are said to be statically indeterminate.

Consider a prismatic bar $AB$, of length $l$, cross-sectional area $A$, and modulus of elasticity $E$, which is attached to rigid supports at $A$ and $B$ before being loaded, Fig. 2.15(a).

Drawing the free-body diagram of the bar, Fig. 2.15(b), we obtain the equilibrium equation:

$$\sum F_y = 0: \quad R_A - R_B - F = 0 \quad (2.12)$$

Since this equation is not sufficient to determine the two unknown reactions $R_A$ and $R_B$, the problem is statically indeterminate to the first degree.

We observe from the geometry that the total deformation $\delta$ of the bar must be zero. Denoting by $\delta_1$ and $\delta_2$, respectively, the deformations of the portions $AC$ and $BC$, we have

$$\delta = \delta_1 + \delta_2 = 0 \quad (2.13)$$

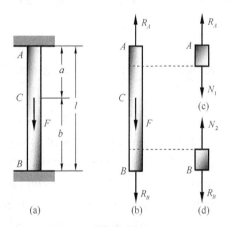

Fig. 2.15

Expressing $\delta_1$ and $\delta_2$ in terms of the corresponding internal forces $N_1$ and $N_2$, we have

$$\delta_1 = \frac{N_1 a}{EA}, \quad \delta_2 = \frac{N_2 b}{EA} \quad (2.14)$$

We note from the free-body diagrams shown in Fig. 2.15(c) and Fig. 2.15(d) that $N_1 = R_A$ and $N_2 = R_B$. Carrying these values into Eq. (2.14), and using Eq. (2.13), we have

$$R_A a + R_B b = 0 \tag{2.15}$$

Eq. (2.12) and Eq. (2.15) can be solved simultaneously for $R_A$ and $R_B$, we obtain

$$R_A = \frac{Fb}{l}, \quad R_B = -\frac{Fa}{l} \tag{2.16}$$

## 2.7 Design of Tensile and Compressive Bar

In engineering applications, stresses are used in the design of structures that will safely and economically perform a specified function.

The largest force which can be applied to a bar is called the ultimate load for this bar and is denoted by $F_u$. Since the applied load is centric, we may divide the ultimate load by the original cross-sectional area of the bar to obtain the ultimate strength of the bar. This stress, also known as the ultimate strength, can be written as

$$\sigma_u = \frac{F_u}{A} \tag{2.17}$$

The working load that a bar will be allowed to carry under normal conditions of utilization is smaller than the ultimate load. This maximum working load which can be applied to a bar is referred to as the allowable load, denoted by $F_{allow}$. Thus, only a fraction of the ultimate-load capacity of the bar is utilized when the allowable load is applied. The remaining portion of the load-carrying capacity of the bar is kept in reserve to assure its safe performance. The ratio of the ultimate load to the allowable load is used to define the factor of safety which can be expressed as

$$n = \frac{F_u}{F_{allow}} \tag{2.18}$$

or, an alternative definition of the factor of safety is based on the use of stresses:

$$n = \frac{\sigma_u}{\sigma_{allow}} \tag{2.19}$$

The selection of the factor of safety is one of the most important engineering tasks. On one hand, if a factor of safety is chosen too small, the possibility of failure becomes unacceptably large; on the other hand, if a factor of safety is chosen unacceptably large, the result is an uneconomical design. The choice of the factor of safety that is appropriate for a given design requires engineering judgment based on many considerations.

## Problems

2.1 Two solid cylindrical rods $AB$ and $BC$ are welded together at $B$ and loaded as shown in

Fig. 2.16. Knowing that $d_1$=50 mm and $d_2$=30 mm, find the normal stress in the midsection of rod *AB* rod *BC*.

2.2  A solid cylindrical rod *AB* of diameter *d* is subjected to two axial loads, $F_1$ at *C* and $F_2$ at *B*, Fig. 2.17. Knowing that *d*=50 mm, $F_1$=200 kN and $F_2$=50 kN, determine the normal stresses in portions *AC* and *BC*.

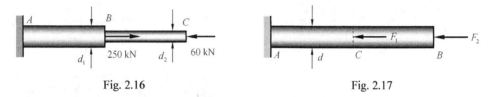

Fig. 2.16　　　　　　　　　　　　Fig. 2.17

2.3  Link *AB*, of width *b*=50 mm, and thickness *t*=6 mm, is used to support the end of a horizontal beam, Fig. 2.18. Knowing that the average normal stress in the link is 138 MPa and that the average shearing stress in the pin is 82 MPa, determine the diameter *d* of the pin, and the average bearing stress in the link.

2.4  Each of the two vertical links has a 16 mm×26 mm uniform rectangular cross section and each of the three pins has a 6 mm diameter, Fig. 2.19. Determine (1) the average shearing stress in the pin at *B*, (2) the average bearing stress at *B* in link *BD*, (3) the average bearing stress at *B* in beam *ABC*, knowing that this beam has a 10 mm×50 mm uniform rectangular cross section.

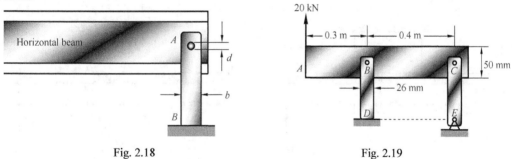

Fig. 2.18　　　　　　　　　　　　Fig. 2.19

2.5  An axial load is applied at end *C* of the steel rod *ABC*, Fig. 2.20. Knowing that *E*=200 GPa, determine the diameter $d_2$ of portion *BC* for which the deflection of point *C* will be 3 mm.

2.6  Considering a rod *AB* as shown in Fig. 2.21, made of steel (*E*=200 GPa) and subjected to an axial load *F*, If the stress in the rod must not exceed 120 MPa and the maximum change in length of the rod must not exceed 0.001 times the length of the rod, determine the smallest diameter of the rod that can be used.

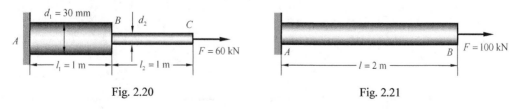

Fig. 2.20　　　　　　　　　　　　Fig. 2.21

2.7  Both portions of the rod $ABC$ are made of a brass for which $E=105$ GPa, Fig. 2.22. Knowing that the diameter of portion $BC$ is $d_2=10$ mm, determine the largest load $F_{max}$ that can be applied if $\sigma_{allow}=100$ MPa and the corresponding deflection at point $C$ is not to exceed 4 mm.

2.8  Each of the links $AB$ and $CD$ is made of aluminum ($E=75$ GPa) and has a cross-sectional area of 125 mm$^2$, Fig. 2.23. Knowing that they are used to support a rigid beam $AC$, determine the deflection of point $E$.

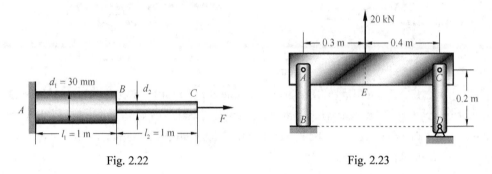

Fig. 2.22                                    Fig. 2.23

2.9  The rigid bar $AD$ is supported by two steel wires of 1.5 mm diameter ($E=200$ GPa) and a pin support at $A$, Fig. 2.24. Knowing that the wires were initially taut, determine (1) the additional tension in each wire when a 1.0 kN load $F$ is applied at point $D$, (2) the corresponding deflection of point $D$.

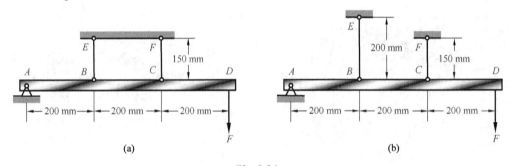

Fig. 2.24

# Chapter 3  Torsion

Consider a bar of circular cross section, loaded at its ends by a pair of equal and opposite external couples of magnitude $M_e$ applied in two planes perpendicular to the longitudinal axis of the bar as shown in Fig. 3.1. Such a bar is said to be a shaft in torsion.

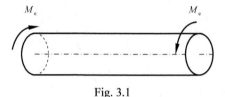

Fig. 3.1

## 3.1  Torsional Moment

Under the action of a pair of external couples, Fig. 3.1, distributed internal forces are set up on the cross section of the shaft. The resultant moment of these distributed internal forces about the longitudinal axis of the shaft is called the torsional moment (twisting moment or torque) of the shaft, denoted by $T$, Fig. 3.2.

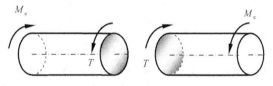

Fig. 3.2

## 3.2  Hooke's Law in Shear

Consider an element in pure shear, Fig. 3.3(a), subjected to shearing stresses $\tau_{xy}$ and $\tau_{yx}$ applied to four faces of the element respectively perpendicular to the $x$ and $y$ axes. According to the equilibrium of the element, we can obtain

$$\tau_{xy} = \tau_{yx} \tag{3.1}$$

This relation is called the reciprocal theorem of shearing stresses.

The element is observed to deform into a rhomboid, Fig. 3.3(b), under pure shearing stresses. Two of the angles formed by four faces under the shearing stresses are reduced from $\frac{1}{2}\pi$ to $\frac{1}{2}\pi - \gamma_{xy}$, while the other two are increased from $\frac{1}{2}\pi$ to $\frac{1}{2}\pi + \gamma_{xy}$, where $\gamma_{xy}$ is the shearing strain. For the shearing stress which does not exceed the proportional limit in shear, we have for any homogeneous isotropic material

$$\tau_{xy} = G\gamma_{xy} \tag{3.2}$$

This relation is known as Hooke's law in shear, and the constant $G$, expressed in the same units as $\tau_{xy}$, is called the shear modulus of elasticity of the material.

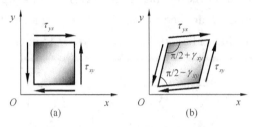

Fig. 3.3

The moduli of elasticity in tension and shear for a homogeneous isotropic material are related by the following equation:

$$G = \frac{E}{2(1+\mu)} \tag{3.3}$$

where $\mu$ is Poisson's ratio.

## 3.3 Shearing Stress on Cross Section

**1. Equilibrium Condition**

Considering a circular shaft $AB$ subjected to torsion, we cut the shaft through a section perpendicular to the longitudinal axis of the shaft at some arbitrary point $C$, Fig. 3.4(a). The section $C$ of the shaft must include the elementary shearing forces $dF$, Fig. 3.4(b), perpendicular to the radius of the shaft, which portion $BC$ exerts on portion $AC$ as the shaft is twisted.

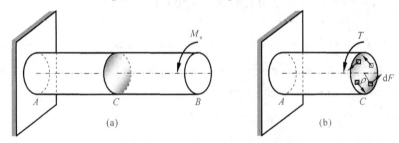

Fig. 3.4

The condition of static equilibrium requires that resultant moment of these elementary forces about the longitudinal centroidal axis of the shaft be equivalent to the torsional moment $T$ on the cross section of the shaft. Denoting by $\rho$ the perpendicular distance from the force $dF$ to the longitudinal centroidal axis of the shaft, then we have

$$T = \int \rho dF = \int \rho \tau dA \tag{3.4}$$

where $\tau$ is the shearing stress on the element of area $dA$.

**2. Deformation Condition**

When the circular shaft $AB$ is subjected to an external couple $M_e$, Fig. 3.5(a), the shaft will twist and every cross section remains plane and undistorted. Consider a segment $CD$ with a length $dx$, which has been twisted through a relative angle of twist $d\varphi$, Fig. 3.5(b).

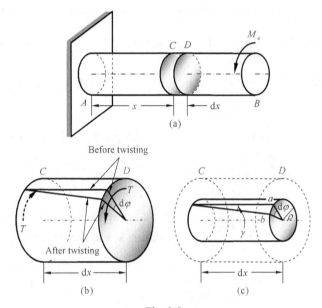

Fig. 3.5

Detach from the segment $CD$ a cylinder of radius $\rho$, Fig. 3.5(c). For small deformation, we can express the length of the arc $ab$ as $l_{ab} = \gamma dx$. We also have $l_{ab} = \rho d\varphi$. It follows that $\gamma dx = \rho d\varphi$, or

$$\gamma = \frac{d\varphi}{dx} \rho \tag{3.5}$$

where $\gamma$ and $d\varphi$ are expressed in radians. Eq. (3.5) obtained shows that the shearing strain in the circular shaft varies linearly with the distance $\rho$ from the longitudinal centroidal axis of the shaft.

**3. Stress-Strain Relation**

Assume that the stresses in the shaft remain below the proportional limit. From Hooke's law in shear, $\tau = G\gamma$, we have

$$\tau = G \frac{d\varphi}{dx} \rho \tag{3.6}$$

where $G$ is the modulus of elasticity in shear. The equation obtained shows that, as long as the proportional limit is not exceeded in any point of a circular shaft, the shearing stress in the circular shaft varies linearly with the distance $\rho$ from the longitudinal centroidal axis of the shaft. Fig. 3.6(a) shows the shearing stress distribution in a solid circular shaft and Fig. 3.6(b) in a hollow circular shaft.

Substituting $\tau$ from Eq. (3.6) into Eq. (3.4), we obtain

$$T = \int \rho \tau \mathrm{d}A = G\frac{\mathrm{d}\varphi}{\mathrm{d}x}\int \rho^2 \mathrm{d}A = GI_\mathrm{p}\frac{\mathrm{d}\varphi}{\mathrm{d}x} \tag{3.7}$$

where $I_\mathrm{p} = \int \rho^2 \mathrm{d}A$ represents the polar moment of inertia of the cross section with respect to its centroid $O$. For a solid circular shaft of diameter $d$, $I_\mathrm{p} = \frac{1}{32}\pi d^4$; and for a hollow circular shaft of inner diameter $d$ and outer diameter $D$, $I_\mathrm{p} = \frac{1}{32}\pi(D^4 - d^4) = \frac{1}{32}\pi D^4(1-\alpha^4)$, where $\alpha = d/D$.

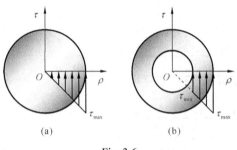

(a)    (b)

Fig. 3.6

Using Eq. (3.6) and Eq. (3.7), we have

$$\tau = \frac{T\rho}{I_\mathrm{p}} \tag{3.8}$$

It can be seen from Eq.(3.8) that the maximum shearing stress can be written as

$$\tau_{max} = \frac{T_{max}\rho_{max}}{I_\mathrm{p}} \tag{3.9}$$

We note that the ratio $I_\mathrm{p}/\rho_{max}$ depends only upon the geometry of the cross section. This ratio is called the polar modulus of section and is denoted by $S_\mathrm{p}$. From Eq. (3.9), we write

$$\tau_{max} = \frac{T_{max}}{S_\mathrm{p}} \tag{3.10}$$

For a solid circular shaft of diameter $d$, $S_\mathrm{p} = \frac{1}{16}\pi d^3$, and for a hollow circular shaft of inner diameter $d$ and outer diameter $D$, $S_\mathrm{p} = \frac{1}{16}\pi D^3(1-\alpha^4)$, where $\alpha = d/D$.

**Example 3.1** Two external couples $M_B$ and $M_C$ are applied to the sections $B$ and $C$ of a variable-section solid shaft $ABC$, Fig. 3.7. Knowing that $D$=46 mm, $d$=30 mm, $M_B$=400 N·m, and $M_C$=300 N·m, determine the maximum shearing stress in shaft $AB$ and in shaft $BC$.

**Solution** Using the method of sections, we obtain the torsional moments:

$$T_{AB} = M_B + M_C = 700 \text{ N}\cdot\text{m}, \qquad T_{BC} = M_C = 300 \text{ N}\cdot\text{m}$$

For shaft $AB$, from Eq. (3.10), we obtain

$$(\tau_{max})_{AB} = \frac{T_{AB}}{(S_p)_{AB}} = \frac{T_{AB}}{\frac{1}{16}\pi D^3} = 36.6 \text{ MPa}$$

Similarly, for shaft $BC$, we have

$$(\tau_{max})_{BC} = \frac{T_{BC}}{(S_p)_{BC}} = \frac{T_{BC}}{\frac{1}{16}\pi d^3} = 56.6 \text{ MPa}$$

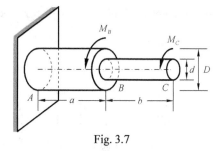

Fig. 3.7

## 3.4  Normal and Shearing Stresses on Oblique Section

Up to this time, the analysis of stresses in a shaft has been limited to shearing stresses on the cross section. We will now consider the stresses on an oblique section. We choose three elements $a$, $b$ and $c$ located on the front surface of the circular shaft subjected to torsion, Fig. 3.8. Since the faces of element $a$ are respectively parallel and perpendicular to the longitudinal axis of the shaft, the only stresses on the element will be the shearing stresses and the shearing stress is maximum. The faces of element $b$, which form an arbitrary angle with the longitudinal axis of the shaft, will be subjected to a combination of normal and shearing stresses. The element $c$ at $45°$ to the longitudinal axis of the shaft, hence the only stresses on element $c$ will be the normal stresses.

Ductile materials are weak in shear. Therefore, when subjected to torsion, a specimen made of a ductile material breaks along the cross section of the specimen, Fig. 3.9(a). Brittle materials are weaker in tension than in shear. Thus when subjected to torsion, a specimen made of a brittle material will break along the plane perpendicular to the direction of the maximum tensile stress, i.e., along the plane forming a $45°$ angle with the cross section of the specimen, Fig. 3.9(b).

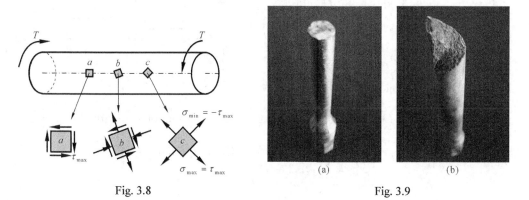

Fig. 3.8    Fig. 3.9

## 3.5  Angle of Twist

Considering a circular shaft $AB$ subjected to an external couple $M_e$ at its free end, Fig. 3.10, and assuming that the corresponding torsional moment produced in the shaft is denoted by $T$ and that the entire shaft remains linearly elastic, we recall from Eq. (3.7) that the angle of twist $d\varphi$ of the segment $CD$ can be written as

$$d\varphi = \frac{Tdx}{GI_p} \tag{3.11}$$

where $GI_p$ is the torsional rigidity of the shaft. The above equation shows that, within the linearly elastic range, the angle of twist $d\varphi$ is proportional to the torsional moment $T$ in the shaft. Integrating in $x$ from 0 to $l$, we obtain the total angle of twist of the shaft:

$$\varphi = \int \frac{Tdx}{GI_p} \tag{3.12}$$

For a circular shaft of a uniform cross section subjected to a constant torsional moment in the shaft, then Eq. (3.12) can be simplified as

$$\varphi = \frac{Tl}{GI_p} \tag{3.13}$$

**Example 3.2** Two external couples $M_B$ and $M_C$ are applied to the sections $B$ and $C$ of a variable-section solid shaft $ABC$, made of aluminum ($G$=77 GPa). Knowing that $D$=46 mm, $d$=30 mm, $a$=750 mm, $b$=900 mm, $M_B$=400 N·m, and $M_C$=300 N·m, determine the angle of twist between the sections $B$ and $C$ and of the section $C$, Fig. 3.11.

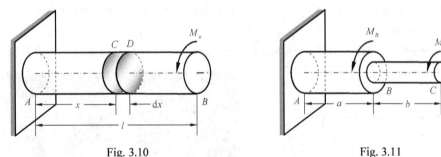

Fig. 3.10  Fig. 3.11

**Solution** Using Eq. (3.13), we obtain the angle of twist of section $C$ relative to section $B$:

$$\varphi_{C/B} = \frac{T_{BC}l_{BC}}{G(I_p)_{BC}} = \frac{T_{BC}l_{BC}}{G\left(\dfrac{1}{32}\pi d^4\right)} = 4.41\times 10^{-2} \text{ rad} = 2.53°$$

and the absolute angle of twist of section $C$:

$$\varphi_C = \varphi_{C/A} = \varphi_{C/B} + \varphi_{B/A} = \frac{T_{BC}l_{BC}}{G\left(\dfrac{1}{32}\pi d^4\right)} + \frac{T_{AB}l_{AB}}{G\left(\dfrac{1}{32}\pi D^4\right)} = 5.96\times 10^{-2} \text{ rad} = 3.41°$$

## 3.6 Statically Indeterminate Shaft

Consider now a circular shaft $AB$, which is fixed at ends $A$ and $B$ and subjected to an external couple $M_e$ at the section $C$, Fig. 3.12(a). We note that the reactions involve two unknowns, while only one equilibrium condition is available. Therefore, the circular shaft is statically indeterminate to the first degree.

We consider the reaction at the support $B$ as redundant and eliminate the corresponding support $B$, Fig. 3.12(b). The redundant reaction is then treated as an unknown load $M_B$ that, together with the other load $M_e$, must produce deformation which is compatible with the original support $B$. The redundant reaction $M_B$ will be determined from the condition that the total angle of twist at $B$ of the circular shaft must be zero.

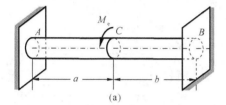 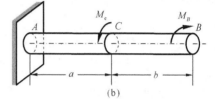

(a)            (b)

Fig. 3.12

The solution is carried out by considering separately the angle of twist $(\varphi_B)_{M_B}$ caused at $B$ by the redundant reaction $M_B$, and the angle of twist $(\varphi_B)_{M_e}$ produced at the same section by the external couple $M_e$. From Eq. (3.13), we find that $(\varphi_B)_{M_B} = \dfrac{M_B(a+b)}{GI_p}$ and $(\varphi_B)_{M_e} = (\varphi_C)_{M_e} = \dfrac{M_e a}{GI_p}$. The total angle of twist at $B$ is the sum of these two quantities and must be zero, i.e., $\varphi_B = (\varphi_B)_{M_B} + (\varphi_B)_{M_e} = 0$, or $\dfrac{M_B(a+b)}{GI_p} + \dfrac{M_e a}{GI_p} = 0$. Solving for $M_B$, we have $M_B = -\dfrac{M_e a}{a+b}$.

## 3.7 Design of Torsional Shaft

The principal specifications to be met in the design of a shaft are the power to be transmitted and the speed of rotation of the shaft. We need to select the material and the dimensions of the cross section of the shaft, so that the allowable shearing stress will not be exceeded when the shaft is transmitting the required power at the specified speed.

The torsional moment $T$ produced in the rotating shaft to transmit a power $P$ (kW) at a rotating speed $\omega$ (rad/s) or $n$ (r/min) can be written as

$$T(\text{N}\cdot\text{m}) = \frac{P(\text{W})}{\omega(\text{rad/s})} \quad \text{or} \quad T(\text{N}\cdot\text{m}) = 9549\frac{P(\text{kW})}{n(\text{r/min})} \qquad (3.14)$$

After having determined the torsional moment $T$ that will be applied to the shaft and having selected the material to be used, we will carry the allowable shearing stress into Eq. (3.9) and the allowable angle of twist into Eq. (3.12). We have

$$\tau_{max} = \frac{T_{max}\rho_{max}}{I_p} \leqslant \tau_{allow}, \qquad \varphi = \int \frac{Tdx}{GI_p} \leqslant \varphi_{allow} \qquad (3.15)$$

where $\tau_{allow}$ is the allowable shearing stress, and $\varphi_{allow}$ is the allowable angle of twist.

**Example 3.3** A 1.8 m long tubular shaft of 45 mm outer diameter having a hollow circular section is to be made of steel for which $\tau_{allow}$ =75 MPa and $G$ =77 GPa, Fig. 3.13. Knowing that the angle of twist of the shaft must not exceed 5° when the shaft is subjected to an external couple of $M_e$= 900 N·m, determine the largest inner diameter $d$ which can be used in the design.

**Solution** Using $\tau_{max} = \frac{T\rho_{max}}{I_p} \leqslant \tau_{allow}$, we have

$$(I_p)_\tau \geqslant \frac{T\rho_{max}}{\tau_{allow}} = 2.70 \times 10^{-7} \text{ m}^4$$

Using $\varphi = \frac{Tl}{GI_p} \leqslant \varphi_{allow}$, we have

$$(I_p)_\varphi \geqslant \frac{Tl}{G\varphi_{allow}} = 2.41 \times 10^{-7} \text{ m}^4$$

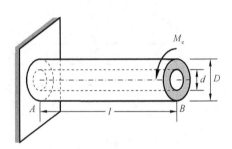

Fig. 3.13

Larger value of $I_p$ should be used in the design, thus we have

$$I_p = \max[(I_p)_\tau, (I_p)_\varphi] \geqslant 2.70 \times 10^{-7} \text{ m}^4$$

For a hollow circular shaft of inner diameter $d$ and outer diameter $D$, using $I_p = \frac{1}{32}\pi(D^4 - d^4)$, we have

$$d \leqslant \sqrt[4]{D^4 - \frac{32I_p}{\pi}} = 34 \text{ mm}$$

Therefore, the largest inner diameter $d$ which can be used in the design is equal to 34 mm.

# Problems

3.1  Knowing that $d$=30 mm and $D$=40 mm, Fig. 3.14, determine the external couple $M_e$ which causes a maximum shearing stress of 52 MPa in the hollow shaft.

3.2  Two external couples $M_B$ and $M_C$ are applied to the sections $B$ and $C$ of a variable-section solid shaft $ABC$, Fig. 3.15. Knowing that $D$=46 mm, $d$=30 mm, $M_B$=400 N·m, and $M_C$=300 N·m, in order to reduce the total mass of the shaft $ABC$, determine the smallest diameter of shaft $AB$ for which the largest shearing stress in the shaft is not increased.

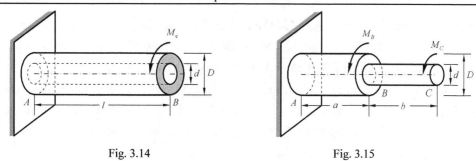

Fig. 3.14　　　　　　　　　　Fig. 3.15

3.3　The pipe AB has an outer diameter of 90 mm and a wall thickness of 6 mm. The solid rod BC has a diameter of 60 mm, Fig. 3.16. Knowing that both the pipe and the rod are made of steel for which the allowable shearing stress is 95 MPa, determine the largest external torque $M_e$ which may be applied at C.

3.4　The allowable stress is 25 MPa in the aluminum rod AB and 50 MPa in the brass rod BC, Fig. 3.17. Knowing that $M_e$=1.5 kN·m, determine the required diameter of rod AB and rod BC.

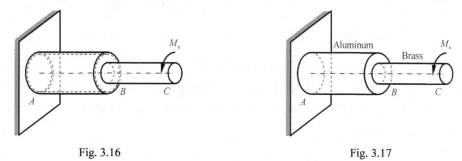

Fig. 3.16　　　　　　　　　　Fig. 3.17

3.5　For the aluminum shaft shown in Fig. 3.18, knowing that G=27 GPa, l=1.1 m, d=10 mm, and D=20 mm, determine (1) the external couple $M_e$ which causes an angle of twist of 5°, (2) the angle of twist caused by the same external couple $M_e$ in a solid cylindrical shaft of the same length and cross-sectional area.

3.6　While a steel shaft of the cross section rotates at 120 r/min, a stroboscopic measurement indicates that angle of twist is 2° in a 4m length, Fig. 3.19. Using G=77 GPa, determine the power being transmitted.

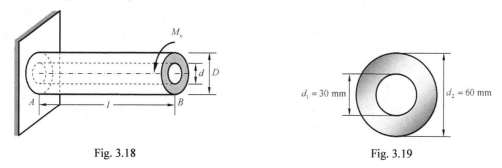

Fig. 3.18　　　　　　　　　　Fig. 3.19

3.7　A 1.8 m long tubular shaft of 45 mm inner diameter having a hollow circular section is to be made of steel for which $\tau_{allow}$=75 MPa and $G$=77 GPa, Fig. 3.20. Knowing that the angle of twist of the shaft must not exceed 5° when the shaft is subjected to an external couple of $M_e$=900 N·m, determine the smallest outer diameter $D$ which can be used in the design.

3.8  The design of a machine element requires a $D$=38 mm outer diameter shaft to transmit 45 kW, Fig. 3.21. (1) If the speed of rotation is 800 r/min, determine the maximum shearing stress in the solid shaft, Fig. 3.21(a). (2) If the speed of rotation can be increased 50% to 1200 r/min, determine the largest inner diameter of the hollow shaft, Fig. 3.21(b), for which the maximum shearing stress will be the same in each shaft.

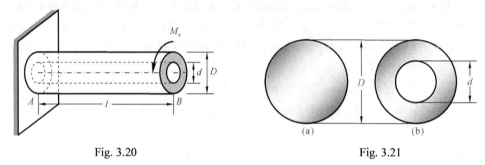

Fig. 3.20  Fig. 3.21

3.9  A 1.5 m long tubular steel shaft ($G$=77 GPa) of 40 mm outer diameter $D$ and 30 mm inner diameter $d$ is to transmit 120 kW, Fig. 3.22. Knowing that the allowable shearing stress is 65 MPa and that the angle of twist must not exceed 3°, determine the minimum rotating speed at which the shaft may rotate.

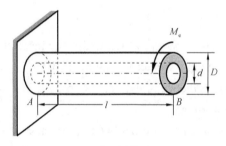

Fig. 3.22

# Chapter 4  Bending Internal Forces

A bar subjected to external forces that are perpendicular to the longitudinal axis of the bar, or external couples that lie in a plane containing the longitudinal axis of the bar, is called a beam in bending.

Beams are classified according to the way in which they are supported. Several types of frequently-used beams are shown in Fig. 4.1. A simply supported beam is pin-supported at one end and roller-supported at the other, Fig. 4.1(a). An overhanging beam is supported at two points and has one or both ends extended beyond the supports, Fig. 4.1(b). A cantilevered beam is fixed at one end and free at the other, Fig. 4.1(c). External loads commonly applied to beams may consist of concentrated forces, Fig. 4.1(a), distributed forces, Fig. 4.1(b), and concentrated couples, Fig. 4.1(c).

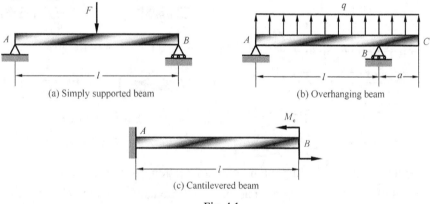

Fig. 4.1

The reactions at the supports of the above beams involve a total of only three unknowns and, therefore, can be determined by the equations of static equilibrium. Such beams are said to be statically determinate. If the reactions at the supports of the beams involve more than three unknowns and cannot be determined by the equations of static equilibrium alone, such beams are said to be statically indeterminate. The beam shown in Fig. 4.2 is statically indeterminate to the first degree.

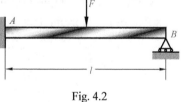

Fig. 4.2

## 4.1  Shearing-Force and Bending-Moment Diagrams

When external loads are applied to a beam, the beam will develop shearing forces and bending moments that, in general, vary from section to section along the axis of the beam, Fig. 4.3. Therefore, the shearing forces and bending moments are functions of the arbitrary position $x$ along the axis of the beam, and can be expressed as

$$V = V(x), \quad M = M(x) \tag{4.1}$$

These two equations are respectively called the shearing-force and bending-moment equations. According to the above equations, the shearing forces and bending moments along the axis of the beam can be plotted and represented by graphs, which are called shearing-force and bending-moment diagrams.

To draw the shearing-force and bending-moment diagrams of the beam, it is first necessary to establish a sign convention for the shearing forces and bending moments. Although the choice of a sign convention is arbitrary, the sign convention often used will be adopted for convenience. The shearing force that causes a clockwise rotation of the beam segment is said to be positive, Fig. 4.4(a). the bending moment that causes compression in the upper part of the beam segment is said to be positive, Fig. 4.4(b).

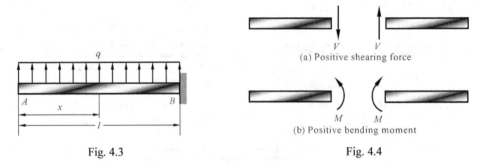

Fig. 4.3

Fig. 4.4

The shearing force and bending moment at any section of the beam can be determined by using the method of sections, i.e., by cutting the beam through the section $C$ where the shearing forces and bending moments are to be determined, Fig. 4.5(a), and considering the equilibrium of the beam segment located on the left side of the section, Fig. 4.5(b), or the right side of the section, Fig. 4.5(c).

Considering the equilibrium of the left portion $AC$ of the beam, Fig. 4.5(b), and using the equations of equilibrium, then we have

$$\sum F_y = 0: \quad qx - V = 0 \quad (0 \leqslant x < l)$$
$$\sum M_C = 0: \quad M - \frac{1}{2}qx^2 = 0 \quad (0 \leqslant x < l) \tag{4.2}$$

Solving Eq.(4.2) for $V$ and $M$, we obtain

$$V = qx \quad (0 \leqslant x < l)$$
$$M = \frac{1}{2}qx^2 \quad (0 \leqslant x < l) \tag{4.3}$$

After finding the shearing-force and bending-moment equations for the entire beam, next we can draw the shearing-force and bending-moment diagrams based on the equations obtained. The shearing-force and bending-moment diagrams for the beam shown in Fig. 4.5(a) are given in Fig. 4.6(b) and Fig. 4.6(c).

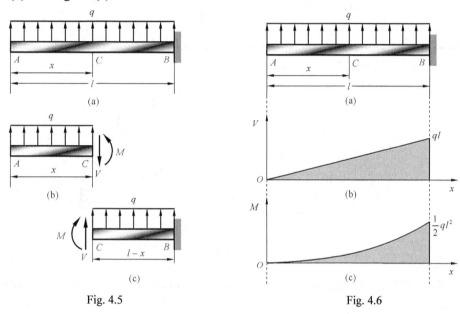

Fig. 4.5  Fig. 4.6

It can be seen from Fig. 4.6(b) and Fig. 4.6(c) that the section $B$ is a critical section (or dangerous section) of the beam due to the fact that this section is subjected to the maximum values of shearing forces and bending moments.

**Example 4.1** Draw the bending-moment diagram for the structure shown in Fig. 4.7(a) and determine the maximum bending moment.

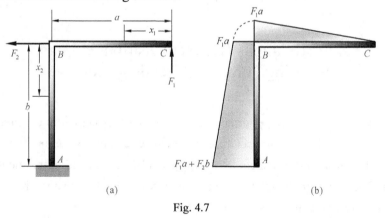

Fig. 4.7

**Solution** Using the method of sections and the condition of equilibrium, we can obtain the bending-moment equations:

$$M_1 = F_1 x_1 \quad (0 \leqslant x_1 \leqslant a)$$
$$M_2 = F_1 a + F_2 x_2 \quad (0 \leqslant x_2 < b)$$

It can also be seen from the bending-moment equations obtained that the bending-moment diagram is an oblique straight-line for each portion of the structure. The bending-moment diagram for the entire structure is shown in Fig. 4.7(b), and the maximum bending moment, which is located at section $A$, is equal to

$$M_{max} = F_1 a + F_2 b$$

**Example 4.2** Draw the bending-moment diagram for the structure shown in Fig. 4.8(a) and determine the maximum bending moment.

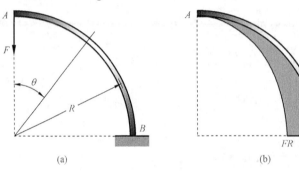

Fig. 4.8

**Solution** Using the method of sections and the condition of equilibrium, we can obtain the bending-moment equation:

$$M = FR\sin\theta \qquad \left(0 \leqslant \theta \leqslant \frac{1}{2}\pi\right)$$

The bending-moment diagram for the structure is shown in Fig. 4.8(b). The maximum bending moment at the fixed end is

$$M_{max} = FR$$

## 4.2 Relations Between Distributed Load, Shearing Force, and Bending Moment

When a beam carries several external loads, the method discussed in the preceding section for drawing shearing force and bending moment diagrams is quite cumbersome. The construction of the shearing-force diagram and, especially, of the bending-moment diagram will be greatly facilitated if certain relations existing between load intensity, shearing force, and bending moment are taken into consideration.

Consider a simply supported beam $AB$ subjected to a distributed load $q$ ($q$ is the intensity of distributed load, i.e., the force per unit length along the axis of the beam), Fig. 4.9(a), and let $C$ and $D$ be two points of the beam at a distance $dx$ from each other. We now detach the portion of the beam $CD$ and draw its free-body diagram, Fig. 4.9(b).

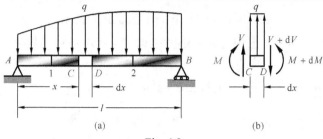

Fig. 4.9

### 1. Relations Between Distributed Load and Shearing Force

According to the condition of static equilibrium of forces in the vertical direction, we have

$$\sum F_y = 0: V - (V + dV) + q dx = 0 \tag{4.4}$$

i.e.,

$$dV = q dx$$

Dividing both members of Eq. (4.4) by $dx$, we obtain

$$\frac{dV}{dx} = q \tag{4.5}$$

Integrating Eq. (4.5) between points 1 and 2, we have

$$V_2 = V_1 + \int_{x_1}^{x_2} q dx \tag{4.6}$$

where $\int_{x_1}^{x_2} q dx$ is the resultant force of the distributed load between points 1 and 2.

### 2. Relations Between Distributed Load and Bending Moment

From the condition of static equilibrium of moments about $D$, we write

$$\sum M_D = 0: (M + dM) - M - V dx - \frac{1}{2} q (dx)^2 = 0 \tag{4.7}$$

i.e.,

$$dM = V dx + \frac{1}{2} q (dx)^2$$

Neglecting the second-order infinitesimal $(dx)^2$ in Eq. (4.7), and dividing both members of Eq. (4.7) by $dx$, we obtain

$$\frac{dM}{dx} = V \tag{4.8}$$

Integrating Eq. (4.8) between points 1 and 2, we write

$$M_2 = M_1 + \int_{x_1}^{x_2} V dx \tag{4.9}$$

where $\int_{x_1}^{x_2} V dx$ is the area of the shearing-force diagram between points 1 and 2.

Eq. (4.5), Eq. (4.6), Eq. (4.8), and Eq. (4.9) provide a convenient means for quickly drawing the shearing-force and bending-moment diagrams of a beam subjected to distributed loads.

**Example 4.3** Draw the shearing-force and bending-moment diagrams for the simply supported beam shown in Fig. 4.10(a) and determine the maximum values of the shearing force and bending moment.

**Solution** (1) Reactions at $A$ and $B$ are equal to $R_A = R_B = \dfrac{1}{2}ql$.

(2) Just to the right of the section $A$ of the beam, the shearing force is equal to $V_A^R = R_A = \dfrac{1}{2}ql$. Using Eq. (4.6), we can obtain the shearing force at any distance $x$ from $A$:

$$V = V_A^R + \int_0^x (-q)\mathrm{d}x = \frac{1}{2}ql - qx \quad (0 < x < l)$$

It can be seen that the shearing-force diagram is an oblique straight line, Fig. 4.10(b), and that the end $A$ or $B$ of the beam has a maximum absolute value of shearing forces.

(3) Using $M_A = 0$, the bending moment at any distance $x$ from $A$ can be obtained from Eq. (4.9):

$$M = M_A + \int_0^x V\mathrm{d}x = \frac{1}{2}qlx - \frac{1}{2}qx^2 \quad (0 \leqslant x \leqslant l)$$

It can also be seen that the bending-moment diagram is a parabola, Fig. 4.10(c), and that the midsection of the beam is subjected to a maximum positive bending moment.

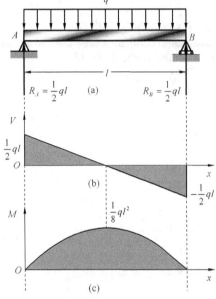

Fig. 4.10

## 4.3 Relations Between Concentrated Load, Shearing Force, and Bending Moment

Consider a simply supported beam $AB$ subjected to a concentrated force $F$ and a concentrated couple $M_e$, Fig. 4.11(a), and let $C$ and $D$ be two points of the beam at a distance $\mathrm{d}x$ from each other. Assuming that the concentrated force and couple are located at the midpoint of the segment $CD$, then we detach the segment $CD$ and draw its free-body diagram, Fig. 4.11(b).

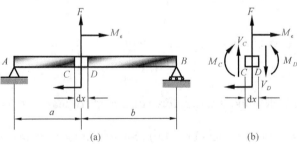

Fig. 4.11

## 1. Relations Between Concentrated Load and Shearing Force

According to the condition of static equilibrium of forces in the vertical direction, we have

$$\sum F_y = 0: \quad V_C - V_D + F = 0 \tag{4.10}$$

i.e.,
$$V_D = V_C + F \tag{4.11}$$

It can be seen that a sudden change of the shearing-force diagram will happen on the section where a concentrated force is applied.

## 2. Relations Between Concentrated Load and Bending Moment

From the condition of static equilibrium of moments about $D$, we write

$$\sum M_D = 0: \quad M_D - M_C - V_C dx - M_e - F\left(\frac{1}{2}dx\right) = 0 \tag{4.12}$$

Neglecting the infinitesimal $dx$ in Eq. (4.12), we obtain

$$M_D = M_C + M_e \tag{4.13}$$

It can also be seen that the bending-moment diagram will have a sudden change at the section subjected to a concentrated couple.

Eq. (4.11) and Eq. (4.13) also provide a convenient means for quickly drawing the shearing-force and bending-moment diagrams of a beam under the action of concentrated forces and/or couples.

**Example 4.4** Draw the shearing-force and bending-moment diagrams for the simply supported beam shown in Fig. 4.12(a) and determine the maximum values of the shearing force and bending moment.

**Solution** (1) Using the equilibrium condition of the entire beam, we have

$$R_A = \frac{3}{4}F, \quad R_B = \frac{1}{4}F$$

(2) The shearing force, close to the right of the section $A$ of the beam, is equal to $V_A^R = R_A = \frac{3}{4}F$. Since no any external load is applied to the segment $AC$, the shearing force over the segment $AC$ is a constant, and the corresponding shearing-force diagram is a horizontal straight line. Using $V_C^L = V_A^R = \frac{3}{4}F$ and Eq. (4.11), we can obtain the shearing force close to the right of the section $C$ of the beam:

$$V_C^R = V_C^L + (-F) = -\frac{1}{4}F$$

Similarly, we can obtain that the shearing force over the segment $BC$ is a constant, and that the shearing-force diagram is also a horizontal straight line.

The shearing-force diagram for the entire beam is shown in Fig. 4.12(b), and the maximum

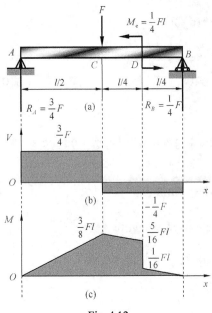

Fig. 4.12

shearing forces is equal to $V_{max} = \dfrac{3}{4}F$ over the segment $AC$.

(3) Using $M_A = 0$, the bending moment at the section $C$ can be obtained from Eq. (4.9):

$$M_C = M_A + \int_{x_A}^{x_C} V dx = \dfrac{3}{4}F \cdot \dfrac{1}{2}l = \dfrac{3}{8}Fl$$

Since the shearing force over the segment $AC$ is a constant, the corresponding bending-moment diagram is an oblique straight line. Similarly, the bending-moment diagram over the segment $CD$ is also an oblique straight line, and the bending moment just to the left of the section $D$ is equal to

$$M_D^L = M_C + \int_{x_C}^{x_D} V dx = \dfrac{5}{16}Fl$$

There is a concentrated couple on the section $D$, and hence the bending-moment diagram will have a sudden change at that section. Using Eq. (4.13), we have

$$M_D^R = M_D^L + (-M_e) = \dfrac{1}{16}Fl$$

Likewise, we can obtain the bending-moment diagram over the segment $DB$ and the bending moment on the section $B$:

$$M_B = M_D^R + \int_{x_D}^{x_B} V dx = 0$$

The bending-moment diagram for the entire beam is shown in Fig. 4.12(c), and the maximum bending moment is located on the section $C$ and equal to $M_{max} = \dfrac{3}{8}Fl$.

## Problems

4.1   Draw the shearing-force and bending-moment diagrams for the beam and loading shown in Fig. 4.13.

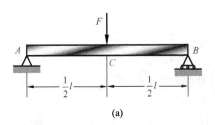

(a)

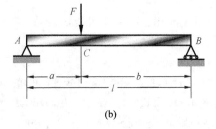

(b)

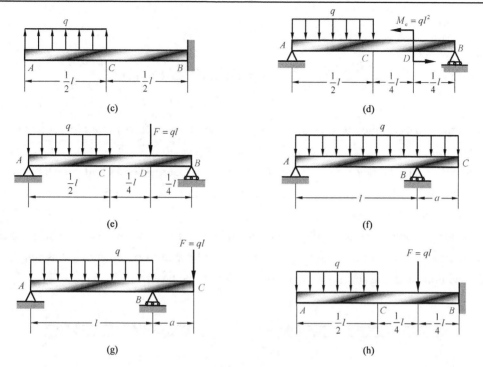

Fig. 4.13

# Chapter 5  Bending Stresses

External loads applied to a beam usually include external forces that are perpendicular to the longitudinal axis of the beam and external couples that lie in a plane containing the longitudinal axis of the beam. One of the effects of these external forces and couples acting on a beam is to produce both normal and shearing stresses on the cross section perpendicular to the longitudinal axis of the beam.

If external couples are applied to the beam and no external forces act on the beam, then the bending is called pure bending, Fig. 5.1(a). A beam subjected to pure bending has only normal stresses without shearing stresses on the cross section of the beam.

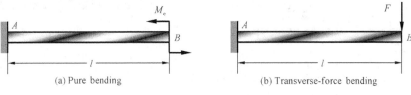

(a) Pure bending  (b) Transverse-force bending

Fig. 5.1

Bending produced by the external forces that do not form couples is called transverse-force bending, ordinary bending, or nonuniform bending, Fig. 5.1(b). A beam subjected to transverse-force bending has both normal and shearing stresses on the cross section of the beam.

## 5.1  Normal Stresses on Cross Section in Pure Bending

Consider a prismatic beam $AB$ possessing a longitudinal plane of symmetry and subjected to equal and opposite external couples of magnitude $M_e$ acting in the longitudinal plane of symmetry, Fig. 5.2.

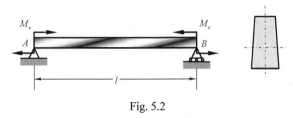

Fig. 5.2

Suppose that the beam is divided into a large number of small elements with faces respectively perpendicular to the three coordinate planes, Fig. 5.3(a). These elements will be deformed as shown in Fig. 5.3(b) when the beam is subjected to pure bending.

The only nonzero stress component exerted on each of the elements is the normal stress on the faces perpendicular to the longitudinal axis of the beam. Thus, each point of a beam in pure bending is in a state of uniaxial stress. For a positive bending moment, the top and bottom surfaces of the beam are observed, respectively, to decrease and increase in length. Therefore, the normal strain and the normal stress are negative in the upper portion of the beam (compression) and positive in the lower portion (tension).

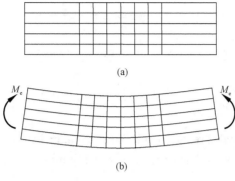

Fig. 5.3

It follows from the above analysis that there must exist a surface parallel to the top and bottom surfaces of the beam, where the normal strain and the normal stress are zero. This surface is called the neutral surface, Fig. 5.4(a), and intersects a transverse section along a straight line called the neutral axis of the section, Fig. 5.4(b). The origin of coordinates will be selected on the neutral surface so that the distance from any point to the neutral surface will be measured by its coordinate $y$.

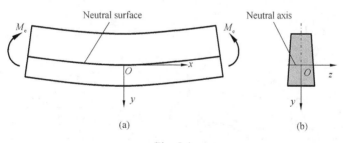

Fig. 5.4

### 1. Deformation Condition

Denoting by $\rho$ the radius of the neutral surface $aa$ and by $\theta$ the central angle corresponding to the neutral surface $aa$, Fig. 5.5.

Observing that the length of the neutral surface $aa$ after deformation is equal to the length $l$ of the undeformed neutral surface, then we have

$$l = \rho\theta \tag{5.1}$$

Considering now the arc $bb$ located at a distance $y$ below the neutral surface, we note that its length after deformation is

$$l'_{bb} = (\rho + y)\theta \tag{5.2}$$

Since the original length of the arc $bb$ was equal to $l_{bb} = l$, the deformation of the arc $bb$ is

$$\delta = l'_{bb} - l_{bb} = y\theta \tag{5.3}$$

The longitudinal normal strain $\varepsilon$ along the arc $bb$ can be obtained by dividing $\delta$ by the original length $l_{bb}$ of the arc $bb$. Therefore, we obtain

$$\varepsilon = \frac{\delta}{l_{bb}} = \frac{y}{\rho} \tag{5.4}$$

Eq. (5.4) shows that the longitudinal normal strain $\varepsilon$ varies linearly with the distance $y$ from the neutral surface.

### 2. Stress-Strain Relation

Assume that the normal stresses in the beam remain below the proportional limit, and Hooke's law for uniaxial stress applies, i.e., $\sigma = E\varepsilon$. From Eq. (5.4), we have

$$\sigma = \frac{E}{\rho} y \tag{5.5}$$

where $E$ is the modulus of elasticity of the material. The above equation shows that, in the elastic range, the normal stress varies linearly with the distance from the neutral surface, Fig. 5.6.

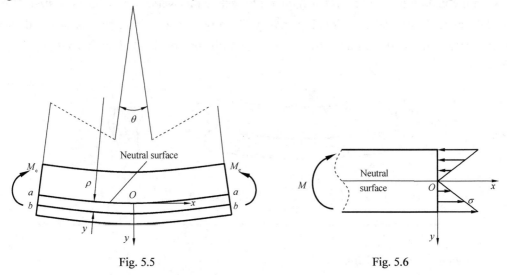

Fig. 5.5        Fig. 5.6

### 3. Equilibrium Condition

According to the condition of static equilibrium in $x$ direction, we have

$$\sum F_x = 0: \quad N = \int \sigma dA = \frac{E}{\rho} \int y dA = 0 \tag{5.6}$$

from which it follows that

$$\int y dA = 0 \tag{5.7}$$

This equation shows that, for a beam subjected to pure bending, as long as the stresses remain in the linearly elastic range, the neutral axis passes through the centroid of the section.

Based on the condition of static equilibrium about the neutral axis, we have

$$\sum M_z = 0: M = \int y\sigma dA = \frac{E}{\rho}\int y^2 dA \tag{5.8}$$

from which it follows that

$$\frac{1}{\rho} = \frac{M}{E\int y^2 dA} = \frac{M}{EI} \tag{5.9}$$

where $I = \int y^2 dA$ is the moment of inertia of the cross section with respect to the centroidal axis. Substituting $1/\rho$ from Eq. (5.9) into Eq (5.5), we obtain the normal stress $\sigma$ at any distance $y$ from the neutral axis:

$$\sigma = \frac{My}{I} \tag{5.10}$$

The normal stress is compressive ($\sigma < 0$) above the neutral axis ($y < 0$) when the bending moment $M$ is positive, and tensile ($\sigma > 0$) when $M$ is negative.

According to Eq. (5.10), the maximum normal stress is equal to $\sigma_{max} = \frac{My_{max}}{I}$. We note that the ratio $I/y_{max}$ depends only upon the geometry of the cross section. This ratio is called the modulus of section and is denoted by $S$. From Eq. (5.10), we have

$$\sigma_{max} = \frac{My_{max}}{I} = \frac{M}{S} \tag{5.11}$$

For a beam with a rectangular cross section of width $b$ and depth $h$, we have $S = \frac{1}{6}bh^2$.

**Example 5.1** Knowing that the external couple acts in a vertical plane, Fig. 5.7, determine the stress at point $P$ and point $Q$ on the midsection $C$ of the beam.

**Solution** Sectioning the beam at the midsection $C$ into two parts and considering the equilibrium of the right part, we obtain the bending moment equal to the external couple applied to the section $B$, i.e., $M=M_e$.

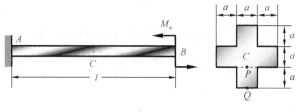

Fig. 5.7

For the section given in Fig. 5.7, the moment of inertia of the cross section about the neutral axis can be given as

$$I = \frac{1}{12}a^4 + \frac{1}{12}a(3a)^3 + \frac{1}{12}a^4 = \frac{29}{12}a^4$$

Using Eq. (5.10), the stresses at points $P$ and $Q$ can be obtained by

$$\sigma_P = \frac{My_P}{I} = \frac{M_e\left(\frac{1}{2}a\right)}{\frac{29}{12}a^4} = \frac{6M_e}{29a^3}, \quad \sigma_Q = \frac{My_Q}{I} = \frac{M_e\left(\frac{3}{2}a\right)}{\frac{29}{12}a^4} = \frac{18M_e}{29a^3}$$

## 5.2 Normal and Shearing Stresses on Cross Section in Transverse-Force Bending

A transverse loading applied to a beam will result in normal and shearing stresses in each transverse section of the beam. The normal stresses are created by the bending moment in that section and the shearing stresses by the shearing force.

**1. Normal Stresses**

The formula for determining normal stresses on the cross section of a beam in pure bending is still valid for a beam subjected to transverse loadings as long as the beam under transverse loads is slender, Fig. 5.8, and the normal stresses in the beam subjected to transverse-force bending can be expressed as

$$\sigma = \frac{My}{I} \qquad (5.12)$$

where the bending moment $M$ will vary from section to section, i.e., $M = M(x)$ is a function of the position $x$ under the action of transverse loads.

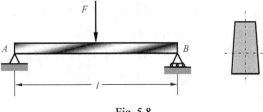

Fig. 5.8

The maximum normal stress for a beam in transverse-force bending can be written as

$$\sigma_{max} = \frac{M_{max} y_{max}}{I} = \frac{M_{max}}{S} \qquad (5.13)$$

**2. Shearing Stresses**

Since the dominant criterion in the design of a beam for strength is the maximum value of the normal stress in the beam, the determination of the normal stresses is quite important in the design of a beam. Shearing stresses, however, can also be important in the design of a short, stubby beam.

Considering a prismatic beam *AB* with a vertical plane of symmetry that supports external loads (a concentrated force *F*, a concentrated couple $M_e$, and a distributed load *q*). At a distance *x* from end *A* we detach from the beam an element *abcd* of length d*x* extending across the width of the beam from the bottom surface of the beam to a horizontal plane located at a distance *y* from the neutral axis, Fig. 5.9.

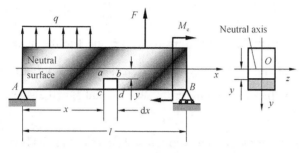

Fig. 5.9

When d*x* approaches zero, the shearing stresses on the left and right faces of the element are equal in magnitude. As long as the width of the cross section of the beam remains small compared to its depth, the shearing stress can be considered to be a constant in width. Assuming that the shearing stress at the points located a distance *y* from the neutral axis of the cross section, on the left or right face of the element is denoted by $\tau$, then the shearing stress on the top face of the element is also equal to $\tau$ according to the reciprocal theorem of shearing stresses, Fig. 5.10(a).

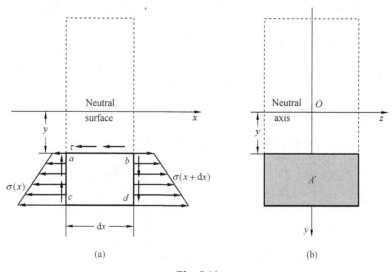

Fig. 5.10

Considering the equilibrium of the element *abcd* in the horizontal direction, then we obtain

$$\sum F_x = 0: \quad \int [\sigma(x+dx) - \sigma(x)]dA' - \tau(bdx) = 0 \qquad (5.14)$$

where $b$ is the width of the cross section of the beam and $A'$ is the shaded area of the section located below the line with a distance $y$ from the neutral axis, Fig. 5.10(b).

Using $\sigma(x+dx) - \sigma(x) = \dfrac{M(x+dx)y'}{I} - \dfrac{M(x)y'}{I} = \dfrac{dMy'}{I} = \dfrac{(Vdx)y'}{I}$, and solving Eq. (5.14) for $\tau$, we have

$$\tau = \frac{V}{Ib}\int y'dA' = \frac{VQ'}{Ib} \qquad (5.15)$$

where $Q' = \int y'dA'$ represents the first moment with respect to the neutral axis of the area $A'$ that is located below or above the line with a distance $y$ from the neutral axis, $I$ is the centroidal moment of inertia of the entire cross section, and $V$ is the shearing force on the cross section.

For a beam of rectangular section of width $b$ and depth $h$, Fig. 5.11(a), $I = \dfrac{1}{12}bh^3$, and

$$Q' = A'\bar{y}' = \left[b\left(\frac{1}{2}h - y\right)\right]\left[\frac{1}{2}\left(\frac{1}{2}h + y\right)\right] = \frac{1}{8}b(h^2 - 4y^2) \qquad (5.16)$$

then we have

$$\tau = \frac{3V}{2bh}\left[1 - \left(\frac{2y}{h}\right)^2\right] \qquad (5.17)$$

Eq. (5.17) shows that the distribution of shearing stresses in the transverse section of a rectangular beam is parabolic, Fig. 5.11(b). The shearing stresses are zero at the top and bottom surfaces of the cross section ($y = \pm\dfrac{1}{2}h$). Making $y = 0$, we obtain the maximum value of the shearing stress on the cross section of a rectangular beam:

$$\tau_{max} = \frac{3V}{2bh} = \frac{3V}{2A} \qquad (5.18)$$

where $A$ is the cross-sectional area of the rectangular beam.

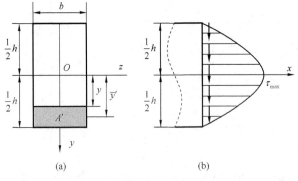

Fig. 5.11

For a beam of I-shaped section (I-beam for short), Fig. 5.12, the shearing stresses on the web vary only very slightly along vertical direction of the section, and the entire shearing force is almost carried by the web. Therefore, the maximum value of the shearing stresses in the cross section can be obtained by dividing the shearing force by the cross-sectional area of the web:

$$\tau_{max} = \frac{V}{A_w} = \frac{V}{b_0 h_0} \tag{5.19}$$

where $b_0$ and $h_0$ are the width and depth of the web of the I-beam.

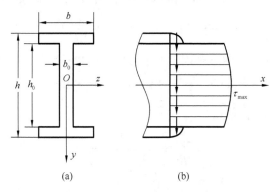

Fig. 5.12

For a beam of solid circular section, Fig. 5.13, the maximum shearing stress is located at the neutral axis of the cross section, and can be expressed as

$$\tau_{max} = \frac{4V}{3A} \tag{5.20}$$

where $A$ is the cross-sectional area of the solid circular beam.

For a hollow circular beam, Fig. 5.14, the maximum shearing stress is also located at the neutral axis of the cross section, and can be expressed as

$$\tau_{max} = \frac{2V}{A} \tag{5.21}$$

where $A$ is the cross-sectional area of the hollow circular beam.

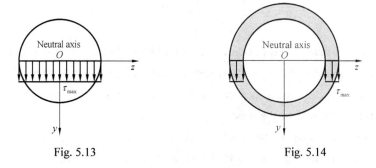

Fig. 5.13  Fig. 5.14

**Example 5.2** Knowing that the external force acts in a vertical plane, Fig. 5.15, determine the shearing stress at point $P$ on the midsection $C$ of the beam.

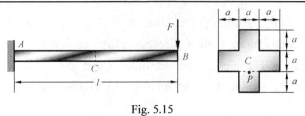

Fig. 5.15

**Solution** Using the method of sections, the shearing force is equal to the external force applied to the section B, i.e., $V=F$. For the given section, the moment of inertia of the entire section about the neutral axis is equal to $I = \frac{29}{12}a^4$, and for the point P, the first moment with respect to the neutral axis of the area that is located below the horizontal line with a distance $\frac{1}{2}a$ from the neutral axis is equal to $Q' = A'\bar{y}' = a^3$. Therefore, the shearing stress at point P on the cross section can be obtained by

$$\tau_P = \frac{VQ'}{Ib} = \frac{Fa^3}{\left(\frac{29}{12}a^4\right)a} = \frac{12F}{29a^2}$$

## 5.3  Design of Bending Beam

The design of a bending beam is usually controlled by the maximum absolute value $|M|_{max}$ of the bending moment that will occur in the beam. The largest normal stress $\sigma_{max}$ in the beam is found at the top or bottom surface of the beam in the critical section where $|M|_{max}$ occurs and can be obtained by

$$\sigma_{max} = \frac{|M|_{max} y_{max}}{I} = \frac{|M|_{max}}{S} \tag{5.22}$$

A safe design requires that $\sigma_{max} \leq \sigma_{allow}$, where $\sigma_{allow}$ is the allowable stress for the material used. Substituting $\sigma_{allow}$ for $\sigma_{max}$ in Eq. (5.22), and solving for S, we obtain

$$S \geq \frac{|M|_{max}}{\sigma_{allow}} \tag{5.23}$$

In the design of a beam, a proper procedure should lead to the most economical design. This means that, among beams of the same type and the same material, and other things being equal, the beam with the smallest mass per unit length/the smallest cross-sectional area should be selected, since this beam will be the least expensive.

**Example 5.3**  Knowing that the allowable normal stress for the steel used is 165 MPa, select an I-steel beam to support the external load as shown in Fig. 5.16.

**Solution**  The bending moment is maximum at section C and $M_{max} = \frac{1}{4}Fl = 125 \text{ kN} \cdot \text{m}$.

Thus the minimum modulus of section is

$$S_{min} = \frac{|M|_{max}}{\sigma_{allow}} = 757.58 \text{ cm}^3$$

Referring to the table of Shape Steels in Appendix II, we choose a group of I-steel beams having a modulus of section at least as large as $S_{min}$, Tab. 5.1.

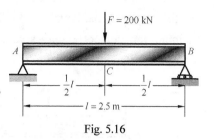

Fig. 5.16

Tab. 5.1

| Designation | Mass per unit length/(kg/m) | Section modulus/cm³ |
| --- | --- | --- |
| 32c | 62.765 | 760 |
| 36a | 60.037 | 875 |
| 36b | 65.689 | 919 |

From Tab. 5.1, it can be seen that both 32c and 36a I-steel beams can be chosen for a safe design. However, the most economical design is to choose 36a I-steel beam since its mass per unit length has only 60.037 kg/m even though it has a large modulus of section, compared to 32c I-steel beam.

The total weight of the beam chosen is 1.47 kN. This weight is small compared to the 200 kN external load and can be neglected in the design.

# Problems

**5.1** Knowing that the external couple acts in a vertical plane, Fig. 5.17, determine the stress at point $P_1$ point $P_2$ on the midsection $C$ of the beam.

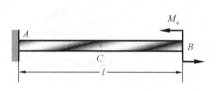 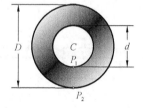

Fig. 5.17

**5.2** Knowing that the I-beam is made of steel for which the allowable stress $\sigma_{allow}$=150 MPa, Fig. 5.18, determine the largest couple that can be applied to the beam when it is bent about the neutral axis. Neglect the effect of fillets.

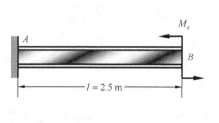 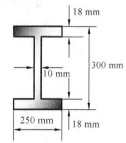

Fig. 5.18

**5.3** An external couple is applied to a T-section beam shown in Fig. 5.19. Determine the maximum tensile and compressive stresses in the beam when it is bent about the neutral axis. Neglect the effect of fillets.

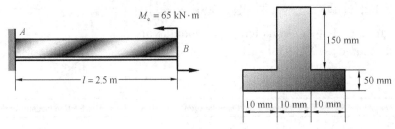

Fig. 5.19

**5.4** Knowing that the external force acts in a vertical plane, Fig. 5.20, determine the shearing stress at point $P$ on the midsection $C$ of the beam.

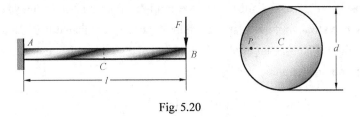

Fig. 5.20

**5.5** Knowing that the external force acts in a vertical plane, Fig. 5.21, determine the shearing stress at point $P$ on the midsection $C$ of the beam.

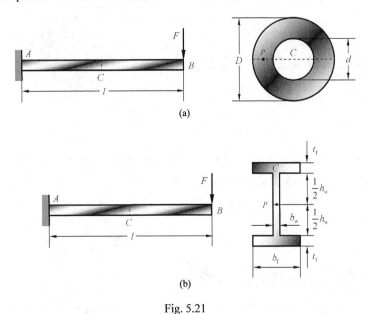

Fig. 5.21

**5.6** Select an I-beam to support the external load as shown in Fig. 5.22. The allowable normal stress for the steel used is 165 MPa.

5.7 Knowing that the allowable normal stress for the steel used is 165 MPa, select the most economical I-beam to support the external loads as shown in Fig. 5.23.

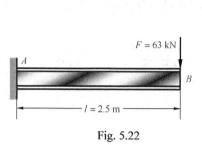

Fig. 5.22

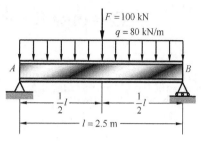

Fig. 5.23

# Chapter 6  Bending Deformation

A prismatic beam subjected to pure bending, Fig. 6.1, is bent into an arc of circle and that, within the linearly elastic range, the curvature of the neutral surface can be expressed as

$$\frac{1}{\rho} = \frac{M}{EI} \tag{6.1}$$

where the bending moment $M$ is a constant, and $EI$ is the flexural/bending rigidity of the beam.

When a slender beam is subjected to a transverse loading, Fig. 6.2, Eq. (6.1) remains valid for any given transverse section. However, both the bending moment and the curvature of the neutral surface will vary from section to section. Denoting by $x$ the distance of the section from the left end of the beam, we have

$$\frac{1}{\rho} = \frac{M(x)}{EI} \tag{6.2}$$

where the bending moment $M(x)$ is a function of $x$.

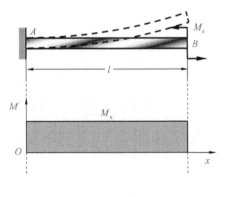

Fig. 6.1

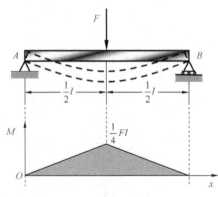

Fig. 6.2

The curvature of the neutral surface of the beam after deformation can be expressed mathematically as

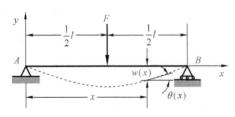

Fig. 6.3

$$\frac{1}{\rho} = \frac{w''}{[1+(w')^2]^{3/2}} \tag{6.3}$$

where $w = w(x)$ is the deflection function, Fig. 6.3, and $w'$ and $w''$ are the first and second derivatives of the function $w = w(x)$. For a small deformation beam, the slope $w'$ is very small, and its square is negligible compared to unity. We have, therefore,

$$\frac{1}{\rho} = w'' \tag{6.4}$$

Substituting $\rho$ from Eq. (6.4) into Eq. (6.2), we obtain

$$w'' = \frac{M(x)}{EI} \tag{6.5}$$

Eq. (6.5) is called the differential equation of the deflection curve for a beam in bending.

## 6.1 Method of Integration

Integrating Eq. (6.5) in $x$, we have

$$w' = \int \frac{M(x)}{EI} dx + C_1 \tag{6.6}$$

where $C_1$ is a constant of integration. Denoting by $\theta(x)$ the angle, measured in radians, that the tangent to the neutral surface after deformation forms with the horizontal, Fig. 6.3, and recalling that this angle is very small, we have

$$w' = \tan\theta \approx \theta \tag{6.7}$$

Thus, we write Eq. (6.6) in the alternative form:

$$\theta = \int \frac{M(x)}{EI} dx + C_1 \tag{6.8}$$

Integrating Eq. (6.8) in $x$, we have

$$w = \int \left( \int \frac{M(x)}{EI} dx \right) dx + C_1 x + C_2 \tag{6.9}$$

where $C_2$ is a second constant.

The constants $C_1$ and $C_2$ are determined from the boundary conditions. Three types of supports often used and their boundary conditions are given in Fig. 6.4.

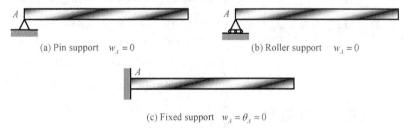

(a) Pin support  $w_A = 0$

(b) Roller support  $w_A = 0$

(c) Fixed support  $w_A = \theta_A = 0$

Fig. 6.4

**Example 6.1** For the loading shown in Fig. 6.5, determine (1) the equation of the deflection curve for the cantilevered beam $AB$, (2) the deflection and slope at the free end.

**Solution** Considering the $x$ section of the beam,

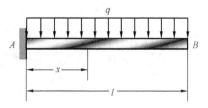

Fig. 6.5

Fig. 6.5, then we have bending moment $M(x) = -\frac{1}{2}q(l-x)^2$ and the differential equation of the deflection curve $w'' = -\frac{q(l-x)^2}{2EI}$. Using the method of integration, we obtain

$$\theta = \int \left[-\frac{q(l-x)^2}{2EI}\right]dx + C_1 = -\frac{q}{6EI}(3l^2x - 3lx^2 + x^3) + C_1$$

$$w = \int\left\{\int\left[-\frac{q(l-x)^2}{2EI}\right]dx\right\}dx + C_1x + C_2 = -\frac{q}{24EI}(6l^2x^2 - 4lx^3 + x^4) + C_1x + C_2$$

Using the boundary conditions $\theta = w = 0$ at $x = 0$, we obtain $C_1 = 0$ and $C_2 = 0$. Thus the equation of the deflection curve can be given as

$$w = -\frac{q}{24EI}(6l^2x^2 - 4lx^3 + x^4)$$

From the above equation, we can obtain the deflection and slope at the free end of the beam:

$$w_B = w_{x=l} = -\frac{ql^4}{8EI}, \quad \theta_B = \theta_{x=l} = -\frac{ql^3}{6EI}$$

## 6.2 Method of Superposition

When a beam is subjected to several loads, Fig. 6.6(a), it is often convenient to compute separately the deflection and slope caused by each of the given loads. The deflection and slope due to the combined loads are then obtained by applying the principle of superposition and adding the values of the deflection and slope corresponding to the various loads. Most structural and mechanical engineering handbooks include tables giving the deflections and slopes of beams for various loadings and supports.

Consider a simple supported beam subjected to two loads, Fig. 6.6(a). Using the principle of superposition, the deflection and slope at any section of the beam can be obtained by superposing the deflections and slopes caused respectively by the concentrated load, Fig. 6.6(b), and by the distributed load, Fig. 6.6(c). Therefore, the deflection and slope at any section of the beam subjected to combined loads can be expressed as

$$w = w_F + w_q, \quad \theta = \theta_F + \theta_q \qquad (6.10)$$

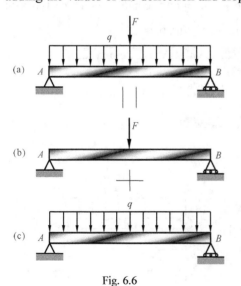

Fig. 6.6

**Example 6.2** For the beam and loading shown in Fig. 6.7, determine the deflection at the midsection $C$, and the slope at the end $B$.

**Solution** For the concentrated load $F$ and from Appendix III, we have

$$(w_C)_F = -\frac{Fl^3}{48EI}, \quad (\theta_B)_F = \frac{Fl^2}{16EI}$$

Similarly, for the distributed load $q$ and from Appendix III, we have

$$(w_C)_q = -\frac{5ql^4}{384EI}, \quad (\theta_B)_q = \frac{ql^3}{24EI}$$

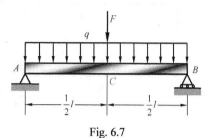

Fig. 6.7

Using the principle of superposition, then we obtain the deflection at the midsection $C$ and the slope at the end $B$:

$$w_C = (w_C)_F + (w_C)_q = -\left(\frac{Fl^3}{48EI} + \frac{5ql^4}{384EI}\right), \quad \theta_B = (\theta_B)_F + (\theta_B)_q = \frac{Fl^2}{16EI} + \frac{ql^3}{24EI}$$

## 6.3 Statically Indeterminate Beam

Consider a prismatic beam $AB$, Fig. 6.8, which has a fixed end at $A$ and is supported by a roller at $B$. We note that the reactions involve four unknowns, while only three equilibrium conditions are available. Therefore, the beam is statically indeterminate to the first degree.

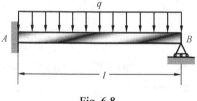

Fig. 6.8

The method of superposition can be used to determine the reactions at the supports of a statically indeterminate beam. For the statically indeterminate beam to the first degree shown in Fig. 6.8, we designate one of the reactions, say the reaction at $B$, as redundant and eliminate the corresponding support. The redundant reaction is then treated as an unknown load that, together with the other loads, must produce deformation which is compatible with the original support. The deflection and slope at the point where the support has been eliminated are obtained by computing separately the deflection and slope caused by the given loads and by the redundant reaction, and by superposing the results obtained.

**Example 6.3** For the beam and loading shown in Fig. 6.9(a), determine the reaction at the roller support.

**Solution** Considering the reaction at $B$ as redundant and releasing the beam from support $B$, Fig. 6.9(a), then the reaction $R_B$ is considered as a redundant reaction and will be determined from the condition that the deflection of the beam at $B$ must be zero.

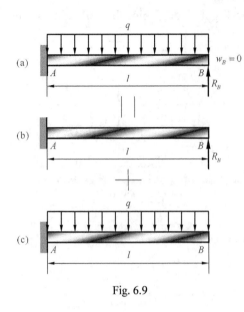

Fig. 6.9

From Appendix III, the deflection caused at $B$ by the redundant reaction $R_B$, Fig. 6.9(b), can be given as

$$(w_B)_{R_B} = \frac{R_B l^3}{3EI}$$

Similarly, the deflection produced at $B$ by the uniformly distributed load $q$, Fig. 6.9(c), can be expressed as

$$(w_B)_q = -\frac{ql^4}{8EI}$$

Using the principle of superposition, the total deflection at $B$ can be written as

$$w_B = (w_B)_{R_B} + (w_B)_q = \frac{R_B l^3}{3EI} - \frac{ql^4}{8EI}$$

Since the total deflection of the beam at $B$ must be zero, then we have

$$\frac{R_B l^3}{3EI} - \frac{ql^4}{8EI} = 0$$

Solving for $R_B$, we obtain the reaction at the roller support $B$:

$$R_B = \frac{3}{8}ql$$

## Problems

6.1 For the loading shown in Fig. 6.10, determine (1) the equation of the deflection curve for the cantilevered beam $AB$, (2) the deflection and slope at the free end.

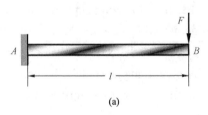
(a)

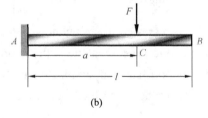
(b)

Fig. 6.10

6.2 For the loading shown in Fig. 6.11, determine (1) the equation of the deflection curve for the simple supported beam $AB$, (2) the deflection at the midsection $C$ and slope at the section $B$.

6.3 For the beam and loading shown in Fig. 6.12, determine (1) the deflection at the midsection $C$, (2) the slope at the end $B$.

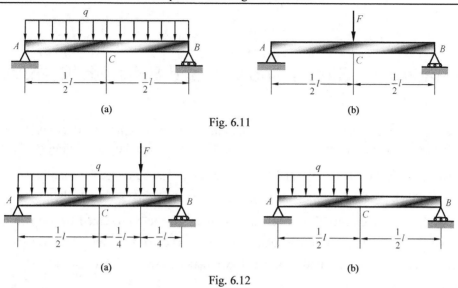

Fig. 6.11

Fig. 6.12

6.4  For the cantilever beam and loading shown in Fig. 6.13, determine the deflection and slope at point $B$.

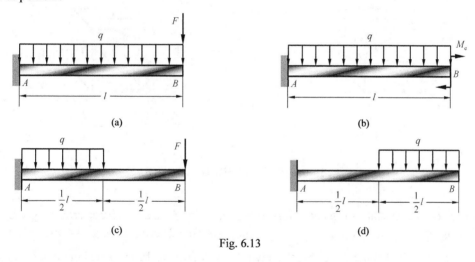

Fig. 6.13

6.5  For the beam and loading shown in Fig. 6.14, determine the reaction at the roller support $B$.

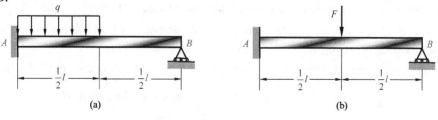

Fig. 6.14

# Chapter 7  Stress Analysis and Strength Theories

The most general state of stress at a given point $P$ may be represented by six components. Three of these components, $\sigma_x$, $\sigma_y$, and $\sigma_z$, define the normal stresses acting on the faces of a small cubic element centered at $P$ and having the same orientation as the coordinate axes, Fig. 7.1, and the other three, $\tau_{xy}(\tau_{yx}=\tau_{xy})$, $\tau_{yz}(\tau_{zy}=\tau_{yz})$, and $\tau_{zx}(\tau_{xz}=\tau_{zx})$, define the shearing stresses on the same element. The same state of stress can be represented by a different set of components if the coordinate axes are rotated. The discussion of the stress state will mainly deal with plane stress, i.e., with a situation in which two parallel faces of the cubic element are free of any stress, Fig. 7.2.

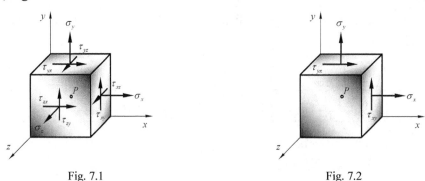

Fig. 7.1      Fig. 7.2

## 7.1   Stress Transformation

Consider that a state of plane stress exists at point $P$ and is characterized by the stress components $\sigma_x$, $\sigma_y$, and $\tau_x(\tau_y=\tau_x)$, Fig. 7.3(a). We now determine the stress components $\sigma_{x'}$, $\sigma_{y'}$, and $\tau_{x'}(\tau_{y'}=\tau_{x'})$ after the element shown in Fig. 7.3(a) has been rotated through an angle $\theta$ counterclockwise about the $z$ axis, Fig. 7.3(b).

Consider an element with faces respectively perpendicular to the $x$, $y$, and $x'$ axes, Fig. 7.4. We observe that, if the area of the oblique face is denoted by $\Delta A$, the areas of the vertical and horizontal faces are respectively equal to $\Delta A \cos\theta$ and $\Delta A \sin\theta$.

Using the equations of static equilibrium along the $x'$ and $y'$ axes, respectively, we have

$$\sum F_{x'} = 0: \quad \sigma_{x'}\Delta A - \sigma_x(\Delta A\cos\theta)\cos\theta + \tau_x(\Delta A\cos\theta)\sin\theta$$
$$-\sigma_y(\Delta A\sin\theta)\sin\theta + \tau_y(\Delta A\sin\theta)\cos\theta = 0$$

$$\sum F_{y'} = 0: \quad -\tau_{x'}\Delta A + \sigma_x(\Delta A\cos\theta)\sin\theta + \tau_x(\Delta A\cos\theta)\cos\theta$$
$$-\sigma_y(\Delta A\sin\theta)\cos\theta - \tau_y(\Delta A\sin\theta)\sin\theta = 0$$

(7.1)

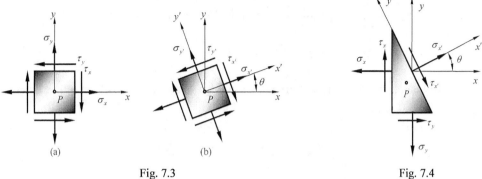

Fig. 7.3    Fig. 7.4

Solving for $\sigma_{x'}$ and $\tau_{x'}$, and using $\tau_y = \tau_x$, $\cos^2\theta = \frac{1}{2}[1+\cos(2\theta)]$, $\sin^2\theta = \frac{1}{2}[1-\cos(2\theta)]$, and $2\sin\theta\cos\theta = \sin(2\theta)$, we obtain

$$\sigma_{x'} = \frac{1}{2}(\sigma_x + \sigma_y) + \frac{1}{2}(\sigma_x - \sigma_y)\cos(2\theta) - \tau_x \sin(2\theta)$$

$$\tau_{x'} = \frac{1}{2}(\sigma_x - \sigma_y)\sin(2\theta) + \tau_x \cos(2\theta)$$

(7.2)

If $\sigma_{y'}$ is needed, it can be obtained by substituting $\theta + \frac{1}{2}\pi$ for $\theta$ into Eq. (7.2). This yields

$$\sigma_{y'} = \frac{1}{2}(\sigma_x + \sigma_y) - \frac{1}{2}(\sigma_x - \sigma_y)\cos(2\theta) + \tau_x \sin(2\theta)$$

(7.3)

**Example 7.1**  For the given stress state shown is Fig. 7.5(a), determine the normal and shearing stresses after the element has been rotated through 20° clockwise.

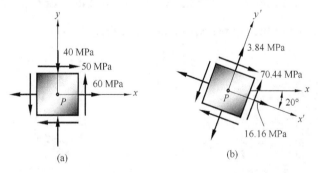

Fig. 7.5

**Solution**  According to the reference system given in Fig. 7.5(a), then we have

$\sigma_x = 60$ MPa,    $\sigma_y = -40$ MPa,    $\tau_x = -50$ MPa,    $\theta = -20°$

Using Eq. (7.2) and Eq. (7.3), we obtain

$$\sigma_{x'} = \frac{1}{2}(\sigma_x + \sigma_y) + \frac{1}{2}(\sigma_x - \sigma_y)\cos(2\theta) - \tau_x \sin(2\theta) = 16.16 \text{ MPa}$$

$$\sigma_{y'} = \frac{1}{2}(\sigma_x + \sigma_y) - \frac{1}{2}(\sigma_x - \sigma_y)\cos(2\theta) + \tau_x \sin(2\theta) = 3.84 \text{ MPa}$$

$$\tau_{x'} = \frac{1}{2}(\sigma_x - \sigma_y)\sin(2\theta) + \tau_x \cos(2\theta) = -70.44 \text{ MPa}$$

The normal and shearing stresses are shown in Fig. 7.5(b) after the element has been rotated through 20° clockwise.

## 7.2 Principal Stresses

Setting $\sigma_{\text{avg}} = \frac{1}{2}(\sigma_x + \sigma_y)$, $R = \sqrt{\left[\frac{1}{2}(\sigma_x - \sigma_y)\right]^2 + \tau_x^2}$, and using Eq. (7.2), we have

$$(\sigma_{x'} - \sigma_{\text{avg}})^2 + \tau_{x'}^2 = R^2 \tag{7.4}$$

which is the equation of a circle of radius $R$ centered at the point $C$ of abscissa $\sigma_{\text{avg}}$ and ordinate 0, Fig. 7.6. This circle is called Mohr's circle or the stress circle.

The two points $A$ and $B$ where the circle of Fig. 7.6 intersects the horizontal axis are of special interest: point $A$ corresponds to the maximum value $\sigma_{\max}$ of the normal stress, while point $B$ corresponds to the minimum value $\sigma_{\min}$. Besides, these both points correspond to zero value of the shearing stress. The values of the parameter $\theta$ which correspond to points $A$ and $B$ can be obtained by setting $\tau_{x'} = 0$ in Eq. (7.2), i.e.,

$$\tan(2\theta_p) = \frac{-2\tau_x}{\sigma_x - \sigma_y} \tag{7.5}$$

This equation defines two values of $\theta_p$ that are 90° apart. Either of these values can be used to determine the orientation of the corresponding element, Fig. 7.7.

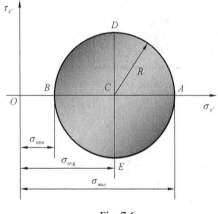
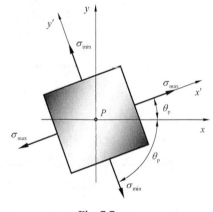

Fig. 7.6     Fig. 7.7

The planes containing the faces of the element obtained in this way are called the principal planes at point $P$, and the corresponding values $\sigma_{\max}$ and $\sigma_{\min}$ are called the in-plane principal stresses at point $P$. It is clear that no shearing stress is exerted on the principal planes. According to Eq. (7.4), we have

$$\sigma_{max,min} = \sigma_{avg} \pm R = \frac{1}{2}(\sigma_x + \sigma_y) \pm \sqrt{\left[\frac{1}{2}(\sigma_x - \sigma_y)\right]^2 + \tau_x^2} \tag{7.6}$$

There exist three perpendicular principal stresses at any point within a homogenous isotropic material subjected to plane stress. If $\sigma_{min} \geq 0$ in Eq. (7.6), then three principal stresses can be given as $\sigma_1 = \sigma_{max}$, $\sigma_2 = \sigma_{min}$, and $\sigma_3 = 0$; if $\sigma_{max} \geq 0 \geq \sigma_{min}$, three principal stresses can be expressed as $\sigma_1 = \sigma_{max}$, $\sigma_2 = 0$, and $\sigma_3 = \sigma_{min}$; if $\sigma_{max} \leq 0$, three principal stresses are $\sigma_1 = 0$, $\sigma_2 = \sigma_{max}$, and $\sigma_3 = \sigma_{min}$.

**Example 7.2** For the given stress state, Fig. 7.8(a), determine the principal stresses and the corresponding principal planes.

**Solution** Referring to the coordinate system given in Fig. 7.8(a), we have

$$\sigma_x = 60 \text{ MPa}, \quad \sigma_y = -40 \text{ MPa}, \quad \tau_x = -50 \text{ MPa}$$

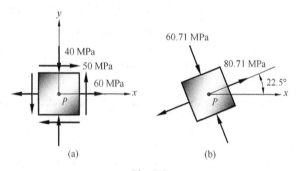

Fig. 7.8

Using Eq. (7.6), we obtain the in-plane principal stresses as follows:

$$\sigma_{max} = \frac{1}{2}(\sigma_x + \sigma_y) + \sqrt{\left[\frac{1}{2}(\sigma_x - \sigma_y)\right]^2 + \tau_x^2} = 80.71 \text{ MPa}$$

$$\sigma_{min} = \frac{1}{2}(\sigma_x + \sigma_y) - \sqrt{\left[\frac{1}{2}(\sigma_x - \sigma_y)\right]^2 + \tau_x^2} = -60.71 \text{ MPa}$$

Using Eq. (7.5), we can obtain the orientations of the in-plane principal stresses:

$$\theta_p = \frac{1}{2}\arctan\frac{-2\tau_x}{\sigma_x - \sigma_y} = \begin{matrix} 22.5° & \text{(corresponding to } \sigma_{max}) \\ -67.5° & \text{(corresponding to } \sigma_{min}) \end{matrix}$$

The principal stresses and the corresponding principal planes are shown in Fig. 7.8(b). Therefore, three principal stresses for the given state of plane stress are equal to $\sigma_1 = 80.71$ MPa, $\sigma_2 = 0$, and $\sigma_3 = -60.71$ MPa.

## 7.3 Maximum Shearing Stress

Referring to the circle of Fig. 7.6, the points $D$ and $E$ located on the vertical diameter of the circle correspond to the largest and smallest values of the shearing stress $\tau_{x'}$. Since the abscissa of points $D$ and $E$ is $\sigma_{avg} = \frac{1}{2}(\sigma_x + \sigma_y)$, the values of the parameter $\theta$ corresponding to these points are obtained by setting $\sigma_{x'} = \frac{1}{2}(\sigma_x + \sigma_y)$ in Eq. (7.2). Thus, we have

$$\tan(2\theta_s) = \frac{\sigma_x - \sigma_y}{2\tau_x} \tag{7.7}$$

This equation defines two values of $\theta_s$ which are 90° apart. Either of these values can be used to determine the orientation of the element corresponding to the maximum in-plane shearing stress, Fig. 7.9.

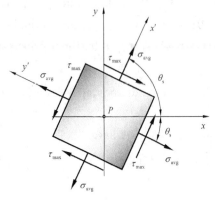

Fig. 7.9

From Eq. (7.4), we have

$$(\tau_{max})_{in\text{-}plane} = R = \sqrt{\left[\frac{1}{2}(\sigma_x - \sigma_y)\right]^2 + \tau_x^2} \tag{7.8}$$

and the normal stress corresponding to the maximum in-plane shearing stress is

$$\sigma_{avg} = \frac{1}{2}(\sigma_x + \sigma_y) \tag{7.9}$$

Comparing to Eq. (7.5) and Eq. (7.7), we note that $\tan(2\theta_s)$ is the negative reciprocal of $\tan(2\theta_p)$, i.e., $\tan(2\theta_s) \cdot \tan(2\theta_p) = -1$. This means that the angles are 45° apart. We thus conclude that the planes of maximum in-plane shearing stress are at 45° to the principal planes.

We should note that the shearing stresses $\tau_{max} = \frac{1}{2}(\sigma_1 - \sigma_3)$ may be larger than the shearing stress $(\tau_{max})_{in\text{-}plane}$ defined by Eq. (7.8). This occurs when the principal stresses defined by Eq. (7.6) have the same sign, i.e., when they are either both tensile or both compressive.

**Example 7.3** For the given stress state, Fig. 7.10(a), determine (1) the maximum in-plane shearing stress, (2) the orientation of the planes of maximum in-plane shearing stress, (3) the normal stress on the planes of maximum in-plane shearing stress.

**Solution** For the coordinate system given in Fig. 7.10(a), we have

$$\sigma_x = 60 \text{ MPa}, \quad \sigma_y = -40 \text{ MPa}, \quad \tau_x = -50 \text{ MPa}$$

(1) Using Eq. (7.8), we obtain the maximum in-plane shearing stress:

$$(\tau_{max})_{in\text{-}plane} = \sqrt{\left[\frac{1}{2}(\sigma_x - \sigma_y)\right]^2 + \tau_x^2} = 70.71 \text{ MPa}$$

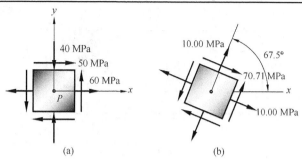

Fig. 7.10

(2) Using Eq. (7.7), we obtain the orientation of the planes of maximum in-plane shearing stress:

$$\theta_s = \frac{1}{2}\arctan\frac{\sigma_x - \sigma_y}{2\tau_x} = \frac{67.5°}{-22.5°}$$

(3) Using Eq. (7.9), we obtain the normal stress on the planes of maximum in-plane shearing stress:

$$\sigma_{avg} = \frac{1}{2}(\sigma_x + \sigma_y) = 10.00 \text{ MPa}$$

The maximum in-plane shearing stress and the normal stress on the planes of maximum in-plane shearing stress, as well as the orientation of these stresses, are shown in Fig. 7.10(b).

## 7.4 Pressure Vessels

Thin-walled pressure vessels provide an important application of plane stress. The analysis of stresses in thin-walled pressure vessels will be limited to the two types of vessels most frequently encountered: cylindrical pressure vessel and spherical pressure vessel.

**1. Cylindrical Pressure Vessel**

Consider a cylindrical vessel of inner diameter $d$ and wall thickness $t$ subjected to pressure $p$, Fig. 7.11. We now determine the stresses exerted on a small element of wall with sides respectively parallel and perpendicular to the longitudinal axis of the cylinder.

It is clear that no shearing stress is exerted on the element. The normal stresses $\sigma_1$ and $\sigma_2$ shown in Fig. 7.11 are therefore principal stresses. The stress $\sigma_1$ is known as the hoop or circumferential stress, and the stress $\sigma_2$ is called the longitudinal or axial stress.

To determine the longitudinal stress $\sigma_2$, considering the equilibrium of the free body shown in Fig. 7.12, we have

$$\sum F_x = 0: \quad \sigma_2(\pi dt) - p\left(\frac{1}{4}\pi d^2\right) = 0 \qquad (7.10)$$

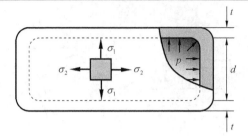

Fig. 7.11

Solving this equation for $\sigma_2$, then we obtain

$$\sigma_2 = \frac{pd}{4t} \tag{7.11}$$

In order to determine the hoop stress $\sigma_1$, we detach a portion of the vessel, Fig. 7.13. Using the condition of equilibrium, we have

$$\sum F_z = 0: \ \sigma_1(2t\Delta x) - p(d\Delta x) = 0 \tag{7.12}$$

Solving for $\sigma_1$, then we obtain the hoop stress:

$$\sigma_1 = \frac{pd}{2t} \tag{7.13}$$

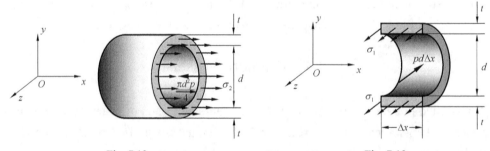

Fig. 7.12　　　　　　　　　　Fig. 7.13

Using Eq. (7.11) and Eq. (7.13), we have the maximum shearing stresses as follows:

$$(\tau_{max})_{\text{in-plane}} = \frac{1}{2}(\sigma_1 - \sigma_2) = \frac{pd}{8t}, \quad \tau_{max} = \frac{1}{2}(\sigma_1 - \sigma_3) = \frac{pd}{4t} \tag{7.14}$$

## 2. Spherical Pressure Vessel

We now consider a spherical vessel of inner diameter $d$ and wall thickness $t$ under pressure $p$, Fig. 7.14(a). Using the symmetry of vessel, the two in-plane principal stresses must be equal, i.e., $\sigma_1 = \sigma_2$.

To determine the stress $\sigma_1$ (or $\sigma_2$), we consider the free body shown in Fig. 7.14(b). From the condition of static equilibrium of the free body in the vertical direction, we have

$$\sigma_1(\pi dt) - p\left(\frac{1}{4}\pi d^2\right) = 0 \tag{7.15}$$

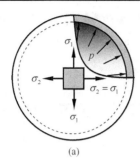 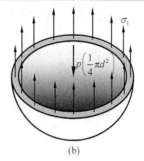

Fig. 7.14

Solving the above equation for $\sigma_1$, we have

$$\sigma_1 = \sigma_2 = \frac{pd}{4t} \tag{7.16}$$

Using Eq. (7.16), we obtain the maximum shearing stresses:

$$(\tau_{max})_{in\text{-}plane} = \frac{1}{2}(\sigma_1 - \sigma_2) = 0, \quad \tau_{max} = \frac{1}{2}(\sigma_1 - \sigma_3) = \frac{pd}{8t} \tag{7.17}$$

## 7.5  Generalized Hooke's Law

Consider an element with side lengths $dx$, $dy$, and $dz$, Fig. 7.15(a). When this element is subjected to principal stresses $\sigma_1$, $\sigma_2$, and $\sigma_3$, it will deform into a new element having side lengths $(1+\varepsilon_1)dx$, $(1+\varepsilon_2)dy$, and $(1+\varepsilon_3)dz$, Fig. 7.15(b), where $\varepsilon_1$, $\varepsilon_2$, and $\varepsilon_3$ are the normal strains corresponding to the principal stresses $\sigma_1$, $\sigma_2$, and $\sigma_3$, respectively.

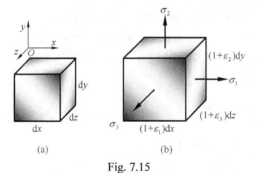

Fig. 7.15

Assuming that the material is homogenous and isotropic and behaves in a linear-elastic manner, then we can determine the relations between the stresses and strains by using the principle of superposition.

We first consider the normal strain of the element in the $x$ direction, caused by each normal stress. When $\sigma_1$ is applied, the element elongates in the $x$ direction and the normal strain $\varepsilon_1'$ in this direction, caused by $\sigma_1$, is

$$\varepsilon_1' = \frac{\sigma_1}{E} \tag{7.18}$$

Application of $\sigma_2$ causes the element to contract with a strain $\varepsilon_1''$ in the x direction and $\varepsilon_1''$ can be expressed as

$$\varepsilon_1'' = -\mu \frac{\sigma_2}{E} \tag{7.19}$$

Similarly, $\sigma_3$ will cause a contraction of the element in the x direction and the normal strain $\varepsilon_1'''$ along this direction can be written as

$$\varepsilon_1''' = -\mu \frac{\sigma_3}{E} \tag{7.20}$$

Using the principle of superposition, the normal strain $\varepsilon_1$ in the x direction, caused by $\sigma_1$, $\sigma_2$, and $\sigma_3$, can be obtained by

$$\varepsilon_1 = \varepsilon_1' + \varepsilon_1'' + \varepsilon_1''' = \frac{1}{E}[\sigma_1 - \mu(\sigma_2 + \sigma_3)] \tag{7.21}$$

Likewise, the normal strains in the y and z directions can also be obtained by using the principle of superposition. Thus the relations between the stresses and strains for a stress state subjected to three principal stresses can be expressed as

$$\varepsilon_1 = \frac{1}{E}[\sigma_1 - \mu(\sigma_2 + \sigma_3)], \quad \varepsilon_2 = \frac{1}{E}[\sigma_2 - \mu(\sigma_3 + \sigma_1)], \quad \varepsilon_3 = \frac{1}{E}[\sigma_3 - \mu(\sigma_1 + \sigma_2)] \tag{7.22}$$

The above relations are referred to as the generalized Hooke's law for the state of principal stresses of a homogeneous isotropic material. The results obtained are valid only as long as the stresses do not exceed the proportional limit, and as long as the deformations involved remain small.

**Example 7.4** A 40 mm square is scribed on the side of a large steel pressure vessel. After pressurization the biaxial stress condition at the square is as shown in Fig. 7.16. Knowing that E=200 GPa and $\mu$=0.30, determine the change in length of side AB, side BC, and diagonal AC.

**Solution** Knowing that $\sigma_1 = 60$ MPa, $\sigma_2 = 30$ MPa, $\sigma_3 = 0$, then from generalized Hooke's law, we have

$$\varepsilon_1 = \frac{1}{E}[\sigma_1 - \mu(\sigma_2 + \sigma_3)] = 2.55 \times 10^{-4}, \quad \varepsilon_2 = \frac{1}{E}[\sigma_2 - \mu(\sigma_3 + \sigma_1)] = 6.0 \times 10^{-5}$$

Using $\varepsilon_{AB} = \delta_{AB}/l_{AB} = \varepsilon_1$, we have

$$\delta_{AB} = \varepsilon_1 l_{AB} = 1.02 \times 10^{-2} \text{ mm}$$

Similarly, we can obtain

$$\delta_{BC} = \varepsilon_2 l_{BC} = 2.4 \times 10^{-3} \text{ mm}$$

Using $\delta_{AC} = \sqrt{(l_{AB} + \delta_{AB})^2 + (l_{BC} + \delta_{BC})^2} - \sqrt{l_{AB}^2 + l_{BC}^2}$, we have

$$\delta_{AC} = 8.9 \times 10^{-3} \text{ mm}$$

For a general stress state shown in Fig. 7.17, shearing stresses $\tau_{xy}$, $\tau_{yz}$, and $\tau_{zx}$ will tend to deform a cubic element into an oblique parallelepiped. For values of the stress which do not exceed the proportional limit, applying the principle of superposition we can obtain the following the generalized Hooke's law for a homogeneous isotropic material under the most general stress state:

$$\varepsilon_x = \frac{1}{E}[\sigma_x - \mu(\sigma_y + \sigma_z)], \quad \varepsilon_y = \frac{1}{E}[\sigma_y - \mu(\sigma_z + \sigma_x)], \quad \varepsilon_z = \frac{1}{E}[\sigma_z - \mu(\sigma_x + \sigma_y)]$$
$$\gamma_{xy} = \frac{\tau_{xy}}{G}, \quad \gamma_{yz} = \frac{\tau_{yz}}{G}, \quad \gamma_{zx} = \frac{\tau_{zx}}{G}$$
(7.23)

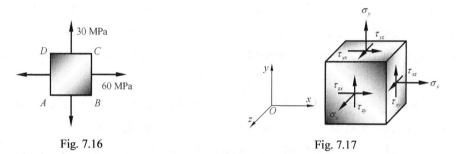

Fig. 7.16　　　　　　　　　　　Fig. 7.17

## 7.6　Strength Theories

A structural member made of a ductile or brittle material is usually designed so that that material will not fail under expected loadings. A failure of materials is usually defined as either yielding or fracture. If the material is ductile, the failure is usually specified by yielding, whereas if the material is brittle, it is specified by fracture.

A member subjected to a uniaxial stress is safe as long as $\sigma < \sigma_s$ for a ductile material, where $\sigma_s$ is the yield strength of the material, or $\sigma < \sigma_u$ for a brittle material, where $\sigma_u$ is the ultimate strength of the material, Fig. 7.18.

When a member is in a state of plane stress or biaxial stress, it is impossible to predict whether or not the member under investigation will fail by using the test, Fig. 7.19. Thus, for a complex stress state, some theory regarding the failure of materials must be established. Such theory is called strength theory or failure theory.

Two types of strength theory are available: one applicable to brittle fracture and the other suitable for ductile yielding.

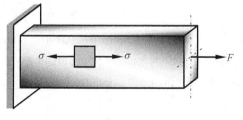

Fig. 7.18

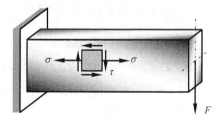

Fig. 7.19

## 1. Maximum Normal Stress Theory

This theory states that the failure of materials subjected to a biaxial or triaxial stress occurs when the maximum normal stress reaches the limit value at which the failure occurs in a tension test on the same material. This theory is based on the fact that the brittle fracture is caused by the maximum normal stress.

Assume that three principal stresses are represented by $\sigma_1$, $\sigma_2$ and $\sigma_3$, and sorted from large to small, i.e. $\sigma_1 \geqslant \sigma_2 \geqslant \sigma_3$. According to the maximum-normal-stress theory, a given member fails when the maximum normal stress in the member reaches the ultimate strength obtained from the tensile test on the same material, thus the failure criterion can be expressed as

$$\sigma_{eq1} = \sigma_1 = \sigma_u \qquad (7.24)$$

where $\sigma_{eq1}$ is the equivalent stress corresponding to the maximum-normal-stress theory, and $\sigma_u$ is the ultimate strength of the material obtained from the tensile test.

Assuming that the factor of safety is denoted by $n$, then the strength condition for the material subjected to a biaxial or triaxial stress can be expressed as

$$\sigma_{eq1} = \sigma_1 \leqslant \sigma_{allow} = \frac{\sigma_u}{n} \qquad (7.25)$$

where $\sigma_{allow}$ is the allowable stress, and $n$ is the factor of safety. The maximum-normal-stress theory is in good agreement with experimental evidence on brittle fracture.

**Example 7.5** The state of plane stress occurs in a member made of cast-iron with $\sigma_{ut}$=160 MPa and $\sigma_{uc}$=320 MPa, Fig. 7.20. Using the maximum-normal-stress theory, determine whether fracture occurs. If fracture does not occur, determine the corresponding factor of safety.

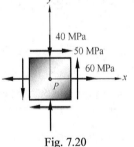

Fig. 7.20

**Solution** Referring to Example 7.2, we have

$$\sigma_1 = 80.71 \text{ MPa}, \quad \sigma_2 = 0, \quad \sigma_3 = -60.71 \text{ MPa}$$

Since $\sigma_{eq1} = \sigma_1 = 80.71 \text{ MPa} < \sigma_{ut} = 160 \text{ MPa}$, then fracture does not occur. The factor of safety is equal to

$$n = \frac{\sigma_{ut}}{\sigma_{eq1}} = 1.98$$

## 2. Maximum Normal Strain Theory

This theory states that the failure of materials subjected to a biaxial or triaxial stress occurs when the maximum normal strain reaches the limit value at which the failure occurs in a tension test on the same material. This theory is based on the fact that the brittle fracture is caused by the maximum normal strain.

According to the maximum-normal-strain theory, a given member fails when the maximum normal strain in the member reaches the ultimate tensile strain obtained from the tensile test on the same material, thus the failure criterion can be expressed as

$$\varepsilon_1 = \frac{1}{E}[\sigma_1 - \mu(\sigma_2 - \sigma_3)] = \frac{1}{E}\sigma_b \tag{7.26}$$

where $\varepsilon_1$ is the largest normal strain, and $\mu$ is the Poisson's ratio. Alternatively, this criterion can also be expressed as

$$\sigma_{eq2} = \sigma_1 - \mu(\sigma_2 - \sigma_3) = \sigma_b \tag{7.27}$$

where $\sigma_{eq2}$ is the equivalent stress corresponding to the maximum-normal-strain theory. Therefore, the strength condition for the material subjected to a biaxial or triaxial stress can be expressed as

$$\sigma_{eq2} = \sigma_1 - \mu(\sigma_2 - \sigma_3) \leqslant \sigma_{allow} = \frac{\sigma_b}{n} \tag{7.28}$$

where $\sigma_{allow}$ is the allowable stress, and $n$ is the factor of safety. At present, the maximum-normal-strain theory has been seldom used.

**3. Maximum Shearing Stress Theory**

This theory states that the failure of materials subjected to a biaxial or triaxial stress occurs when the maximum shearing stress reaches the limit value at which the failure occurs in a tension test on the same material. This theory is based on the fact that ductile yielding is caused by the maximum shearing stress.

According to the maximum-shearing-stress theory, a given member fails when the maximum shearing stress in the member reaches the yield strength obtained from the tensile test on the same material, so the failure criterion can be expressed as

$$\tau_{max} = \frac{1}{2}(\sigma_1 - \sigma_3) = \frac{1}{2}\sigma_s \tag{7.29}$$

where $\tau_{max}$ is the largest shearing stress, and $\sigma_s$ is the yield strength of the material obtained from the tensile test. Alternatively, this criterion can also be expressed as

$$\sigma_{eq3} = \sigma_1 - \sigma_3 = \sigma_s \tag{7.30}$$

where $\sigma_{eq3}$ is the equivalent stress corresponding to the maximum-shearing-stress theory. Therefore, the strength condition for the material subjected to a biaxial or triaxial stress can be expressed as

$$\sigma_{eq3} = \sigma_1 - \sigma_3 \leqslant \sigma_{allow} = \frac{\sigma_s}{n} \tag{7.31}$$

where $\sigma_{allow}$ is the allowable stress, and $n$ is the factor of safety. The maximum-shearing-stress theory is widely used for ductile yielding.

**Example 7.6** The state of plane stress occurs in a member made of steel with $\sigma_s$=310 MPa,

Fig. 7.21. Using the maximum-shearing-stress theory, determine whether yielding occurs. If yielding does not occur, determine the corresponding factor of safety.

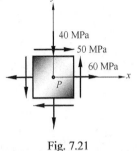

Fig. 7.21

**Solution**  Referring to Example 7.2, we have
$$\sigma_1 = 80.71 \text{ MPa}, \quad \sigma_2 = 0, \quad \sigma_3 = -60.71 \text{ MPa}$$
Since $\sigma_{eq3} = \sigma_1 - \sigma_3 = 141.42 \text{ MPa} < \sigma_s = 310 \text{ MPa}$, then yielding does not occur. The factor of safety is equal to
$$n = \frac{\sigma_s}{\sigma_{eq3}} = 2.19$$

### 4. Maximum Distortion Energy Theory

This theory states that the failure of materials subjected to a biaxial or triaxial stress occurs when the distortion-energy density (distortion energy per unit volume) reaches the limit value at which the failure occurs in a tension test on the same material. This theory is based on the fact that ductile yielding is caused by the maximum distortion energy associated with change in shape of the material.

According to the maximum-distortion-energy theory, a given member fails when the maximum distortion energy in the member reaches the ultimate distortion energy obtained from the tensile test on the same material, thus the failure criterion can be expressed as

$$u_d = \frac{(1+\mu)}{6E}[(\sigma_1-\sigma_2)^2+(\sigma_2-\sigma_3)^2+(\sigma_3-\sigma_1)^2] = \frac{(1+\mu)}{6E}2\sigma_s^2 \qquad (7.32)$$

where $u_d$ is the distortion-energy density corresponding to a biaxial or triaxial stress, and $E$ is the modulus of elasticity. Alternatively, this criterion can also be expressed as

$$\sigma_{eq4} = \sqrt{\frac{1}{2}[(\sigma_1-\sigma_2)^2+(\sigma_2-\sigma_3)^2+(\sigma_3-\sigma_1)^2]} = \sigma_s \qquad (7.33)$$

where $\sigma_{eq4}$ is the equivalent stress corresponding to the maximum-distortion-energy theory. Therefore, the strength condition for the material subjected to a biaxial or triaxial stress can be expressed as

$$\sigma_{eq4} = \sqrt{\frac{1}{2}[(\sigma_1-\sigma_2)^2+(\sigma_2-\sigma_3)^2+(\sigma_3-\sigma_1)^2]} \leq \sigma_{allow} = \frac{\sigma_s}{n} \qquad (7.34)$$

where $\sigma_{allow}$ is the allowable stress, and $n$ is the factor of safety. The maximum-distortion-energy theory is excellent agreement with experiments on ductile yielding.

**Example 7.7**  The state of plane stress occurs in a member made of steel with $\sigma_s$=310 MPa, Fig. 7.22. Using the maximum-distortion-energy theory, determine whether yielding occurs. If yielding does not occur, determine the corresponding factor of safety.

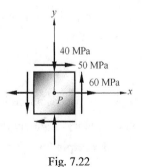

Fig. 7.22

**Solution** Referring to Example 7.2, we have

$$\sigma_1 = 80.71 \text{ MPa}, \quad \sigma_2 = 0, \quad \sigma_3 = -60.71 \text{ MPa}$$

Since $\sigma_{eq4} = \sqrt{\frac{1}{2}[(\sigma_1 - \sigma_2)^2 + (\sigma_2 - \sigma_3)^2 + (\sigma_3 - \sigma_1)^2]} = 122.88 \text{ MPa} < \sigma_s$, then yielding does not occur. The factor of safety is equal to

$$n = \frac{\sigma_s}{\sigma_{eq4}} = 2.52$$

# Problems

**7.1** For the given stress state, determine the normal and shearing stresses after the element shown in Fig. 7.23 has been rotated through 30° anticlockwise.

**7.2** For the given stress state, determine the normal and shearing stresses after the element shown in Fig. 7.24 has been rotated through 15° clockwise.

**7.3** For the given stress state, determine the normal and shearing stresses after the element shown in Fig. 7.25 has been rotated through 55° clockwise.

**7.4** For the given stress state, determine the normal and shearing stresses after the element shown in Fig. 7.26 has been rotated through 45° anticlockwise.

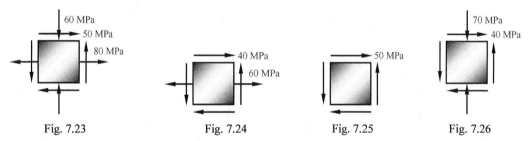

Fig. 7.23　　　　　Fig. 7.24　　　　　Fig. 7.25　　　　　Fig. 7.26

**7.5** For the given stress state, Fig. 7.27, determine the principal stresses and the corresponding principal planes.

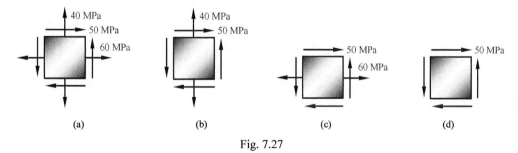

(a)　　　　　(b)　　　　　(c)　　　　　(d)

Fig. 7.27

**7.6** For the given stress state, Fig. 7.28, determine (1) the maximum in-plane shearing stress, (2) the orientation of the planes of maximum in-plane shearing stress, (3) the normal stress on the planes of maximum in-plane shearing stress.

7.7 A 40 mm square plate is subjected to plane stress shown in Fig. 7.29. Knowing that $E=200$ GPa and $\mu=0.30$, determine the change in length of side $AB$, side $BC$, and diagonal $AC$.

7.8 The state of plane stress occurs in a member made of steel with $\sigma_s=310$ MPa, Fig. 7.30. Using the maximum-shearing-stress theory, determine whether yielding occurs. If yielding does not occur, determine the corresponding factor of safety.

7.9 The state of plane stress occurs in a member made of steel with $\sigma_s=310$ MPa, Fig. 7.31. Using the maximum-distortion-energy theory, determine whether yielding occurs. If yielding does not occur, determine the corresponding factor of safety.

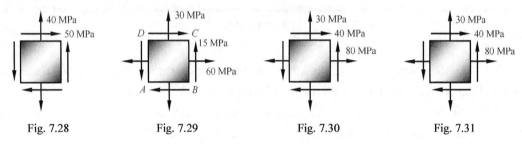

Fig. 7.28  Fig. 7.29  Fig. 7.30  Fig. 7.31

# Chapter 8  Combined Loadings

In preceding chapters, we discussed members subjected to a single type of internal forces. For instance, we analyzed bars in tension or compression, shafts in torsion, and beams in bending. We also developed methods for determining stresses and strains for each type of internal forces.

However, in many structures a member is often subjected to two or more types of internal forces. These internal forces applied instantaneously to a member are called combined loadings. For example, the segment *AD* of the beam *ADB* in Fig. 8.1 is subjected to the simultaneous action of a bending moment and an axial force, and the portion *AB* of the crank *ABC* in Fig. 8.2 carries instantaneously torsional and bending moments.

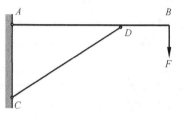
Fig. 8.1

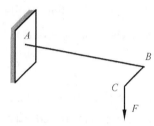
Fig. 8.2

The stress distribution within the member subjected to combined loadings can be determined by using the principle of superposition. The stress distribution due to each internal force is first determined, and then these distributed stresses are superimposed to determine the resultant distributed stress. The principle of superposition can be used provided a linear relation exists between the stresses and the applied loads. Also, the geometry of the member should not undergo significant change when the loads are applied. This is to ensure that the stress produced by one load is not related to the stress produced by any other load. Most ordinary structures satisfy these two conditions, and therefore the principle of superposition is extensively used in engineering field.

## 8.1  Bar in Eccentric Tension or Compression

The distribution of stress in the cross section of a bar under an axial loading can be assumed uniform only if the line of action of the external load passes through the centroid of the cross section. Such a loading is said to be centric.

Consider now the distribution of stress when the line of action of the external load does not pass through the centroid of the cross section, i.e., when the loading is eccentric, Fig. 8.3.

Assuming that the external load is applied in a plane of symmetry of the bar and that the eccentric distance is $e$, then the internal forces acting on any cross section include an axial force $N = F$ applied at the centroid of the cross section and a bending moment $M = Fe$ acting in the plane of symmetry of the bar, Fig. 8.4.

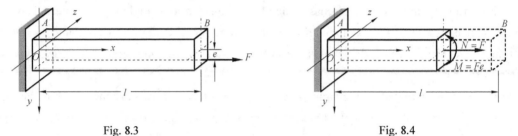

Fig. 8.3          Fig. 8.4

If the material remains linearly elastic and is subjected to a small deformation, then the stress distribution due to the eccentric loading can be obtained by superposing the uniform stress distribution $\sigma_N$ corresponding to the axial force $N$ and the linear stress distribution $\sigma_M$ corresponding to the bending moment $M$. Therefore, we have

$$\sigma = \sigma_N + \sigma_M = \frac{N}{A} + \frac{My}{I} \tag{8.1}$$

where $A$ is the cross-sectional area, $I$ is the moment of inertia of the cross section about the centroidal axis, and $y$ is measured from the centroidal axis of the cross section.

Depending upon the geometry of the cross section and the eccentricity of the external load, the resultant stress may be positive and negative, or some may have the same sign. From Eq. (8.1), we know that there is a line in the cross section along which $\sigma = 0$. This line represents the neutral axis of the cross section. We note that the neutral axis does not coincide with the centroidal axis of the cross section.

The resultant normal stress has a maximum value $\sigma_{max}$ at the bottom and a minimum value $\sigma_{min}$ at the top, which can be expressed as

$$\sigma_{max,min} = \frac{N}{A} + \frac{My_{max}}{I} = \frac{N}{A} \pm \frac{M}{S} \tag{8.2}$$

where $y_{max}$ is the perpendicular distance from the points at the top or bottom to the centroidal axis, and $S = I/y_{max}$ is the modulus of section about the centroidal axis.

From the analysis given above, the points located at the bottom or top of the cross section will have a maximum or minimum stress. These points are called the critical points (or dangerous points). Assuming that the allowable stress of the material used is denoted by $\sigma_{allow}$, then the strength condition for the bar subjected to an eccentric tension or compression can be expressed as

$$\left|\sigma_{max,min}\right| = \left|\frac{N}{A} \pm \frac{M}{S}\right| \leqslant \sigma_{allow} \tag{8.3}$$

**Example 8.1** Two external loads of magnitude $F=50$ kN are applied to the end of a 20a I-steel beam, Fig. 8.5. Knowing that $e=80$ mm, determine the stress on the bottom surface of the beam.

**Solution** The axial force and bending moment at any section of the beam can be expressed as

$$N = 2F = 100 \text{ kN}, \qquad M = Fe = 4 \text{ kN} \cdot \text{m}$$

Referring to the table of shape steels in Appendix II, we have the cross-sectional area and section modulus of the 20a I-steel beam as follows:

$$A = 35.578 \text{ cm}^2, \qquad S = 237 \text{ cm}^3$$

Using Eq. (8.2), we have

$$\sigma = \sigma_N + \sigma_M = \frac{N}{A} + \frac{M}{S} = 44.98 \text{ MPa}$$

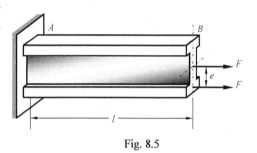

Fig. 8.5

## 8.2 I-Beam in Transverse-Force Bending

Depending on the shape of the cross section of a beam and the value of the shearing force at the critical section (or dangerous section) where the bending moment has a maximum value, it may happen that the largest stress will not occur at the top or bottom of the section, but at some other point within the section. A combination of large values of the normal and shearing stresses near the junction of the web and the flanges of an I-beam can result in a value of the stress that is larger than the value on the surfaces of the I-beam.

Consider an I-beam $AB$ subjected to a transverse loading, Fig. 8.6. The shearing force remains constant over the entire beam, while the bending moment will reach a maximum value at the fixed end. Therefore, the critical section is located at the fixed end and the internal forces at the critical section can be expressed as

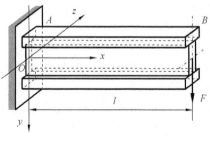

Fig. 8.6

$$\begin{aligned} V &= F \\ M_{max} &= Fl \end{aligned} \qquad (8.4)$$

The shearing stress at the critical section caused by the shearing force will reach a maximum value on the neutral axis of the I-beam, and the maximum shearing stress can be expressed as

$$\tau_{max} = \frac{VQ'_{max}}{Ib} = \frac{V}{A_w} \qquad (8.5)$$

where $I$ is the moment of inertia of the I-shaped section about the neutral axis, $Q'_{max}$ is the first moment of the area above the neutral axis about the neutral axis, $b$ is the width of the web, and $A_w$ is the cross-sectional area of the web.

The normal stress at the critical section caused by the bending moment has a maximum stress on the top or bottom surface, which can be written as

$$\sigma_{max} = \frac{M_{max} y_{max}}{I} = \frac{M_{max}}{S} \tag{8.6}$$

where $y_{max}$ is the perpendicular distance from the points at the top or bottom surface to the neutral axis, and $S = I/y_{max}$ is the modulus of section about the neutral axis.

At any other point at the critical section, the material is subjected simultaneously to the normal stress:

$$\sigma = \frac{M_{max} y}{I} \tag{8.7}$$

where $y$ is the distance from the neutral axis, and to the shearing stress:

$$\tau = \frac{VQ'}{Ib} \tag{8.8}$$

where $Q'$ is the first moment about the neutral axis of the portion of the cross-sectional area located below or above the point where the stress is computed, and $b$ is the width of the cross section at that point.

Using stress transformation, we can obtain the principal stress at any point of the critical section. For an I-beam, the large shearing stresses will occur at the junctions of the web with the flanges of the beam, where the normal stresses are also large, thus the principal stresses at such points are larger than the stress at the top or bottom surface and on the neutral axis. We should be particularly aware of this possibility when selecting I-beams, and calculate the principal stress at the junctions $a$ and $b$ of the web with the flanges of the beam, Fig. 8.7.

The principal stresses at the point $a$ and point $b$ on the critical section can, by using stress transformation, be expressed as, respectively

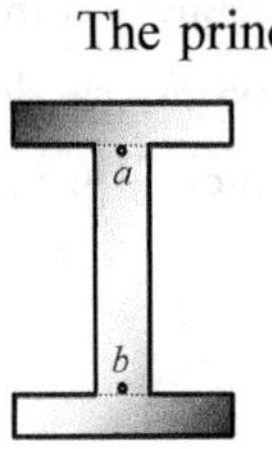

$$\sigma_{1,3} = \frac{\sigma}{2} \pm \sqrt{\left(\frac{\sigma}{2}\right)^2 + \tau^2} \quad \text{(at } a\text{)}$$

$$\sigma_{1,3} = -\frac{\sigma}{2} \pm \sqrt{\left(\frac{\sigma}{2}\right)^2 + \tau^2} \quad \text{(at } b\text{)} \tag{8.9}$$

Fig. 8.7

If the material is known to be ductile, then the maximum-shearing-stress theory or the maximum-distortion-energy theory should be used for design of the I-beam.

If the maximum-shearing-stress theory is applied, then the equivalent stress at point $a$ or $b$ can be written as

$$\sigma_{eq3} = \sigma_1 - \sigma_3 = \sqrt{\sigma^2 + 4\tau^2} \tag{8.10}$$

where $\sigma_{eq3}$ is the equivalent stress corresponding to the maximum-shearing-stress theory. Assuming that the allowable stress of the material used is $\sigma_{allow}$, then the maximum-shearing-stress strength condition at the junction points of the I-beam subjected to transverse-force bending can be expressed as

$$\sigma_{eq3} = \sqrt{\sigma^2 + 4\tau^2} \leq \sigma_{allow} \tag{8.11}$$

If the maximum-distortion-energy theory is applied, then the strength condition at the junction points of the I-beam subjected to transverse-force bending can be given as

$$\sigma_{eq4} = \sqrt{\frac{1}{2}[(\sigma_1 - \sigma_2)^2 + (\sigma_2 - \sigma_3)^2 + (\sigma_3 - \sigma_1)^2]} = \sqrt{\sigma^2 + 3\tau^2} \leq \sigma_{allow} \tag{8.12}$$

where $\sigma_{eq4}$ is the equivalent stress corresponding to the maximum-distortion-energy theory.

**Example 8.2**  An external load, $F$=150 kN, is applied to the free end of an I-beam made of steel for which the yield strength $\sigma_s$=235 MPa, Fig. 8.8. Using the maximum-shearing-stress theory, determine whether yielding occurs. If yielding does not occur, determine the factor of safety. Neglect the effect of fillets.

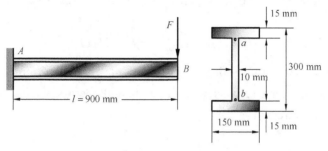

Fig. 8.8

**Solution**  The critical section is located at the fixed end $A$ of the beam, where the bending moment and shearing force can be expressed as

$$M_{max} = Fl = 135 \text{ kN} \cdot \text{m}, \qquad V = F = 150 \text{ kN}$$

The moment of inertia of the cross section is equal to

$$I = \frac{1}{12} \times (150) \times (300)^3 - \frac{1}{12} \times (140) \times (270)^3 = 1.07865 \times 10^8 \text{ (mm}^4\text{)}$$

The maximum normal and shearing stresses on the cross section at the fixed end are

$$\sigma_{max} = \frac{M_{max} y_{max}}{I} = 187.73 \text{ MPa}, \qquad \tau_{max} = \frac{V}{A_w} = 55.56 \text{ MPa}$$

The normal and shearing stresses at point $a$ (or $b$) on the cross section at the fixed end are

$$\sigma_a = \frac{y_a}{y_{max}} \sigma_{max} = 168.96 \text{ MPa}, \qquad \tau_a \approx \tau_{max} = \frac{V}{A_w} = 55.56 \text{ MPa}$$

The equivalent stress corresponding to the maximum-shearing-stress theory can be expressed as

$$\sigma_{eq3} = \sqrt{\sigma_a^2 + 4\tau_a^2} = 202.23 \text{ MPa}$$

The fact that $\sigma_{eq3} > \sigma_{max}$ has shown that the critical point is located at point $a$ (or $b$). Since $\sigma_{eq3} < \sigma_s$, no yielding occurs. The factor of safety is equal to

$$n = \frac{\sigma_s}{\sigma_{eq3}} = 1.16$$

## 8.3 Beam in Bending and Tension/Compression

Consider a rectangular beam $AB$, which is fixed at the left end and subjected to two external loads, $F_1$ and $F_2$, at the free end, Fig. 8.9. Provided the material used remains linearly elastic, and is only subjected to a small deformation, then we can use the principle of superposition to determine the resultant stress distribution in the beam due to combined loadings.

The critical section is located at the fixed end and the internal forces at the critical section can be expressed as

$$V = F_2, \quad M_{max} = F_2 l, \quad N = F_1 \qquad (8.13)$$

The shearing stress at the critical section caused by the shearing force will reach a maximum value on the centroidal axis of the rectangular cross section, and the maximum shearing stress can be expressed as

$$\tau_{max} = \frac{V Q'_{max}}{I b} = \frac{3V}{2A} \qquad (8.14)$$

where $I$ is the moment of inertia of the cross section about the centroidal axis, $Q'_{max}$ is the first moment of the cross section above the centroidal axis about the centroidal axis, $b$ is the width of the rectangular beam, and $A$ is the cross-sectional area of the rectangular beam. The stress produced by the shearing force generally has a much smaller contribution compared with that produced by the axial force or bending moment, and hence it can be neglected.

The normal stress at the critical section caused by the bending moment has a maximum stress at the top or bottom surface, which can be written as

$$\sigma_M = \frac{M_{max} y_{max}}{I} = \frac{M_{max}}{S} \qquad (8.15)$$

where $y_{max}$ is the perpendicular distance from the points at the top or bottom surface to the centroidal axis, and $S = I/y_{max}$ is the modulus of section about the centroidal axis.

The normal stress at the critical section caused by the axial force has a uniform distribution, and can be expressed as

$$\sigma_N = \frac{N}{A} \qquad (8.16)$$

where $A$ is the cross-sectional area of the rectangular beam.

The points located at the top or bottom of the critical section will have a maximum or minimum stress, Fig. 8.10. The maximum or minimum value can be given as

$$\sigma_{max,min} = \sigma_N \pm \sigma_M = \frac{N}{A} \pm \frac{M_{max}}{S} \qquad (8.17)$$

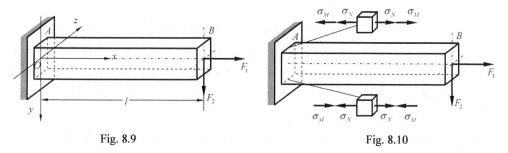

Fig. 8.9　　　　　　　　　　　　　Fig. 8.10

Assuming that the allowable stress of the material used is $\sigma_{allow}$, then the strength condition for the beam subjected to both axial tension/compression and bending is

$$\left|\sigma_{max,min}\right| = \left|\frac{N}{A} \pm \frac{M_{max}}{S}\right| \leqslant \sigma_{allow} \qquad (8.18)$$

**Example 8.3**　Two external loads, $F_1$=100 kN and $F_2$=4 kN, are applied to the free end of a 20a I-steel beam, Fig. 8.11. Knowing that $l$=1 m, determine the maximum tensile stress in the beam.

**Solution**　The maximum tensile stress will develop on the top-surface at the fixed end of the beam. The axial force and bending moment at the fixed end can be expressed as

$N = F_1 = 100 \text{ kN}, \qquad M_{max} = F_2 l = 4 \text{ kN} \cdot \text{m}$

For a 20a I-steel beam, we have

$A = 35.578 \text{ cm}^2, \qquad S = 237 \text{ cm}^3$

Using Eq. (8.17), we obtain

$\sigma_{max}^t = \sigma_N + \sigma_M = \frac{N}{A} + \frac{M_{max}}{S} = 44.98 \text{ MPa}$

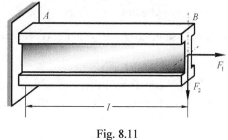

Fig. 8.11

## 8.4　Shaft in Torsion and Bending

Consider a circular shaft $AB$, which is fixed at the left end and subjected to two external loads, $F$ and $M_e$, at the free end, Fig. 8.12. Assuming that the material remains linearly elastic, and that the deformation is small, then we can use the principle of superposition to determine the resultant stress distribution in the shaft due to combined loadings.

The critical section is located at the section $A$ and the internal forces at the critical section can be expressed as

$$V = F, \qquad M_{max} = Fl, \qquad T = M_e \qquad (8.19)$$

The shearing stress at the critical section caused by the shearing force has a maximum value on the neutral axis of the circular cross section, that is

$$\tau_V = \frac{VQ'_{max}}{Ib} = \frac{4V}{3A} \quad (8.20)$$

where $I$ is the moment of inertia of the circular cross section about the neutral axis, $Q'_{max}$ is the first moment of the cross-sectional area above the neutral axis with respect to the neutral axis, $b$ is the diameter of the circular shaft, and $A$ is the cross-sectional area of the circular shaft. This stress is small, and can be neglected. Therefore, we only need to consider the resultant stress caused the torsional and bending moments.

The normal stress on the critical section caused by the bending moment has a maximum absolute value at the top or bottom points of the critical section, that is

$$\sigma_M = \frac{M_{max} y_{max}}{I} = \frac{M_{max}}{S} \quad (8.21)$$

where $y_{max}$ is the perpendicular distance from the points at the top or bottom surface to the neutral axis, and $S = I/y_{max}$ is the modulus of section.

The shearing stress at the critical section caused by the torsional moment reaches a maximum value on the outer surface of the circular shaft, and the maximum value can be expressed as

$$\tau_T = \frac{T_{max} \rho_{max}}{I_p} = \frac{T_{max}}{S_p} \quad (8.22)$$

where $I_p$ is the polar moment of inertia of the circular cross section with respect to the centroid of the cross section, $\rho_{max}$ is the (outer) radius of the circular cross section, and $S_p = I_p/\rho_{max}$ is the polar modulus of section.

It has been shown that the critical point has a state of plane stress and is located at the top point $P_1$ or bottom point $P_2$ of the critical section, Fig. 8.13.

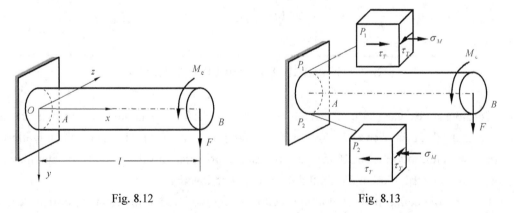

Fig. 8.12　　　　　　　　Fig. 8.13

The principal stresses at the top point $P_1$ and bottom point $P_2$ on the critical section can, by using stress transformation, be expressed as, respectively

$$\sigma_{1,3} = \frac{\sigma_M}{2} \pm \sqrt{\left(\frac{\sigma_M}{2}\right)^2 + \tau_T^2} \quad (\text{at } P_1)$$
$$\sigma_{1,3} = -\frac{\sigma_M}{2} \pm \sqrt{\left(\frac{\sigma_M}{2}\right)^2 + \tau_T^2} \quad (\text{at } P_2)$$
(8.23)

A shaft subjected to both torsion and bending is generally made of a ductile material, hence the maximum-shearing-stress theory or the maximum-distortion-energy theory should be used for design of the shaft.

For a ductile material, if the maximum-shearing-stress theory is applied, then the equivalent stress at point $P_1$ or $P_2$ can be written as

$$\sigma_{eq3} = \sigma_1 - \sigma_3 = \sqrt{\sigma_M^2 + 4\tau_T^2} \tag{8.24}$$

where $\sigma_{eq3}$ is the equivalent stress corresponding to the maximum-shearing-stress theory.

Using Eq. (8.21) and Eq. (8.22), and $S_p = 2S$, Eq. (8.24) can be rewritten as

$$\sigma_{eq3} = \frac{\sqrt{M_{max}^2 + T_{max}^2}}{S} \tag{8.25}$$

where $S = \frac{1}{32}\pi d^3$ for a solid shaft with diameter $d$, and $S = \frac{1}{32}\pi D^3(1-\alpha^4)$, where $\alpha = d/D$, for hollow shaft with outer diameter $D$ and inner diameter $d$.

Assuming that the allowable stress of the material used is $\sigma_{allow}$, then the maximum-shearing-stress strength condition for the shaft subjected to both torsion and bending can be expressed as

$$\sigma_{eq3} = \frac{\sqrt{M_{max}^2 + T_{max}^2}}{S} \leqslant \sigma_{allow} \tag{8.26}$$

For a ductile material, if the maximum-distortion-energy theory is applied, then the strength condition for the shaft subjected to both torsion and bending can be given as

$$\sigma_{eq4} = \sqrt{\frac{1}{2}[(\sigma_1-\sigma_2)^2 + (\sigma_2-\sigma_3)^2 + (\sigma_3-\sigma_1)^2]} = \sqrt{\sigma_M^2 + 3\tau_T^2} = \frac{\sqrt{M_{max}^2 + \frac{3}{4}T_{max}^2}}{S} \leqslant \sigma_{allow}$$
(8.27)

where $\sigma_{eq4}$ is the equivalent stress corresponding to the maximum-distortion-energy theory.

**Example 8.4** The steel pipe $AB$, having $\sigma_s$=325 MPa yield strength, $l$=200 mm length, $D$=80 mm outer diameter, and $t$=5 mm wall thickness, is subjected to two external loads, $F$ = 6 kN and $M_e$=1.5 kN·m, at the free end, Fig. 8.14. Using the maximum-shearing-stress theory, determine whether yielding occurs. If yielding does not occur, determine the factor of safety.

**Solution** The critical point is located at the top or bottom surface at the fixed end of the shaft. The bending and torsional moments at the fixed end can be expressed as

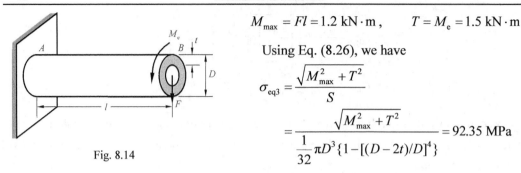

$$M_{max} = Fl = 1.2 \text{ kN} \cdot \text{m}, \quad T = M_e = 1.5 \text{ kN} \cdot \text{m}$$

Using Eq. (8.26), we have

$$\sigma_{eq3} = \frac{\sqrt{M_{max}^2 + T^2}}{S}$$

$$= \frac{\sqrt{M_{max}^2 + T^2}}{\frac{1}{32}\pi D^3\{1-[(D-2t)/D]^4\}} = 92.35 \text{ MPa}$$

Fig. 8.14

Since $\sigma_{eq3} < \sigma_s$, no yielding occurs. Thus the factor of safety is equal to

$$n = \frac{\sigma_s}{\sigma_{eq3}} = 3.52$$

## Problems

8.1 An external load of magnitude $F=80$ kN is applied to the end of a 22a I-steel beam, Fig. 8.15. Knowing that $e=80$ mm, determine the stresses on the top and bottom surfaces of the beam.

8.2 Three external loads of magnitude $F=80$ kN are applied to the end of a 20a I-steel beam, Fig. 8.16. Knowing that $e=80$ mm, determine the maximum tensile and compressive stresses in the beam.

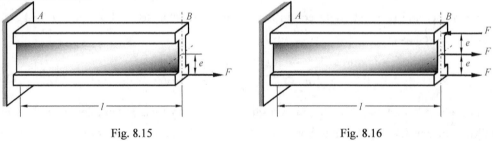

Fig. 8.15      Fig. 8.16

8.3 An external load, $F=180$ kN, is applied to the free end of an H-steel beam for which the yield strength $\sigma_s=235$ MPa, Fig. 8.17. Using the maximum-distortion-energy theory, determine whether yielding occurs. If yielding does not occur, determine the factor of safety. Neglect the effect of fillets.

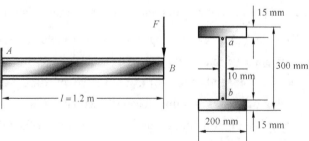

Fig. 8.17

8.4 Two external loads, $F_1=100$ kN and $F_2=20$ kN, are applied to the free end of a 36a I-steel beam for which the yield strength $\sigma_s=235$ MPa, Fig. 8.18. Using the maximum-shearing-stress theory, determine whether yielding occurs. If yielding does not occur, determine the factor of safety.

8.5 The steel pipe $AB$, having $\sigma_s=325$ MPa yield strength, $l=200$ mm length, $D=80$ mm outer diameter, and $t=5$ mm wall thickness, is subjected to two external loads, $F=6$ kN and $M_e=1.5$ kN·m, at the free end, Fig. 8.19. Using the maximum-distortion-energy theory, determine whether yielding occurs. If yielding does not occur, determine the factor of safety.

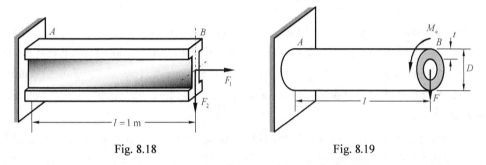

Fig. 8.18                                Fig. 8.19

8.6 A cantilevered I-steel beam supports an external load as shown in Fig. 8.20. Knowing that $F=200$ kN and $\sigma_{\text{allow}}=180$ MPa, determine (1) the maximum value of the normal stress in the beam, (2) the maximum value of the principal stresses at the junction of a flange and the web, (3) whether the specified I-steel beam is acceptable.

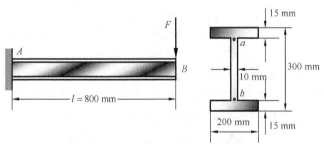

Fig. 8.20

# Chapter 9   Stability of Column

When a member is designed, it is necessary that it satisfy specific strength, rigidity and stability requirements. In the preceding chapters we have discussed the strength and rigidity of members. In this chapter, we will discuss the stability of members, i.e. the ability of members to support a given axial compressive load without experiencing a sudden lateral deflection.

A long slender member subjected to axial compression is called a column. The failure of a column occurs by buckling, i.e., by a sudden lateral deflection. It should be noted that the failure of a short column occurs by yielding or fracturing of the material. The buckling of a column may occur even though the maximum stress in the column is less than the yield strength and ultimate strength of the material.

Quite often the buckling of a column can lead to a sudden failure of a structure, and as a result, special attention must be given to the design of a column so that it can safely support a given load without buckling.

The column to be considered in this chapter is an ideal column, meaning one that is perfectly straight before loading, is made of homogeneous material, and upon which the load is applied through the centroid of the cross section.

## 9.1   Critical Load of Long Column with Pin Supports

The maximum axial compressive force that a column can support when it is on the verge of buckling is called the critical load, denoted by $F_{cr}$. Consider a column $AB$ of length $l$ to support a given load $F$, Fig. 9.1(a). The column is pin-supported at both ends and $F$ is a centric axial load. If the cross-sectional area of the column is selected so that the stress on the cross section is less than the allowable stress for the material used, and if the deformation falls within the given specifications, we might conclude that the column has been properly designed. However, it may happen that, as the load is applied, the column will buckle. Instead of remaining straight, it will suddenly become sharply curved, Fig. 9.1(b). We now determine the critical value $F_{cr}$ of the load for which the position shown in Fig. 9.1(a) cease to be stable and the conditions under which the configuration of Fig. 9.1(b) is possible.

Since a column can be considered as a beam placed in a vertical position and subjected to an axial load. Considering the equilibrium of the free body $AC$, Fig. 9.1(c), we find that the bending moment at the cross section $C$ is $M = -Fw$. Substituting this value for $M$ in $w'' = M/EI$, we have

$$w'' + \frac{F}{EI}w = 0 \qquad (9.1)$$

This is a linear, homogeneous differential equation of the second order with constant coefficients. Setting

$$k^2 = \frac{F}{EI} \qquad (9.2)$$

we write Eq. (9.1) in the form

$$w'' + k^2 w = 0 \qquad (9.3)$$

The general solution of Eq. (9.3) is

$$w = A\sin(kx) + B\cos(kx) \qquad (9.4)$$

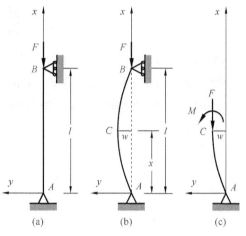

Fig. 9.1

Recalling the boundary conditions that must be satisfied at ends $A$ and $B$ of the column, we first make $x=0$, $w=0$ in Eq. (9.4) and find that $B=0$. Substituting next $x=l$, $w=0$, we obtain

$$A\sin(kl) = 0 \qquad (9.5)$$

This equation is satisfied either if $A=0$, or if $\sin(kx)=0$. If the first of these conditions is satisfied, Eq. (9.5) reduces to $w \equiv 0$ and the column is always straight. For the second condition to be satisfied, we must have $kl = n\pi$ or, using Eq. (9.2) and solving for $F$,

$$F = \frac{n^2 \pi^2 EI}{l^2} \qquad (9.6)$$

The smallest value of $F$ defined by Eq. (9.6) is that corresponding to $n=1$. We thus have

$$F_{cr} = \frac{\pi^2 EI}{l^2} \qquad (9.7)$$

This expression obtained is known as Euler's formula. In the case of a column with a circular or square cross section, the moment of inertia $I$ of the cross section is the same about any centroidal axis, and the column is likely to buckle about any centroidal axis. For other shape of cross section, the critical load should be computed by making $I = I_{min}$ in Eq. (9.7).

## 9.2 Critical Load of Long Column with Other Supports

Euler's formula (9.7) was derived for a column that was pin-supported at both ends. Now the critical load $F_{cr}$ will be determined for columns with different end conditions. In the case of a column with one fixed end $A$ and one free end $B$ supporting a load $F$, Fig. 9.2(a). The column will behave as the upper half of a pin-supported column, Fig. 9.2(b). The critical load for the column of Fig. 9.2(a) is thus the same as for the pin-ended column of Fig. 9.2(b) and can be obtained from Euler's formula (9.7) by using a column length equal to twice the actual length $l$ of the given column. Therefore, the critical load is expressed as

$$F_{cr} = \frac{\pi^2 EI}{(2l)^2} = \frac{\pi^2 EI}{(\mu l)^2} = \frac{\pi^2 EI}{l_e^2} \tag{9.8}$$

where $\mu$ is the effective-length factor, and $l_e = \mu l$ is the effective length. The above expression is an extension of Euler's formula, and thus is also called Euler's formula.

The effective-length factor and effective length corresponding to various end conditions are shown in Fig. 9.3.

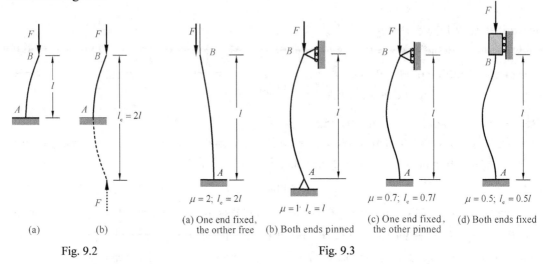

Fig. 9.2  Fig. 9.3

## 9.3 Critical Stress of Long Column

The value of the stress corresponding to the critical load is called the critical stress and is denoted by $\sigma_{cr}$. Recalling Eq. (9.8) and setting $i^2 = I/A$, where $A$ is the cross-sectional area and $i$ is the radius of gyration, we have

$$\sigma_{cr} = \frac{F_{cr}}{A} = \frac{\pi^2 E}{(\mu l/i)^2} = \frac{\pi^2 E}{\lambda^2} \tag{9.9}$$

where $\lambda = \mu l/i$ is called the slenderness ratio of the column. The above equation is also called Euler's formula. Eq. (9.9) shows that the critical stress is proportional to the modulus of elasticity of the material, and inversely proportional to the square of the slenderness ratio of the column.

Euler's formula was derived based on the fact that the critical stress is below the proportional limit of the material, that is,

$$\sigma_{cr} = \frac{\pi^2 E}{\lambda^2} \leqslant \sigma_p \tag{9.10}$$

Defining $\lambda_p = \sqrt{\pi^2 E/\sigma_p}$, then the above equation can be rewritten as

$$\lambda \geqslant \lambda_p \tag{9.11}$$

where $\lambda_p$ is the critical slenderness ratio. The above equation shows that Euler's formula is valid only when $\lambda \geqslant \lambda_p$. For a low-carbon steel (Q235), $E$=206 GPa, and $\sigma_p$=200 MPa, then $\lambda_p$=100.8. For an aluminum alloy, $E$=70 GPa and $\sigma_p$=175 MPa, then $\lambda_p$=62.8.

**Example 9.1** A long column consists of a steel tube that has a 30 mm outer diameter and a 20 mm inner diameter, Fig. 9.4. Knowing that $l$=1.5 m and $E$=200 GPa, determine the critical load and critical stress for the support condition.

**Solution** The critical load for the support condition shown can be determined by using Euler's formula. For the column given, it can be sectioned at the midsection $C$, Fig. 9.4, and divided into two portions, $BC$ and $AC$, each equivalent to a column that is fixed at one end and free at the other. Therefore, the critical load for the original column is equal to that for each of the two portions. Using $\mu_{BC} = \mu_{AC} = 2$ and $l_{BC} = l_{AC} = \frac{1}{2}l$, we obtain the critical load for the support condition shown in Fig. 9.4 as follows:

$$F_{cr} = (F_{cr})_{BC} = (F_{cr})_{AC} = \frac{\pi^2 EI}{(\mu_{AC} l_{AC})^2}$$

$$= \frac{\pi^2 E\left[\frac{1}{64}\pi(D^4 - d^4)\right]}{l^2} = 27.99 \text{ kN}$$

Using Eq. (9.9), the critical stress is given as

$$\sigma_{cr} = \frac{F_{cr}}{A} = \frac{F_{cr}}{\frac{1}{4}\pi(D^2 - d^2)} = 71.28 \text{ MPa}$$

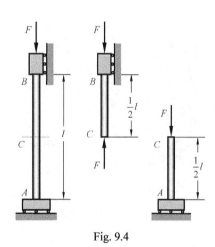

Fig. 9.4

## 9.4 Critical Stress of Intermediate Column

In the preceding sections, we assumed that the column was perfectly straight before loading and made of homogeneous material, and that all stresses remained below the proportional limit. Real columns fall short of such an idealization, and in practice the design of columns is based on empirical formulas.

For a long column, where the slenderness ratio $\lambda$ is large, its failure is closely predicted by Euler's formula, and the value of $\sigma_{cr}$ depends on the modulus of elasticity $E$ of the used material by Euler's formula $\sigma_{cr} = \pi^2 E/\lambda^2$, but not on its yield strength $\sigma_s$ or ultimate strength $\sigma_u$.

For a short column, its failure occurs essentially as a result of yielding or fracture, and we have $\sigma_{cr} = \sigma_s$ for a ductile material or $\sigma_{cr} = \sigma_u$ for a brittle material.

For an intermediate column, its failure is dependent on both $E$ and $\sigma_s$ (or $\sigma_u$). In this range, the column failure is an extremely complex phenomenon and hence empirical formulas are often used for the design of an intermediate column under centric loading. The simplest empirical formula is a linear formula, which can be written as

$$\sigma_{cr} = a - b\lambda \tag{9.12}$$

where $a$ and $b$ are two constants depending on the material used. For low carbon steel, $a = 304$ MPa and $b = 1.12$ MPa. For a cast iron, $a = 332.2$ MPa and $b = 1.454$ MPa.

**Example 9.2** A 20a I-steel column is subjected to a centric compressive load $F$ at the free end, Fig 9.5. Knowing that $\sigma_p$=200 MPa, $E$=206 GPa, $a$=304 MPa, and $b$=1.12, determine the critical load of the column if the length of the column is (1) $l$=1.2 m, (2) $l$=1.0 m.

**Solution** For the column given, the effective-length factor is $\mu = 2.0$, and the critical slenderness ratio equals

$$\lambda_p = \sqrt{\frac{\pi^2 E}{\sigma_p}} = 101$$

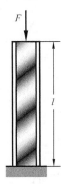

Fig. 9.5

Referring to the table of shape steels in Appendix II, we have the cross-sectional area and minimum moment of inertia for the 20a I-steel column as follows:

$$A = 35.578 \text{ cm}^2, \qquad I_{min} = 158 \text{ cm}^4$$

(1) The slenderness ratio is

$$\lambda = \frac{\mu l}{i} = \frac{\mu l}{\sqrt{I_{min}/A}} = 114$$

Since $\lambda > \lambda_p$, then Euler's formula should be used to determine the critical load, which can be given as

$$F_{cr} = A\sigma_{cr} = A\frac{\pi^2 E}{\lambda^2} = 557 \text{ kN}$$

(2) The slenderness ratio is

$$\lambda = \frac{\mu l}{\sqrt{I_{min}/A}} = 95$$

Since $\lambda < \lambda_p$, then the empirical formula should be used to determine the critical load, which can be given as

$$F_{cr} = A\sigma_{cr} = A(a - b\lambda) = 703 \text{ kN}$$

## 9.5 Design of Column

A straight line empirical formula is often used for an intermediate column, and Euler's formula is used for a long column. Therefore, three formulas of the critical stress used for design of long, intermediate, and short columns can be expressed as

$$\sigma_{cr} = \frac{\pi^2 E}{\lambda^2} \quad (\lambda \geqslant \lambda_p)$$
$$\sigma_{cr} = a - b\lambda \quad (\lambda_s \leqslant \lambda < \lambda_p) \quad (9.13)$$
$$\sigma_{cr} = \sigma_s \quad (\lambda < \lambda_s)$$

where $\lambda_s = (a - \sigma_s)/b$ and $\lambda_p = \sqrt{\pi^2 E / \sigma_p}$.

The ratio of the critical stress $\sigma_{cr}$ to the factor of safety $n$ is used to define the allowable stress $\sigma_{allow}$. The working stress $\sigma$ within a column under centric load should satisfy the following condition if the column is stable:

$$\sigma \leqslant \sigma_{allow} = \frac{\sigma_{cr}}{n} \quad (9.14)$$

**Example 9.3** A hollow circular steel column having a $D$=120 mm outer diameter and a $t$=15 mm wall thickness is subjected to a centric compressive load $F$ at the free end, Fig. 9.6. Determine the critical load if the length of the column is 1.8 m, 1.2 m, and 0.6 m. Use $\sigma_p$=280 MPa, $\sigma_s$=350 MPa, $E$=210 GPa, $a$=461 MPa, and $b$=2.568 MPa.

**Solution** For the column given in Fig. 9.6, the effective-length factor $\mu = 2.0$. And the critical slenderness ratios for the proportional limit and yield strength are equal to

$$\lambda_p = \sqrt{\frac{\pi^2 E}{\sigma_p}} = 86, \quad \lambda_s = \frac{a - \sigma_s}{b} = 43$$

(1) The slenderness ratio is

$$\lambda = \frac{\mu l}{i} = \frac{\mu l}{\sqrt{I/A}} = \frac{\mu l}{\sqrt{\left[\frac{1}{64}\pi(D^4 - d^4)\right] / \left[\frac{1}{4}\pi(D^2 - d^2)\right]}} = \frac{4\mu l}{\sqrt{D^2 + d^2}} = 96$$

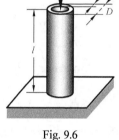

Fig. 9.6

Since $\lambda > \lambda_p$, then Euler's formula should be chosen to determine the critical load. The corresponding critical load can be computed as

$$F_{cr} = A\sigma_{cr} = \frac{1}{4}\pi(D^2 - d^2)\frac{\pi^2 E}{\lambda^2} = 1113 \text{ kN}$$

(2) The slenderness ratio is

$$\lambda = \frac{4\mu l}{\sqrt{D^2 + d^2}} = 64$$

Since $\lambda_s < \lambda < \lambda_p$, then the empirical formula should be chosen to determine the critical load. The corresponding critical load can be given as

$$F_{cr} = A\sigma_{cr} = \frac{1}{4}\pi(D^2 - d^2)(a - b\lambda) = 1468 \text{ kN}$$

(3) The slenderness ratio is

$$\lambda = \frac{4\mu l}{\sqrt{D^2 + d^2}} = 32$$

Since $\lambda < \lambda_s$, then the strength formula should be chosen to determine the critical load. The corresponding critical load can be expressed as

$$F_{cr} = A\sigma_{cr} = \frac{1}{4}\pi(D^2 - d^2)\sigma_s = 1732 \text{ kN}$$

**Example 9.4** A plane structure $ABC$ is composed of three bars, $AB$, $AC$ and $BC$, and connected by three pins at $A$, $B$ and $C$, Fig. 9.7. The bars $AB$, $AC$ and $BC$, each having diameter of 32 mm, have lengths of 1 m, 0.6 m and 0.8 m, respectively. The joint $C$ is subjected to a concentrated load $F$. Knowing that $\sigma_p = 280$ MPa, $\sigma_s = 350$ MPa, $E = 210$ GPa, $a = 461$ MPa, $b = 2.568$ MPa and $0 < \theta < 90°$, determine the critical load $F_{cr}$ which can be applied to the structure and the corresponding critical angle $\theta_{cr}$ (Note: Only in-plane buckling needs to be considered).

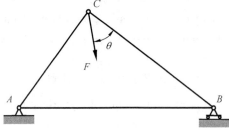

Fig. 9.7

**Solution** $\lambda_p = \sqrt{\dfrac{\pi^2 E}{\sigma_p}} = 86$, $\lambda_s = \dfrac{a - \sigma_s}{b} = 43$, $\lambda_{AC} = \dfrac{4\mu l_{AC}}{d} = 75$, $\lambda_{BC} = \dfrac{4\mu l_{BC}}{d} = 100$

Since $\lambda_s < \lambda_{AC} < \lambda_p < \lambda_{BC}$, i.e. $AC$ and $BC$ are respectively intermediate and long columns, then the empirical and Euler's formulas should be applied to calculating the critical loads of $AC$ and $BC$ respectively, i.e.

$$(F_{AC})_{cr} = \left(\frac{1}{4}\pi d^2\right)(a - b\lambda_{AC}) = 215.86 \text{ kN}, \qquad (F_{BC})_{cr} = \left(\frac{1}{4}\pi d^2\right)\frac{\pi^2 E}{\lambda_{BC}^2} = 166.69 \text{ kN}$$

Thus the critical load and corresponding critical angle can be expressed as

$$F_{cr} = \sqrt{(F_{AC})_{cr}^2 + (F_{BC})_{cr}^2} = 272.73 \text{ kN}, \qquad \theta_{cr} = \arctan\frac{(F_{AC})_{cr}}{(F_{BC})_{cr}} = 52.32°$$

## Problems

**9.1** Each of the four long columns consists of a steel rod that has a 30 mm diameter, Fig. 9.8. Knowing that $l=1.5$ m and $E=200$ GPa, determine the critical load $F_{cr}$ for each support condition.

**9.2** Each of the four long columns consists of an aluminum tube that has a 30 mm outer diameter and a 5 mm wall thickness, Fig. 9.9. Using $l=1.2$ m, $E=70$ GPa and a factor of safety of $n=2.3$, determine the allowable load $F_{\text{allow}}$ for each support condition.

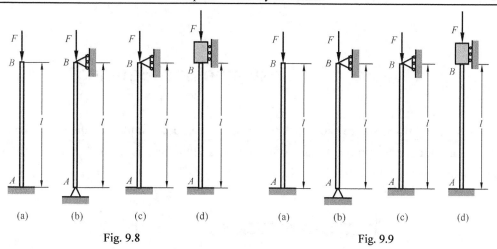

Fig. 9.8  Fig. 9.9

9.3  A solid circular steel column having a $d=150$ mm diameter is subjected to a centric compressive load $F$ at the free end, Fig. 9.10. Determine the critical load if the length of the column is 2.0 m, 1.2 m, and 0.5 m. Use $\sigma_p=280$ MPa, $\sigma_s=350$ MPa, $E=210$ GPa, $a=461$ MPa, and $b=2.568$ MPa.

9.4  A 25a I-steel column is subjected to a centric compressive load $F$ at the free end, Fig. 9.11. Knowing that $\sigma_p=200$ MPa, $E=206$ GPa, $a=304$ MPa, and $b=1.12$ MPa, determine the critical load of the column if the length of the column is $l=1.5$ m, $l=1.0$ m.

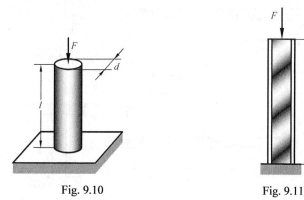

Fig. 9.10  Fig. 9.11

# Chapter 10  Unsymmetrical Bending

The bending discussed in the preceding chapters has been limited to the case where the beam has at least one longitudinal plane of symmetry and the load is applied in the longitudinal plane of symmetry, as shown in Fig. 10.1.

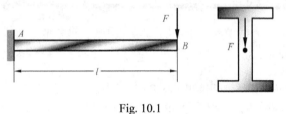

Fig. 10.1

This chapter will discuss two other general situations: the beam does not possess a longitudinal plane of symmetry, as shown in Fig. 10.2, or the load is not applied in any longitudinal plane of symmetry of the beam, as shown in Fig. 10.3.

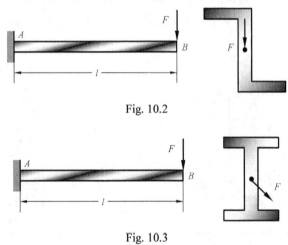

Fig. 10.2

Fig. 10.3

## 10.1  Unsymmetrical Pure Bending

Assume that the external couples are applied in the $xy$ plane, as shown in Fig. 10.4.

Considering an infinitesimal line element $dx$, as shown in Fig. 10.5, which is detached from the beam shown in Fig. 10.4, then the longitudinal normal strain can be given as

$$\varepsilon = \frac{\eta}{\rho} \tag{10.1}$$

where $\eta$ is the distance from the area element $dA$ to the neutral surface, and $\rho$ is the radius of curvature of the neutral surface.

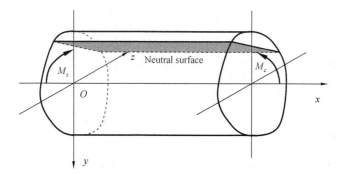

Fig. 10.4

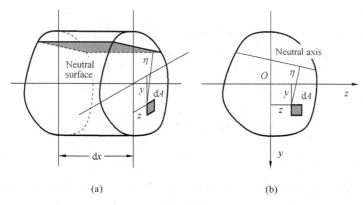

Fig. 10.5

Assume that the normal stresses in the beam remain below the proportional limit, and Hooke's law for uniaxial stress applies, i.e., $\sigma = E\varepsilon$. From Eq. (10.1), we have

$$\sigma = \frac{E}{\rho}\eta \tag{10.2}$$

where $E$ is the modulus of elasticity of the material.

According to the condition of static equilibrium in the $x$ direction, we have

$$\sum F_x = 0: \quad N = \int \sigma dA = \frac{E}{\rho}\int \eta dA = 0 \tag{10.3}$$

i.e.
$$\int \eta dA = 0 \tag{10.4}$$

Eq. (10.4) shows that, for an unsymmetrical beam subjected to pure bending, as long as the stresses remain in the elastic range, the neutral axis passes through the centroid of the section, as shown in Fig. 10.6.

Denoting by $\theta$ the included angle between the neutral axis and the $y$ axis, as shown in Fig. 10.6, then we obtain

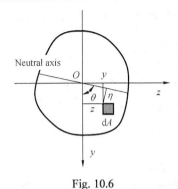

Fig. 10.6

$$\eta = y\sin\theta - z\cos\theta \qquad (10.5)$$

Using the condition of static equilibrium about the $y$ axis, we have

$$\sum M_y = 0: \ M_y = \int z\sigma dA = \frac{E}{\rho}\int z(y\sin\theta - z\cos\theta)dA = 0 \qquad (10.6)$$

or

$$\tan\theta = \frac{I_y}{I_{yz}} \qquad (10.7)$$

where $I_y = \int z^2 dA$ and $I_{yz} = \int yz dA$ are respectively the moment of inertia of the cross section with respect to the $y$ axis and the product of inertia of the cross section with respect to the $y$, $z$ axes.

Based on the condition of static equilibrium about the $z$ axis, we have

$$\sum M_z = 0: \ M_z = \int y\sigma dA = \frac{E}{\rho}\int y(y\sin\theta - z\cos\theta)dA \qquad (10.8)$$

or

$$\frac{1}{\rho} = \frac{M_z}{E\int(y^2\sin\theta - yz\cos\theta)dA} = \frac{M_z}{E(I_z\sin\theta - I_{yz}\cos\theta)} \qquad (10.9)$$

where $I_z = \int y^2 dA$ is the moment of inertia of the cross section with respect to the $z$ axis. Substituting $\eta$ from Eq. (10.5) and $1/\rho$ from Eq. (10.9) into Eq (10.2), we obtain

$$\sigma = \frac{M_z(yI_y - zI_{yz})}{I_y I_z - I_{yz}^2} \qquad (10.10)$$

If the external couples are respectively applied in the $xy$ and $xz$ planes, as shown in Fig. 10.7, then we can obtain

$$\sigma = \frac{M_z(yI_y - zI_{yz})}{I_y I_z - I_{yz}^2} + \frac{M_y(zI_z - yI_{yz})}{I_y I_z - I_{yz}^2} \qquad (10.11)$$

From Eq. (10.11), the orientation of the neutral axis can be determined as

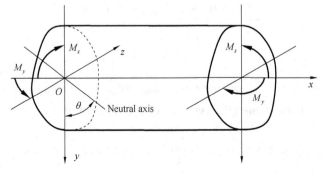

Fig. 10.7

$$\tan\theta = -\frac{M_z I_y - M_y I_{yz}}{M_y I_z - M_z I_{yz}} \tag{10.12}$$

If the external couples are respectively applied in the principal centroidal planes of inertia, i.e., $I_{yz} = 0$, then Eq. (10.11) and Eq. (10.12) can be simplified as

$$\sigma = \frac{M_z y}{I_z} + \frac{M_y z}{I_y}, \qquad \tan\theta = -\frac{M_z I_y}{M_y I_z} \tag{10.13}$$

When an external couple does not act in any principal centroidal plane of inertia, this external couple should first be resolved into components located in the two principal centroidal planes of inertia. Then Eq. (10.13) can be used to determine the normal stress caused by each couple component. Finally, using the principle of superposition, the resultant normal stress at each point can be determined.

If the external couples are only applied in one of the principal centroidal planes of inertia, say the $xy$ plane, then Eq. (10.13) can be rewritten as

$$\sigma = \frac{M_z y}{I_z}, \qquad \theta = 90° \tag{10.14}$$

Eq. (10.14) shows that, when the external couples are applied in the principal centroidal plane of inertia, the deflection distribution is a plane curve located in the principal centroidal plane of inertia. Such bending is called the plane bending.

## 10.2 Unsymmetrical Transverse-Force Bending

The normal stress formula obtained from an unsymmetrical beam in pure bending is still valid for an unsymmetrical beam subjected to transverse-force bending as long as this unsymmetrical beam is slender. However, for an unsymmetrical beam, say a think-walled beam shown in Fig. 10.8, in transverse-force bending, the beam will bend and twist under the action of the external force if the force passes through the centroid of the cross section.

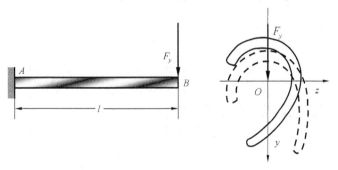

Fig. 10.8

Is it possible to apply an external force so that the beam will bend without twisting? If so, where should the external force be applied? Denoting by $y$ and $z$ axes, as shown in Fig. 10.9, the principal centroidal axes of inertia and assuming that the think-walled beam bends without twisting if the external force is applied at the point $S$, the shearing stress at any point of a given cross section can be given as

$$\tau = \frac{V_y Q_z'}{I_z t} \tag{10.15}$$

where $Q_z'$ is the first moment of the shaded area with respect to the principal centroidal axis $z$, $I_z$ the moment of inertia of the cross section with respect to the principal centroidal axis $z$, $V_y$ the shearing force on the cross section along the $y$ axis, and $t$ the thickness of the think-walled beam.

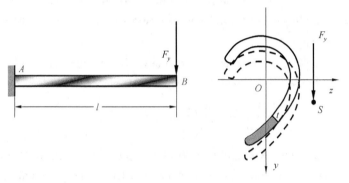

Fig. 10.9

When the external force is applied at the point $S$, the beam bends without twisting. The point $S$ is usually called the shear center or flexural center. The shear center normally does not coincide with the centroid of the cross section for a think-walled beam. Denoting by $a_z$ the distance between the line of action of the external force and an arbitrary reference point $a$ on the cross section, then from Fig. 10.10 we have

$$F_y a_z = \int r\tau dA \tag{10.16}$$

where $r$ is the distance of the line of action of the resultant force on the area element $dA$ from the reference point $a$.

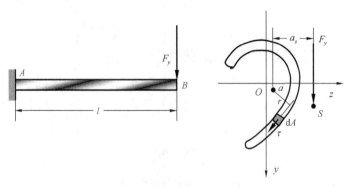

Fig. 10.10

In the case of an oblique external force shown in Fig. 10.11, the think-walled beam will also be free of twist if the force is also applied at the shear center, since this oblique external force can be resolved into two components $F_y$ and $F_z$, neither of which causes the beam to twist.

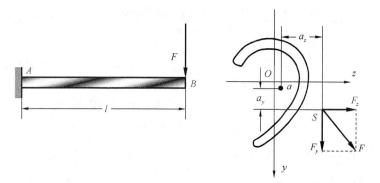

Fig. 10.11

**Example 10.1** As shown in Fig. 10.12, the cantilever made of a 160×160×16 angle steel is subjected to the external force acting on the free end. Knowing that $F = 15$ kN, $l = 1$ m, $E = 200$ GPa, determine the maximum normal stress and the maximum deflection.

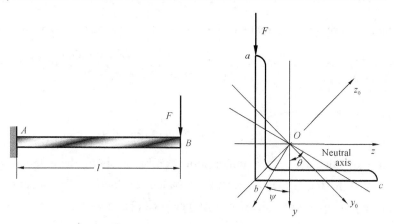

Fig. 10.12

**Solution** (1) From the table of shape steels in Appendix II, we obtain

$$I_y = I_z = 1175.08 \text{ cm}^4, \qquad I_{y_0} = 484.59 \text{ cm}^4, \qquad I_{z_0} = 1865.57 \text{ cm}^4$$

The product of inertia corresponding to the $Oxyz$ coordinate system can be given as

$$I_{yz} = \frac{I_{y_0} - I_{z_0}}{2}\sin(2\alpha) + I_{y_0 z_0}\cos(2\alpha) = 690.49 \text{ cm}^4$$

The orientation of the neutral axis can be determined as

$$\theta = \arctan\frac{I_y}{I_{yz}} = 59.53°$$

It is obvious that the maximum tensile normal stress is located at the point $a$ on the cross section $A$ and the maximum compressive normal stress is located at the point $b$. Using $\sigma = \dfrac{M_z(yI_y - zI_{yz})}{I_y I_z - I_{yz}^2}$, we obtain

$$\sigma_a = \frac{(-15) \times [(-11.45) \times (1175.08) - (-4.55) \times (690.49)]}{1175.08^2 - 690.49^2} = 171.11 \text{(MPa)}$$

$$\sigma_b = \frac{(-15) \times [(4.55) \times (1175.08) - (-4.55) \times (690.49)]}{1175.08^2 - 690.49^2} = -140.84 \text{(MPa)}$$

The above calculation shows that the maximum normal stress is equal to

$$\sigma_{max} = \sigma_a = 171.11 \text{ MPa}$$

(2) The maximum normal stress can also be calculated by resolving the external force into two components located in the principal centroidal planes of inertia. Using $\sigma = \dfrac{M_{z_0} y_0}{I_{z_0}} + \dfrac{M_{y_0} z_0}{I_{y_0}}$, the maximum tensile normal stress at the point $a$ on the cross section $A$ can be expressed as

$$\sigma_a = \frac{Fl}{\sqrt{2}} \left[ -\frac{(y_0)_a}{I_{z_0}} + \frac{(z_0)_a}{I_{y_0}} \right] = \frac{15}{\sqrt{2}} \times \left( -\frac{-11.31}{1865.57} + \frac{4.88}{484.59} \right) = 171.11 \text{(MPa)}$$

and the maximum compressive normal stress at the point $b$ can also be expressed as

$$\sigma_b = \frac{Fl}{\sqrt{2}} \left[ -\frac{(y_0)_b}{I_{z_0}} + \frac{(z_0)_b}{I_{y_0}} \right] = \frac{15}{\sqrt{2}} \times \left( -\frac{0}{1865.57} + \frac{-6.43}{484.59} \right) = -140.74 \text{(MPa)}$$

Hence the maximum normal stress is

$$\sigma_{max} = \sigma_a = 171.11 \text{ MPa}$$

(3) It is easy to calculate the maximum deflection by resolving the external force into two components located in the principal centroidal planes of inertia. The maximum deflection components in the principal centroidal planes of inertia can be rewritten, respectively, as

$$w_{y_0} = \frac{(F/\sqrt{2})l^3}{3EI_{z_0}} = 0.95 \text{ mm}, \quad w_{z_0} = -\frac{(F/\sqrt{2})l^3}{3EI_{y_0}} = -3.65 \text{ mm}$$

It is obvious that the maximum deflection is equal to

$$w_{max} = \sqrt{w_{y_0}^2 + w_{z_0}^2} = 3.77 \text{ mm}$$

and the direction of the maximum deflection can be determined as

$$\psi = \arctan \frac{(w_{y_0}/\sqrt{2}) + (w_{z_0}/\sqrt{2})}{(w_{y_0}/\sqrt{2}) - (w_{z_0}/\sqrt{2})} = -30.41°$$

**Example 10.2** The thin-walled cantilever beam of thickness $t$ ($t \ll h$, $t \ll b$) is subjected to the external force applied at the shear center, as shown in Fig. 10.13. Determine the distribution of shearing stresses and the location of the shear center.

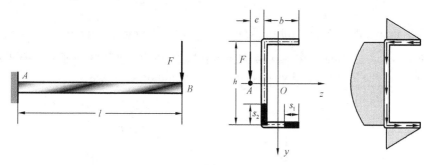

Fig. 10.13

**Solution** (1) The moment of inertia of the cross section with respect to the $z$ axis is given as

$$I_z = \frac{1}{12}th^3 + 2\left[\frac{1}{12}bt^3 + bt\left(\frac{h}{2}\right)^2\right]$$

Neglecting the term containing $t^3$, which is very small, we have

$$I_z = \frac{1}{12}th^2(h+6b)$$

Assuming that the external force is applied at the shear center, then the beam does not twist when it is subjected to a transverse force. The shearing stress distribution in each flange is equal to

$$\tau = \frac{V_y Q_z'}{I_z t} = \frac{6Fs_1}{th(h+6b)}$$

and the shearing stress distribution in the web is

$$\tau = \frac{V_y Q_z'}{I_z t} = \frac{6F(hb + hs_2 - s_2^2)}{th^2(h+6b)}$$

The maximum shearing stress is located on the neutral axis and is equal to

$$\tau_{max} = \frac{3F(h+4b)}{2th(h+6b)}$$

(2) Denoting by $e$ the distance from the shear center to the centerline of the web, then we have

$$Fe = \int_0^b h\tau t ds_1 = \int_0^b \frac{6Fs_1}{h+6b}ds_1 = \frac{3Fb^2}{h+6b}$$

i.e.
$$e = \frac{3b^2}{h+6b}$$

## Problems

**10.1** A thin-walled beam of thickness $t$ ($t \ll a$), simply supported at both ends, is loaded by a force $F$ in the middle, Fig. 10.14. Determine the normal stress distribution under the load as well as the location and value of the maximum normal stress.

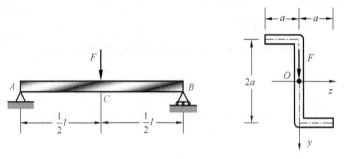

Fig. 10.14

**10.2** A thin-walled cantilever beam of thickness $t$ ($t \ll a$) is subjected to a constant line load $q$, Fig. 10.15. Determine the distribution of the normal stress in the cross section at the support.

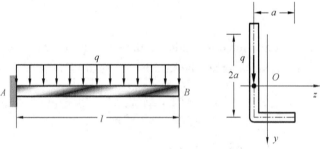

Fig. 10.15

**10.3** A thin-walled beam, produced from an aluminum sheet of $t = 2$ mm thickness, is loaded by a force $F$ in the middle, Fig. 10.16. The beam is. Knowing that $F = 1.5$ kN, $E = 70$ GPa, $l = 1.2$ m, $a = 40$ mm, compute the deflection at the point where the force is applied.

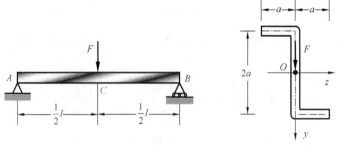

Fig. 10.16

10.4 The thin-walled cantilever beam of uniform thickness $t$ ($t \ll h$, $t \ll b$) is subjected to the force applied at the shear center, Fig. 10.17. Denoting by $E$ the modulus of elasticity, determine the distributions of normal and shearing stresses in the section at the support and the deflection of the beam at the free end.

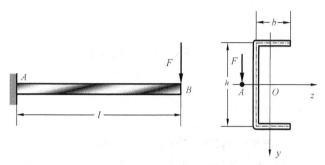

Fig. 10.17

# Chapter 11   Energy Methods

## 11.1   External Work

### 1. Work of External Force

Consider a deformable body supported at $A$ and $B$ shown in Fig. 11.1. When an external force is applied to the body gradually from zero to a final value $F$, the corresponding displacement of the force application point in the force direction is also slowly increased from zero to a final value $\delta$. If the material is within a linearly elastic range, then the work done by the external force applied to the body can be expressed as

$$W_e = \frac{1}{2} F \delta \tag{11.1}$$

### 2. Work of External Couple

When an external couple is applied to the body gradually from zero to a final value $M_e$, the corresponding angular displacement of the couple is slowly increased from zero to a final value $\theta$. If the material is within a linearly elastic range, then the work done by the external couple can be expressed, Fig. 11.2, as

$$W_e = \frac{1}{2} M_e \theta \tag{11.2}$$

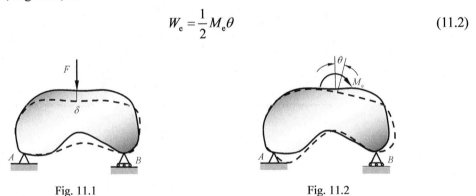

Fig. 11.1                Fig. 11.2

### 3. Work of Multiple External Loads

Consider a deformable body supported at $A$ and $B$ shown in Fig. 11.3, which is subjected to a series of external forces $F_1, F_2, \cdots, F_i, \cdots, F_n$. When these external forces are applied to the body gradually from zero to their final values, the corresponding displacements of the force application points in the force directions are also slowly increased from zero to their final values. If the material is within a linearly elastic range, then the work done by the external forces applied to the body can be expressed as

$$W_e = \sum \frac{1}{2} F_i \Delta_i \qquad (11.3)$$

where $\Delta_i$ is a function of the forces $F_1, F_2, \cdots, F_i, \cdots, F_n$, that is, $\Delta_i = \Delta_i(F_1, F_2, \cdots, F_i, \cdots, F_n)$ $(i = 1, 2, \cdots, n)$. It should be noted that $F_i$ is a generalized force and $\Delta_i$ is the corresponding generalized displacement. If $F_i$ is a couple, then $\Delta_i$ is an angle of rotation corresponding to the couple.

**4. Work of Combined Loadings**

Consider a slender circular member subjected to combined loadings, Fig. 11.4. If the deformation is small, the external work done by the combined loadings can be written as

$$W_e = \sum \frac{1}{2} F_i \delta_i \qquad (11.4)$$

where $\delta_i$ is a function of the force $F_i$, that is, $\delta_i = \delta_i(F_i)$ $(i = 1, 2, 3)$.

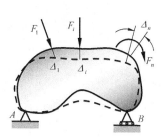
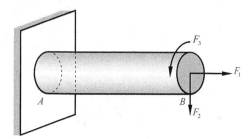

Fig. 11.3　　　　　　　　　　　Fig. 11.4

## 11.2　Stain-Energy Density

**1. Strain-Energy Density for Uniaxial Stress**

When a member is subjected to uniaxial stress, Fig. 11.5, the strain-energy density is equal to the area under the stress-strain curve and can be expressed as

$$u = \int \sigma d\varepsilon \qquad (11.5)$$

where $\varepsilon$ is the normal strain corresponding to $\sigma$. For values of $\sigma$ within the linear elastic range, we have $\sigma = E\varepsilon$, where $E$ is the modulus of elasticity of the material. Substituting $\sigma$ into Eq. (11.5) and performing the integration, we have

$$u = \frac{1}{2} E \varepsilon^2 = \frac{1}{2} \sigma \varepsilon = \frac{\sigma^2}{2E} \qquad (11.6)$$

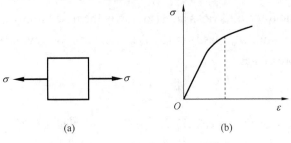

(a)　　　　　　　　　　　　(b)

Fig. 11.5

### 2. Strain-Energy Density for Pure Shearing Stress

When a member is subjected to pure shearing stress, Fig. 11.6, the strain-energy density can be expressed as

$$u = \int \tau \mathrm{d}\gamma \tag{11.7}$$

where $\gamma$ is the shearing strain corresponding to $\tau$. From Eq. (11.7), we can see that the strain-energy density $u$ is equal to the area under the stress-strain curve. For values of $\tau$ within the proportional limit, i.e. $\tau = G\gamma$, where $G$ is the modulus of elasticity in shear, we obtain

$$u = \frac{1}{2}G\gamma^2 = \frac{1}{2}\tau\gamma = \frac{\tau^2}{2G} \tag{11.8}$$

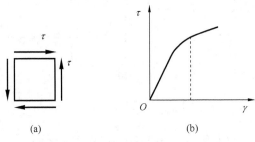

(a)　　　　　　　　　　　　(b)

Fig. 11.6

## 11.3　Strain Energy

### 1. Axially Tensile or Compressive Strain Energy

Consider a prismatic bar in axial tension or compression, which is fixed at the left end and subjected at the right end to a gradually increasing external force $F$ coinciding with the axial line of the bar, Fig. 11.7.

According to Eq. (11.6), the strain energy of the bar can be written as

$$V_\varepsilon = \int \frac{\sigma^2}{2E} A \mathrm{d}x \tag{11.9}$$

Using $\sigma = \dfrac{N}{A}$, where $N$ is the axial force, and $A$ is the cross-sectional area, we have

$$V_\varepsilon = \int \dfrac{N^2}{2EA} dx \tag{11.10}$$

In the case of a bar having a uniform cross-sectional area $A$ and a constant axial force $N$, Eq. (11.10) can be rewritten as

$$V_\varepsilon = \dfrac{N^2 l}{2EA} \tag{11.11}$$

## 2. Torsional Strain Energy

Consider a circular shaft in torsion, which is fixed at the left end and subjected at the right end to a gradually increasing external couple $M_e$ applied in a plane perpendicular to the longitudinal axis of the shaft, Fig. 11.8.

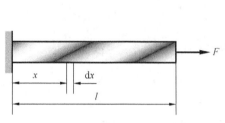
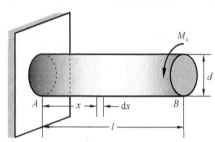

Fig. 11.7          Fig. 11.8

From Eq. (11.8), the strain energy for the shaft can be given as

$$V_\varepsilon = \int \left( \int \dfrac{\tau^2}{2G} dA \right) dx \tag{11.12}$$

Since $\tau = \dfrac{T\rho}{I_p}$, where $T$ is the torsional moment, $I_p$ is the polar moment of inertia of the cross section, and $\rho$ is the distance from the centroid of the cross section, then we have

$$V_\varepsilon = \int \left( \dfrac{T^2}{2GI_p^2} \int \rho^2 dA \right) dx \tag{11.13}$$

Using $I_p = \int \rho^2 dA$, Eq. (11.13) can be rewritten as

$$V_\varepsilon = \int \dfrac{T^2}{2GI_p} dx \tag{11.14}$$

If the shaft has a uniform polar moment of inertia $I_p$ and a constant torsional moment $T$, then Eq. (11.14) can be expressed as

$$V_\varepsilon = \dfrac{T^2 l}{2GI_p} \tag{11.15}$$

## 3. Bending Strain Energy

Consider a prismatic beam in bending, which is fixed at the left end and subjected at the right end to a gradually increasing vertical external force $F$ applied in a symmetric plane containing the longitudinal axis of the beam, Fig. 11.9. Since the strain energy caused by the shearing stress is usually small, compared to that caused by the normal stress, and can be neglected for a slender beam. Therefore, the strain energy for the beam in bending can be expressed as

$$V_\varepsilon = \int \left( \int \frac{\sigma^2}{2E} dA \right) dx \tag{11.16}$$

Using $\sigma = \dfrac{My}{I}$, where $M$ is the bending moment, $I$ is the moment of inertia of the cross section, and $y$ is the perpendicular distance from the neutral axis, we have

$$V_\varepsilon = \int \left( \frac{M^2}{2EI^2} \int y^2 dA \right) dx \tag{11.17}$$

Substituting $I = \int y^2 dA$ into Eq. (11.17), then Eq. (11.17) can be rewritten as

$$V_\varepsilon = \int_0^l \frac{M^2}{2EI} dx \tag{11.18}$$

For a beam in pure bending, Fig. 11.10, the bending moment $M$ is a constant. If the beam has a uniform moment of inertia $I$, then Eq. (11.18) can be expressed as

$$V_\varepsilon = \frac{M^2 l}{2EI} \tag{11.19}$$

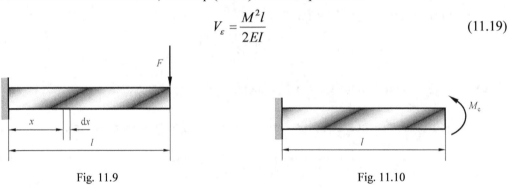

Fig. 11.9        Fig. 11.10

## 4. Combined Loading Strain Energy

Consider a circular member subjected to combined loadings, Fig. 11.11. For a slender member, the strain energy corresponding to the shearing stress caused the shearing force can be neglected, thus the strain energy for the member subjected to combined loadings can be expressed as

$$V_\varepsilon = \int \frac{N^2}{2EA} dx + \int \frac{T^2}{2GI_p} dx + \int \frac{M^2}{2EI} dx \tag{11.20}$$

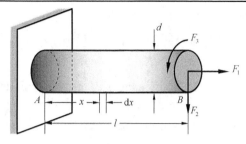

Fig. 11.11

## 11.4 Principle of Work and Energy

If we neglect the loss of energy, then the conservation of energy requires that the work done by the external forces applied to a body be transformed entirely into the strain energy stored in the deformable body. This relation can be stated mathematically as

$$V_\varepsilon = W_e \tag{11.21}$$

where $V_\varepsilon$ is the strain energy and $W_e$ is the external work. Eq. (11.21) is called the principle of work and energy, and can be used to determine the deformation of a structure.

**Example 11.1** Knowing that the bending rigidity is $EI$, Fig. 11.12, use the principle of work and energy to determine the deflection at the free end $B$ of the prismatic cantilever $AB$.

**Solution** The external work and strain energy can be given as

$$W_e = \frac{1}{2} F \delta_B, \quad V_\varepsilon = \int_0^l \frac{M^2}{2EI} dx = \int_0^l \frac{(Fx)^2}{2EI} dx = \frac{F^2 l^3}{6EI}$$

Using the principle of work and energy, i.e., $W_e = V_\varepsilon$, we have

$$\frac{1}{2} F \delta_B = \frac{F^2 l^3}{6EI}, \text{ or } \delta_B = \frac{F l^3}{3EI}$$

**Example 11.2** Using the principle of work and energy, determine the slope at the free end $B$ of the prismatic cantilever $AB$, Fig. 11.13. Know that the bending rigidity is $EI$.

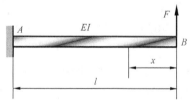
Fig. 11.12

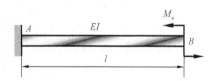
Fig. 11.13

**Solution** $W_e = \frac{1}{2} M_e \theta_B, \quad V_\varepsilon = \int_0^l \frac{M^2}{2EI} dx = \int_0^l \frac{M_e^2}{2EI} dx = \frac{M_e^2 l}{2EI}$

Using $W_e = V_\varepsilon$, we have

$$\frac{1}{2} M_e \theta_B = \frac{M_e^2 l}{2EI}, \text{ or } \theta_B = \frac{M_e l}{EI}$$

## 11.5 Reciprocal Theorem

Consider that a linearly elastic body, supported at $A$ and $B$, is subjected to generalized forces, as shown in Fig. 11.14. Assuming that the generalized forces, $F_1$ and $F_3$, are first applied to the body and then the generalized forces, $F_2$ and $F_4$, are applied to the body again, the strain energy stored in the body can be expressed as

$$V_{\varepsilon 1} = \frac{1}{2}F_1\Delta_1 + \frac{1}{2}F_3\Delta_3 + \frac{1}{2}F_2\Delta_2 + \frac{1}{2}F_4\Delta_4 + F_1\Delta_1' + F_3\Delta_3' \qquad (11.22)$$

where $\Delta_1$ (or $\Delta_3$) is the generalized displacement at the application point of $F_1$ (or $F_3$) in the direction of $F_1$ (or $F_3$), caused by both $F_1$ and $F_3$, $\Delta_2$ (or $\Delta_4$) is the generalized displacement at the application point of $F_2$ (or $F_4$) in the direction of $F_2$ (or $F_4$), caused by both $F_2$ and $F_4$, and $\Delta_1'$ (or $\Delta_3'$) is the generalized displacement at the application point of $F_1$ (or $F_3$) in the direction of $F_1$ (or $F_3$), caused by both $F_2$ and $F_4$.

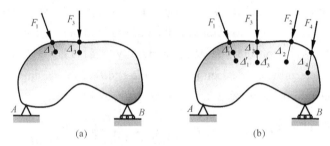

Fig. 11.14

When the generalized forces, $F_2$ and $F_4$, are first applied to the body and then the generalized forces, $F_1$ and $F_3$, are applied to the body again, the strain energy stored in the body can be given as

$$V_{\varepsilon 2} = \frac{1}{2}F_2\Delta_2 + \frac{1}{2}F_4\Delta_4 + \frac{1}{2}F_1\Delta_1 + \frac{1}{2}F_3\Delta_3 + F_2\Delta_2' + F_4\Delta_4' \qquad (11.23)$$

where $\Delta_2'$ (or $\Delta_4'$) is the generalized displacement at the application point of $F_2$ (or $F_4$) in the direction of $F_2$ (or $F_4$), caused by both $F_1$ and $F_3$.

By comparison of Eq. (11.22) and Eq. (11.23) and using $V_{\varepsilon 1} = V_{\varepsilon 2}$, we have

$$F_1\Delta_1' + F_3\Delta_3' = F_2\Delta_2' + F_4\Delta_4' \qquad (11.24)$$

We thus conclude that the work done by the first group of generalized forces on the generalized displacement caused by the second group of generalized forces is equal to the work done by the second group of generalized forces on the generalized displacement caused by the first group of generalized forces. The relation above is called the reciprocal theorem of work.

Assuming that $F_3 = F_4 = 0$ and $F_1 = F_2$, then Eq. (11.24) can be simplified as

$$\Delta_1' = \Delta_2' \qquad (11.25)$$

We thus conclude that, when two generalized forces are equal in magnitude, the generalized displacement at the application point of the first force in the direction of the first force caused by the second force is equal to the generalized displacement at the application point of the second force in the direction of the second force caused by the first force. This relation is called the reciprocal theorem of displacement.

**Example 11.3** A linearly elastic sphere is subjected to two radial tensile loads, equal in magnitude and opposite in direction, as shown in Fig. 11.15(a). Knowing that the sphere has modulus of elasticity $E$, Poisson's ratio $\mu$ and diameter $d$, determine the change in volume of the sphere.

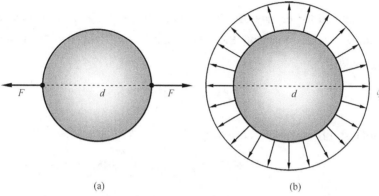

Fig. 11.15

**Solution** Assuming that the sphere is subjected to triaxial stress in uniform tension, as shown in Fig. 11.15(b), then we can obtain

$$\varepsilon_1 = \frac{1}{E}[\sigma_1 - \mu(\sigma_2 + \sigma_3)] = \frac{1-2\mu}{E}q$$

Using the reciprocal theorem of work, we have

$$F(\varepsilon_1 d) = q\Delta V$$

i.e.
$$\Delta V = \frac{F(\varepsilon_1 d)}{q} = \frac{1-2\mu}{E}Fd$$

## 11.6 Castigliano's Theorem

Consider a deformable body supported at $A$ and $B$ shown in Fig. 11.16, which is subjected to a series of external forces $F_1, F_2, \cdots, F_i, \cdots, F_n$. When these external forces are applied to the body gradually from zero to their final values, the corresponding displacements of the force application points in the force directions are also slowly increased from zero to their final values. If the material is within a linearly elastic range, then the work done by the external forces applied to the body can be expressed as

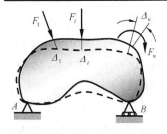

Fig. 11.16

$$W_e = \sum \frac{1}{2} F_i \Delta_i \tag{11.26}$$

where $\Delta_i$ is a linear homogeneous function of the external forces $F_1, F_2, \cdots, F_i, \cdots, F_n$. It can be seen from Eq. (11.26) that the external work $W_e$ is a quadratic homogeneous function of the external forces $F_1, F_2, \cdots, F_i, \cdots, F_n$. Using the principle of work and energy, i.e., $W_e = V_\varepsilon$, then the strain energy is also a quadratic homogeneous function of the external forces $F_1, F_2, \cdots, F_i, \cdots, F_n$ and can be expressed as

$$V_\varepsilon = V_\varepsilon(F_1, F_2, \cdots, F_i, \cdots, F_n) \tag{11.27}$$

If any one of the external forces, say the $i$th external force $F_i$, is increased by $dF_i$ while the other external forces remain constant, then the strain energy will also be increased by $dV_\varepsilon$, which can be denoted as

$$dV_\varepsilon = \frac{\partial V_\varepsilon}{\partial F_i} dF_i \tag{11.28}$$

Therefore, the total strain energy can be written as

$$V_{\varepsilon 1} = V_\varepsilon + dV_\varepsilon = \sum \frac{1}{2} F_i \Delta_i + \frac{\partial V_\varepsilon}{\partial F_i} dF_i \tag{11.29}$$

The total strain energy $V_{\varepsilon 1}$ should be independent of the order in which the external forces are applied to the body. For example, we can apply the external force $dF_i$ to the body first, then apply the external forces $F_1, F_2, \cdots, F_i, \cdots, F_n$. The total strain energy should be the same. When the external force $dF_i$ is applied first, the external work $dW_e'$ done by the external force $dF_i$ can be written as

$$dW_e' = \frac{1}{2} dF_i d\Delta_i \tag{11.30}$$

where $d\Delta_i$ is the displacement corresponding to the external force $dF_i$. When the external forces $F_1, F_2, \cdots, F_i, \cdots, F_n$ are applied, the external work done by the external force $F_1, F_2, \cdots, F_i, \cdots, F_n$ and the external force $dF_i$ can be expressed as

$$W_e' = \sum \frac{1}{2} F_i \Delta_i + dF_i \Delta_i \tag{11.31}$$

Therefore, the total work can be written as

$$W_{e2} = W_e' + dW_e' = \sum \frac{1}{2} F_i \Delta_i + dF_i \Delta_i + \frac{1}{2} dF_i d\Delta_i \tag{11.32}$$

Using the principle of work and energy, the total strain energy can be given as

$$V_{\varepsilon 2} = W_{e2} = \sum \frac{1}{2} F_i \Delta_i + dF_i \Delta_i + \frac{1}{2} dF_i d\Delta_i \tag{11.33}$$

Using $V_{\varepsilon 1} = V_{\varepsilon 2}$, we have

$$\sum \frac{1}{2} F_i \Delta_i + \frac{\partial V_\varepsilon}{\partial F_i} \mathrm{d}F_i = \sum \frac{1}{2} F_i \Delta_i + \mathrm{d}F_i \Delta_i + \frac{1}{2} \mathrm{d}F_i \mathrm{d}\Delta_i \tag{11.34}$$

Neglecting the second-order differential, we obtain

$$\Delta_i = \frac{\partial V_\varepsilon}{\partial F_i} \tag{11.35}$$

where $F_i$ is a generalized force, and $\Delta_i$ is the corresponding generalized displacement. If $F_i$ is a couple, then $\Delta_i$ is an angle of rotation corresponding to the couple. Eq. (11.35) is called Castigliano's second theorem or Castigliano's theorem, which can be stated as follows: the generalized displacement $\Delta_i$ at the application point of $F_i$ in the direction of $F_i$ is equal to the first partial derivative of the strain energy with respect to the generalized force $F_i$.

When Castigliano's theorem is used for a member subjected to combined loadings it can be expressed as

$$\Delta_i = \frac{\partial V_\varepsilon}{\partial F_i} = \int \frac{N}{EA} \frac{\partial N}{\partial F_i} \mathrm{d}x + \int \frac{T}{GI_p} \frac{\partial T}{\partial F_i} \mathrm{d}x + \int \frac{M}{EI} \frac{\partial M}{\partial F_i} \mathrm{d}x \tag{11.36}$$

**Example 11.4** Knowing that the bending rigidity is $EI$, Fig. 11.17, use Castigliano's theorem to determine the deflection at the free end $B$ of the prismatic cantilever $AB$.

**Solution** The strain energy is equal to

$$V_\varepsilon = \int_0^l \frac{M^2}{2EI} \mathrm{d}x = \int_0^l \frac{(Fx)^2}{2EI} \mathrm{d}x = \frac{F^2 l^3}{6EI}$$

Using Castigliano's theorem, we have

$$\delta_B = \frac{\partial V_\varepsilon}{\partial F} = \frac{\partial}{\partial F} \left( \frac{F^2 l^3}{6EI} \right) = \frac{F l^3}{3EI}$$

**Example 11.5** Knowing that the bending rigidity is $EI$, use Castigliano's theorem to determine the slope at the free end $B$ of the prismatic cantilever $AB$, Fig. 11.18.

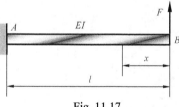

Fig. 11.17

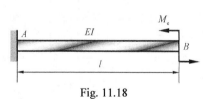

Fig. 11.18

**Solution** The strain energy is

$$V_\varepsilon = \int_0^l \frac{M^2}{2EI} \mathrm{d}x = \int_0^l \frac{M_e^2}{2EI} \mathrm{d}x = \frac{M_e^2 l}{2EI}$$

Using $\Delta_i = \frac{\partial V_\varepsilon}{\partial F_i}$, we have

$$\theta_B = \frac{\partial V_\varepsilon}{\partial M_e} = \frac{\partial}{\partial M_e} \left( \frac{M_e^2 l}{2EI} \right) = \frac{M_e l}{EI}$$

The only displacements that can be determined from Castigliano's theorem are those that correspond to external forces applied to the body. If we want to find a displacement at a point on the body where there is no external force, then a fictitious (or dummy) force corresponding to the displacement to be determined must be applied to the body. We can then determine the displacement by calculating the strain energy and taking the partial derivative with respect to the fictitious force. The result obtained is that displacement produced simultaneousty by the actual forces and the fictitious force. By setting the fictitious force equal to zero, we can then obtain the displacement produced only by the actual forces.

**Example 11.6** Knowing that the bending rigidity is $EI$, use Castigliano's theorem to determine the slope at the free end $B$ of the prismatic cantilever $AB$, Fig. 11.19(a).

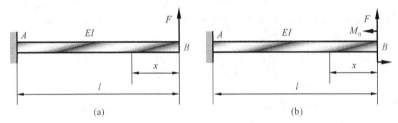

Fig. 11.19

**Solution** Adding a couple $M_0$ at the free end $B$ of the beam, Fig. 11.19(b), then the bending moment at the $x$ section is equal to

$$M = Fx + M_0 \quad (0 \leqslant x < l)$$

Therefore, the corresponding strain energy can be expressed as

$$V_\varepsilon = \int_0^l \frac{M^2}{2EI} dx = \int_0^l \frac{(Fx + M_0)^2}{2EI} dx = \frac{\frac{1}{3}F^2 l^3 + FM_0 l^2 + M_0^2 l}{2EI}$$

Using Castigliano's theorem, i.e., $\varDelta_i = \dfrac{\partial V_\varepsilon}{\partial F_i}$, we have

$$\theta_B = \left(\frac{\partial V_\varepsilon}{\partial M_0}\right)_{M_0=0} = \left(\frac{Fl^2 + 2M_0 l}{2EI}\right)_{M_0=0} = \frac{Fl^2}{2EI}$$

## 11.7 Principle of Virtual Work

Consider that a deformable body supported at $A$ and $B$ is balanced under the action of the loads $F$, $q$ and $M_e$, as shown in Fig. 11.20.

When a virtual deformation is produced in the body due to other factors, such as load change or temperature fluctuation, then the virtual work done by the loads $F$, $q$ and $M_e$ on the virtual deformation can be expressed as

$$W_e = F\Delta_F^* + \int q\Delta_q^* dx + M_e \Delta_{M_e}^* \qquad (11.37)$$

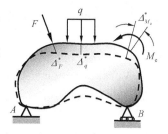

where $\Delta_F^*$, $\Delta_q^*$ and $\Delta_{M_e}^*$ are respectively the corresponding virtual displacements at the application points of $F$, $q$ and $M_e$ in the directions of $F$, $q$ and $M_e$.

When the body is subjected to the virtual deformation, the internal forces caused by the loads $F$, $q$ and $M_e$ applied to the body will do virtual work, which can be given as

Fig. 11.20

$$W_i = \int N d\delta^* + \int M d\theta^* + \int V d\lambda^* + \int T d\varphi^* \qquad (11.38)$$

where $N$, $M$, $V$ and $T$ are respectively the axial force, bending moment, shearing force, and torsional moment caused by the loads $F$, $q$ and $M_e$, and $\delta^*$, $\theta^*$, $\lambda^*$ and $\varphi^*$ are respectively the virtual deformations corresponding to $N$, $M$, $V$ and $T$.

Using $W_i = W_e$, we have

$$\int N d\delta^* + \int M d\theta^* + \int V d\lambda^* + \int T d\varphi^* = F\Delta_F^* + \int q\Delta_q^* dx + M_e \Delta_{M_e}^* \qquad (11.39)$$

This is the principle of virtual work, which can be stated as follows: The virtual work done by the internal forces on the virtual deformation is equal to the virtual work done by the external loads on the virtual displacements. The principle of virtual work can be used for both linearly and nonlinearly elastic bodies.

**Example 11.7** As shown in Fig. 11.21(a), a truss $ABC$ consisting of two rods $AC$ and $BC$, each having a constant cross-sectional area $A$, is subjected to a concentrated force $F$ at joint $C$. The stress-strain relation is $\sigma = E\varepsilon$ for $AC$ and $\sigma^2 = k\varepsilon$ for $BC$, where $E$ and $k$ are constants. Using the principle of virtual work, determine the horizontal and vertical components of displacement at $C$. (Note: The stability of structure does not need to be considered.)

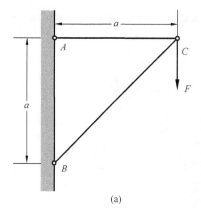
(a)

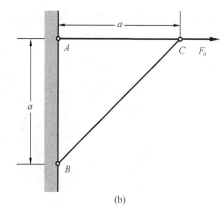
(b)

Fig. 11.21

**Solution** When a horizontal force $F_0$, in the direction to the right, is applied to the truss at $C$, as shown in Fig. 11.21(b), the internal forces produced in $AC$ and $BC$ can be given as

$$N_{AC} = F_0, \quad N_{BC} = 0$$

When the original load $F$ is applied to the structure, the deformations produced in $AC$ and $BC$ can be regarded as virtual deformations, which can be expressed as

$$\delta^*_{AC} = \frac{Fa}{EA}, \quad \delta^*_{BC} = \left[-\frac{(-\sqrt{2}F)^2}{kA^2}\right](\sqrt{2}a) = -\frac{2\sqrt{2}F^2 a}{kA^2}$$

and the displacement produced at $C$ can be regarded as virtual displacement, which is equal to

$$\delta^*_C = \Delta_H$$

where $\Delta_H$ is the horizontal components of displacement at $C$ to be determined.

Using the principle of virtual work, we have

$$F_0 \Delta_H = N_{AC} \frac{Fa}{EA} + N_{BC} \left(-\frac{2\sqrt{2}F^2 a}{kA^2}\right)$$

i.e.

$$\Delta_H = \frac{Fa}{EA}$$

If a vertical force downward is applied to the truss at $C$, in the same way, we can find the vertical component of displacement at $C$, equal to

$$\Delta_V = \frac{Fa}{EA} + (-\sqrt{2})\left(-\frac{2\sqrt{2}F^2 a}{kA^2}\right) = \frac{Fa}{EA} + \frac{4F^2 a}{kA^2}$$

## 11.8 Unit Load Method

Consider that a deformable body supported at $A$ and $B$ is balanced under the action of the loads $F$, $q$ and $M_e$, as shown in Fig. 11.22(a).

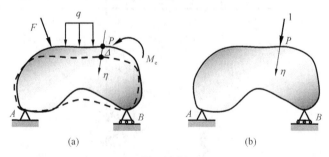

Fig. 11.22

In order to determine the displacement $\Delta$ at $P$ in the direction of $\eta$, a unit load is first applied to the body at $P$ in the direction of $\eta$, as shown in Fig. 11.22(b), and then the original loads are also applied to the body. If the deformations produced by the original loads are regarded as virtual deformations, then from the principle of virtual work we can obtain

$$\Delta = \int \overline{N} \mathrm{d}\delta + \int \overline{M} \mathrm{d}\theta + \int \overline{V} \mathrm{d}\lambda + \int \overline{T} \mathrm{d}\varphi \tag{11.40}$$

where $\overline{N}$, $\overline{M}$, $\overline{V}$, and $\overline{T}$ are respectively the axial force, bending moment, shearing force, and torsional moment produced by the unit load, and $\delta$, $\theta$, $\lambda$, and $\varphi$ are the deformations, respectively corresponding to the axial force, bending moment, shearing force, and torsional moment produced by the loads $F$, $q$ and $M_e$ applied to the body. The above relation, which can be used to determine the displacement at any point in any direction of a structure, is called the unit load method. If an angle of rotation is needed to be determined, then a unit couple is required to be applied at the point where the angle of rotation is to be determined. The unit load method can be used for both linearly and nonlinearly elastic bodies.

For a linearly elastic member subjected to combined loadings, Eq. (11.40) can be rewritten as

$$\Delta = \int \frac{N(x)\overline{N}(x)}{EA} \mathrm{d}x + \int \frac{M(x)\overline{M}(x)}{EI} \mathrm{d}x + \int \frac{T(x)\overline{T}(x)}{GI_p} \mathrm{d}x \tag{11.41}$$

where $N$, $M$, and $T$ are respectively the axial force, bending moment, and torsional moment produced by the combined loadings, and $EA$, $EI$, and $GI_p$ are respectively the axially tensile or compressive, bending, and torsional rigidities of the member subjected to combined loadings.

**Example 11.8** Knowing that the bending rigidity is $EI$, use the unit load method to determine the slope at the free end $B$ of the prismatic cantilever $AB$, Fig. 11.23(a).

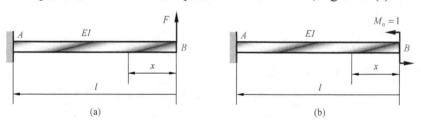

Fig. 11.23

**Solution** The bending moment for the original load, Fig. 11.23(a), can be expressed as

$$M(x) = Fx \quad (0 \leqslant x < l)$$

The bending moment for the unit couple, Fig. 11.23(b), can be expressed as

$$\overline{M}(x) = M_0 = 1 \quad (0 < x < l)$$

From the unit load method, we have

$$\theta_B = \int_0^l \frac{M(x)\overline{M}(x)}{EI} \mathrm{d}x = \frac{Fl^2}{2EI}$$

## 11.9 Applications of Energy Methods

**1. Application to Statically Determinate Structures**

**Example 11.9** Two rods $AB$ and $BC$ of the same flexural rigidity $EI$ are welded together at $B$, Fig. 11.24(a). For the loading shown and using the unit load method, determine the vertical deflection of point $C$, and the slope of section $C$.

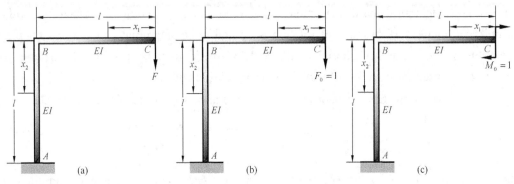

Fig. 11.24

**Solution** The bending moment for the original load, Fig. 11.24(a), can be expressed as
$$M(x_1) = -Fx_1 \ (0 \leqslant x_1 \leqslant l), \qquad M(x_2) = -Fl \ (0 \leqslant x_2 < l)$$

In order to determine the vertical deflection of point $C$, add a downward unit load $F_0 = 1$ at point $C$, Fig. 11.24(b). The bending moment corresponding to the unit load can be expressed as

$$\bar{M}(x_1) = -x_1 \ (0 \leqslant x_1 \leqslant l), \qquad \bar{M}(x_2) = -l \ (0 \leqslant x_2 < l)$$

Using the unit load method, we have
$$(w_C)_y = \int_0^l \frac{M(x_1)\bar{M}(x_1)}{EI} dx_1 + \int_0^l \frac{M(x_2)\bar{M}(x_2)}{EI} dx_2 = \frac{4Fl^3}{3EI}$$

In order to determine the slope of point $C$, add a clockwise unit couple $M_0 = 1$ at point $C$, Fig. 11.24(c). The bending moment corresponding to the unit couple can be expressed as
$$\bar{M}(x_1) = -1 \ (0 \leqslant x_1 \leqslant l), \qquad \bar{M}(x_2) = -1 \ (0 \leqslant x_2 < l)$$

Using the unit load method, we have
$$\theta_C = \int_0^l \frac{M(x_1)\bar{M}(x_1)}{EI} dx_1 + \int_0^l \frac{M(x_2)\bar{M}(x_2)}{EI} dx_2 = \frac{3Fl^2}{2EI}$$

**2. Application to Statically Indeterminate Structures**

**Example 11.10** Two rods $AB$ and $BC$ of the same flexural rigidity $EI$ are welded together at $B$, Fig. 11.25(a), which has a fixed end at $A$ and is supported by a roller at $C$. For the loading and supports shown and using the unit load method, determine the reaction at point $C$.

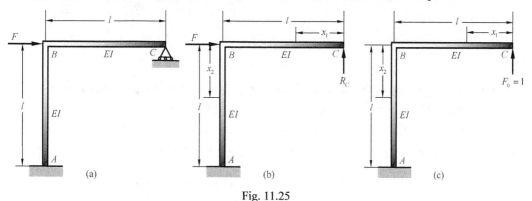

Fig. 11.25

**Solution** The structure shown in Fig. 11.25(a) is statically indeterminate to the first degree. We designate the reaction $R_C$ at point $C$ as redundant and eliminate the corresponding support, Fig. 11.25(b).

Referring to Fig. 11.25(b), the bending moment can be expressed as

$$M(x_1) = R_C x_1 \quad (0 \leqslant x_1 \leqslant l), \quad M(x_2) = R_C l - F x_2 \quad (0 \leqslant x_2 < l)$$

In order to determine the vertical deflection of point $C$, we add a unit load $F_0 = 1$ at point $C$ in the vertical direction, Fig. 11.25(c). The bending moment corresponding to the unit load can be expressed as

$$\bar{M}(x_1) = x_1 \quad (0 \leqslant x_1 \leqslant l), \quad \bar{M}(x_2) = l \quad (0 \leqslant x_2 < l)$$

Using the unit load method, we have

$$(w_C)_y = \int_0^l \frac{M(x_1)\bar{M}(x_1)}{EI} dx_1 + \int_0^l \frac{M(x_2)\bar{M}(x_2)}{EI} dx_2 = \frac{(8R_C - 3F)l^3}{6EI}$$

Since $(w_C)_y = 0$, then we obtain

$$R_C = \frac{3}{8} F$$

## Problems

**11.1** Using the principle of work and energy, determine the deflection at the free end $C$ of the cantilever $ABC$, Fig. 11.26. Know that the bending rigidity is $2EI$ for portion $AB$ and $EI$ for portion $BC$.

**11.2** Using the principle of work and energy, determine the deflection at the midsection $E$ of the simple beam $AB$, Fig. 11.27. Know that the bending rigidity is $2EI$ for portion $CD$ and $EI$ for portions $AC$ and $BD$.

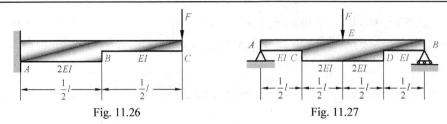

Fig. 11.26                                    Fig. 11.27

11.3 Using Castigliano's theorem, determine the deflection and slope at the free end $B$ of the cantilever $AB$, Fig. 11.28. Know that the bending rigidity is $EI$.

11.4 Using Castigliano's theorem, determine the deflection at the free end $C$ of the cantilever $ABC$, Fig. 11.29. Know that the bending rigidity is $2EI$ for portion $AB$ and $EI$ for portion $BC$.

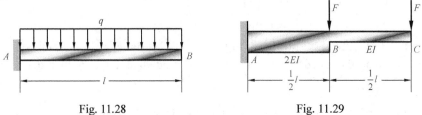

Fig. 11.28                                    Fig. 11.29

11.5 Using Castigliano's theorem, determine the deflection at the midsection $C$ of the simple-supported beam $AB$, Fig. 11.30. Know that the bending rigidity is $EI$.

11.6 Using Castigliano's theorem, determine the deflection and slope at point $B$, Fig. 11.31. Know that the bending rigidity is $EI$.

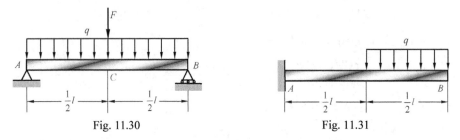

Fig. 11.30                                    Fig. 11.31

11.7 Using the principle of virtual work, determine the deflection and slope at the free end $B$ of the cantilever $AB$ subjected to a uniform load $q$, Fig. 11.32. Know that the stress-strain relation is $\sigma^2 = k\varepsilon$, where $k$ is a constant.

11.8 Using the principle of virtual work, determine the deflection at the midsection $C$ of the simple beam $AB$ subjected to a uniform load $q$, Fig. 11.33. Know that the stress-strain relation is $\sigma^2 = k\varepsilon$, where $k$ is a constant.

11.9 Using the unit load method, determine the deflection and slope at the free end $B$ of the cantilever $AB$, Fig. 11.34. Know that the bending rigidity is $EI$.

11.10 Using the unit load method, determine the deflection and slope at the midsection $B$ of the cantilever $ABC$, Fig. 11.35. Know that the bending rigidity is $2EI$ for portion $AB$ and $EI$ for portion $BC$.

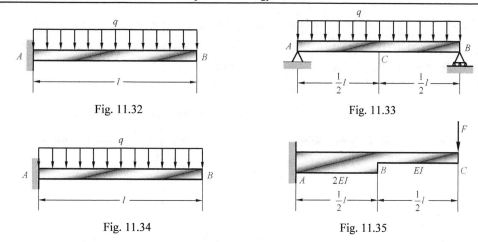

Fig. 11.32  Fig. 11.33  Fig. 11.34  Fig. 11.35

**11.11** Using the unit load method, determine the deflection at the midsection $C$ of the simple-supported beam $AB$, Fig. 11.36. Know that the bending rigidity is $EI$.

**11.12** Using the unit load method, determine the deflection at the midsection $C$ and slope at the section $B$, Fig. 11.37. Know that the bending rigidity is $EI$.

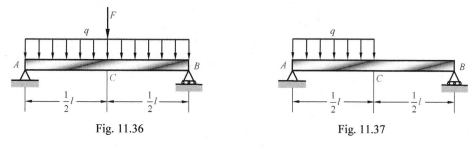

Fig. 11.36  Fig. 11.37

**11.13** Using the unit load method, determine the horizontal and vertical deflections of point $A$, Fig. 11.38. Know that the bending rigidity is $EI$.

**11.14** Two rods $AB$ and $BC$ of the same flexural rigidity $EI$ are welded together at $B$, which has a fixed end at $A$ and is supported by a roller at $C$, Fig. 11.39. For the loading and supports shown and using the unit load method, determine the reaction at point $C$.

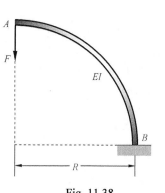

Fig. 11.38

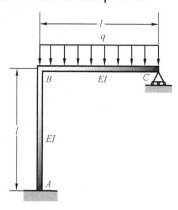

Fig. 11.39

11.15 Two rods *AB* and *BC* of the same flexural rigidity *EI* are welded together at *B*, Fig. 11.40. For the loading shown and using the unit load method, determine the horizontal and vertical deflection of point *C*, and the slope of section *C*.

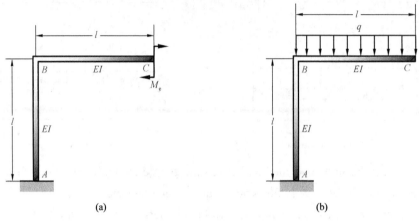

Fig. 11.40

# Chapter 12    Impact Loading

In the preceding chapters, all loadings are static; that is, they are applied gradually to a member and remain constant when they reach a final value. Some loadings, however, are dynamic; that is, they are applied suddenly to a member. These dynamic loadings are called impact loadings. When an object strikes a member, a very large stress will be developed within the member during impact.

## 12.1    Vertical Impact

Consider a block-and-spring system shown in Fig. 12.1. When a block of weight $P$ and kinetic energy $T$ strikes a spring without consideration of mass, it will attack to the spring and move downward. Assume that the block compresses the spring a distance $\Delta_d$ before coming to rest. If there is no dissipation of energy during the impact, then the conservation of energy requires that the kinetic and potential energy of the block be transformed entirely into the strain energy stored in the spring; or in other words, the sum of kinetic and potential energy of the block is equal to the work needed to displace the free end of the spring by an amount $\Delta_d$.

From the conservation of energy, we have

$$T + P\Delta_d = \frac{1}{2} F_d \Delta_d \tag{12.1}$$

where $F_d$ is the force between the block and spring when the spring contraction reaches $\Delta_d$. Assuming that the spring is within linearly elastic range, i.e., $\dfrac{F_d}{\Delta_d} = \dfrac{P}{\Delta_{st}}$ where $\Delta_{st}$ is the static displacement of the free end of the spring when the block is applied statically to the free end of the spring, then we obtain

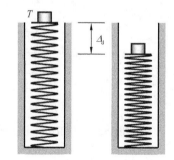

Fig. 12.1

$$\frac{F_d}{P} = \frac{\Delta_d}{\Delta_{st}} = K_d \tag{12.2}$$

where $K_d$ is the impact factor. Substituting Eq. (12.2) into Eq. (12.1), we can obtain

$$\frac{1}{2} K_d^2 - K_d - \frac{T}{P\Delta_{st}} = 0 \tag{12.3}$$

Solving the above equation for $K_d$, we obtain the following positive root:

$$K_d = 1 + \sqrt{1 + \frac{2T}{P\Delta_{st}}} \tag{12.4}$$

Assuming that the block is released from rest, falls a distance $h$ and strikes the spring, then the impact factor can be given as

$$K_d = 1 + \sqrt{1 + \frac{2h}{\Delta_{st}}} \qquad (12.5)$$

Once $K_d$ is determined, the dynamic stress in the spring and the dynamic deformation of the spring can be expressed as

$$\sigma_d = K_d \sigma_{st}, \qquad \delta_d = K_d \delta_{st} \qquad (12.6)$$

where $\sigma_{st}$ and $\delta_{st}$ are respectively the static stress and static deformation when the weight is applied statically to the spring.

If the block is held just above the spring, $h=0$, and released from rest, then, from Eq. (12.4) or Eq. (12.5), the impact factor is $K_d=2$, and the resulting dynamic stress in the spring is $\sigma_d=2\sigma_{st}$; i.e., when the block is dropped suddenly with no initial velocity from the top of the spring (suddenly applied loading), the dynamic stress is twice as large as the static stress caused by the block placed statically on the spring (statically applied load).

**Example 12.1** A block of weight $P$, initially at rest, is dropped from a height $h$ onto the midpoint $C$ of the simple beam $AB$, Fig. 12.2. Knowing that the modulus of elasticity of the beam is $E=105$ GPa, determine (1) the maximum deflection of the beam, and (2) the maximum normal stress in the beam.

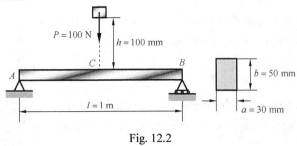

Fig. 12.2

**Solution** (1) Referring to the table of deflection curves in Appendix III, we have

$$\Delta_{st} = \frac{Pl^3}{48EI} = \frac{Pl^3}{48E\left(\frac{1}{12}ab^3\right)} = 6.35 \times 10^{-2} \text{ mm}$$

Using Eq. (12.5), we obtain

$$K_d = 1 + \sqrt{1 + \frac{2h}{\Delta_{st}}} = 57.13$$

Therefore, the maximum dynamic deflection is equal to

$$\Delta_d = K_d \Delta_{st} = 3.63 \text{ mm}$$

(2) Referring to the formula for determining normal stress in Chapter 5, we have the maximum static stress

$$\sigma_{st} = \frac{M}{W} = \frac{\frac{1}{4}Pl^2}{\frac{1}{6}ab^2} = 2.00 \text{ MPa}$$

Using Eq. (12.6), we obtain the maximum dynamic normal stress in the beam:
$$\sigma_d = K_d \sigma_{st} = 114.26 \text{ MPa}$$

**Example 12.2** A collar of mass $m$ is released from rest in the position shown and is stopped by a plate attached at the free end $B$ of the vertical rod $AB$ of diameter $d$, Fig. 12.3. Knowing that $E$=70 GPa, determine (1) the maximum elongation of the rod, and (2) the maximum normal stress in the rod ($g = 9.81$ m/s$^2$).

**Solution** (1) Maximum elongation of the rod:

$$\Delta_{st} = \frac{mgl}{EA} = \frac{mgl}{E\left(\frac{1}{4}\pi d^2\right)} = 8.56 \times 10^{-3} \text{ mm}$$

$$K_d = 1 + \sqrt{1 + \frac{2h}{\Delta_{st}}} = 217.17, \qquad \Delta_d = K_d \Delta_{st} = 1.86 \text{ mm}$$

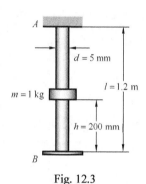

(2) Maximum normal stress in the rod:

$$\sigma_{st} = \frac{mg}{A} = \frac{mg}{\frac{1}{4}\pi d^2} = 0.50 \text{ MPa}, \qquad \sigma_d = K_d \sigma_{st} = 108.58 \text{ MPa}$$

Fig. 12.3

**Example 12.3** A plane mechanism, $ABCD$, is composed of two blocks, $A$ and $D$, two smooth fixed pulleys, $B$ and $C$, and an elastic cable, $ABCD$. The cable has length of 0.6 m and cross-sectional area of 50 mm$^2$. The blocks $A$ and $B$, each having weigh of 120 N, are moving downward and upward with a constant speed of 56 mm/s, Fig. 12.4. Knowing that $E = 10$ GPa, and assuming that the block $D$ is suddenly seized up and stops moving, determine the maximum stress in the cable.

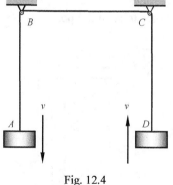

Fig. 12.4

**Solution**

$$\Delta_{st} = \frac{P_A l}{EA} = 144 \text{ μm}, \qquad K_d = 1 + \sqrt{\frac{2T_A}{P_A \Delta_{st}}} = 2.49$$

$$\sigma_{st} = \frac{P_A}{A} = 2.4 \text{ MPa}, \qquad \sigma_d = K_d \sigma_{st} = 5.98 \text{ MPa}$$

## 12.2 Horizontal Impact

Consider a block-and-spring system placed on a smooth horizontal surface, Fig. 12.5. The block of weight $P$ sliding on the surface with kinetic energy $T$ collides with the spring and compresses it a distance $\Delta_d$ before coming to a maximum displacement. If we neglect the loss of energy during impact and the mass of the spring and assume that the spring is within linearly elastic range and that the block sticks to the spring and moves with it, then kinetic energy of the block is transferred entirely to the strain energy of the spring when the block comes to a rest.

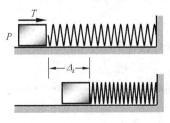

Fig. 12.5

From the conservation of energy, we have

$$T = \frac{1}{2} F_d \Delta_d = \frac{1}{2} P \Delta_{st} K_d^2 \qquad (12.7)$$

The positive root of Eq. (12.7) can be expressed as

$$K_d = \sqrt{\frac{2T}{P \Delta_{st}}} \qquad (12.8)$$

where $\Delta_{st}$ is the static displacement of the free end of the spring when an equivalent force, equal to the weight $P$ of the block, is applied statically to the free end of the spring. Using $T = \frac{1}{2} \frac{P}{g} v^2$, Eq. (12.8) can also be rewritten as

$$K_d = \sqrt{\frac{v^2}{g \Delta_{st}}} \qquad (12.9)$$

From Eq. (12.8) or Eq. (12.9), the dynamic stress in the spring and the dynamic deformation of the spring can be expressed, respectively, as

$$\sigma_d = K_d \sigma_{st}, \qquad \delta_d = K_d \delta_{st} \qquad (12.10)$$

where $\sigma_{st}$ and $\delta_{st}$ are the static stress in the spring and the static deformation of the spring when an equivalent force, equal to the weight $P$ of the block, is applied statically to the spring.

**Example 12.4** A sphere of mass $m$ is moving with a velocity $v$ to the right and hits squarely the free end $A$ of the post $AB$ of diameter $d$, Fig. 12.6. Using $E=200$ GPa, determine (1) the maximum deflection of the post, and (2) the maximum normal stress in the post.

**Solution** (1) Maximum deflection of the post:

$$\Delta_{st} = \frac{mgl^3}{3EI} = \frac{mgl^3}{3E\left(\frac{1}{64}\pi d^4\right)} = 6.58 \text{ mm}$$

$$K_d = \sqrt{\frac{v^2}{g \Delta_{st}}} = 3.94, \qquad \Delta_d = K_d \Delta_{st} = 25.9 \text{ mm}$$

(2) Maximum normal stress in the post:

$$\sigma_{st} = \frac{M}{W} = \frac{mgl}{\frac{1}{32}\pi d^3} = 29.6 \text{ MPa}, \qquad \sigma_d = K_d \sigma_{st} = 117 \text{ MPa}$$

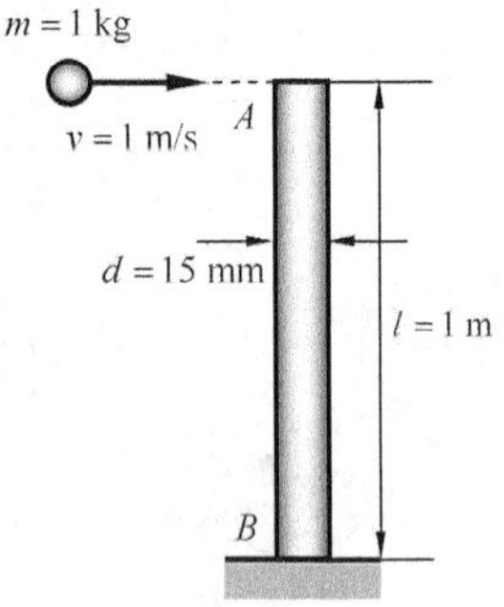

Fig. 12.6

$m = 1$ kg
$v = 1$ m/s
$d = 15$ mm
$l = 1$ m

**Example 12.5** A sphere of mass $m$ moving with a velocity $v$ hits squarely the free end $A$ of the rod $AB$, Fig. 12.7. Knowing that $E=100$ GPa, determine (1) the displacement of point $A$, and (2) the maximum normal stress in the rod.

**Solution** (1) Maximum displacement of point $A$:

$$\Delta_{st} = \frac{mgl}{EA} = \frac{mgl}{E\left(\frac{1}{4}\pi d^2\right)} = 1.25 \times 10^{-6} \text{ m}$$

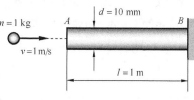

Fig. 12.7

$$K_d = \sqrt{\frac{v^2}{g\Delta_{st}}} = 285.57, \qquad \Delta_d = K_d \Delta_{st} = 0.357 \text{ mm}$$

(2) Maximum normal stress in the rod:

$$\sigma_{st} = \frac{mg}{A} = \frac{mg}{\frac{1}{4}\pi d^2} = 0.125 \text{ MPa}, \qquad \sigma_d = K_d \sigma_{st} = 35.7 \text{ MPa}$$

## Problems

**12.1** A block of mass $m$, initially at rest, falls from a height $h$ onto the free end $B$ of the cantilever beam $AB$, Fig. 12.8. Knowing that $E$=206 GPa, determine (1) the maximum deflection of the beam, and (2) the maximum normal stress in the beam.

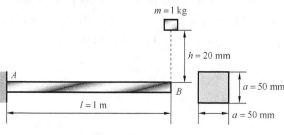

Fig. 12.8

**12.2** A block of mass $m$, initially at rest, falls a distance $h$ before it strikes the midsection $C$ of the cantilever beam $AB$, Fig. 12.9. Knowing that $E$=206 GPa, determine (1) the maximum deflection of the beam, and (2) the maximum normal stress in the beam.

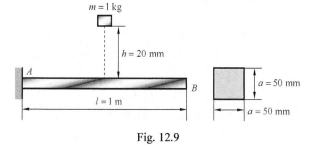

Fig. 12.9

**12.3** A block of mass $m$ falls from a height $h$ onto the free end $C$ of the overhanging beam $ABC$ of diameter $d$, Fig. 12.10. Knowing that $E$=73 GPa, determine (1) the deflection of point $C$, and (2) the maximum normal stress in the beam.

12.4  A cylinder of mass $m$ is released from rest in the position shown and strikes the free end $A$ of the vertical rod $AB$ of diameter $d$, Fig. 12.11. Knowing that $E$=200 GPa, determine the maximum normal stress in the rod.

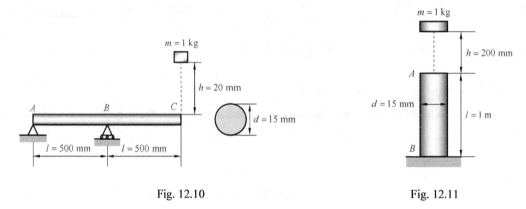

Fig. 12.10  Fig. 12.11

12.5  A sphere of mass $m$ moving with a velocity $v$ strikes the post $AB$ squarely at the midpoint $C$, Fig. 12.12. Using $E$=200 GPa, determine (1) the displacement of point $A$, and (2) the maximum normal stress in the post.

12.6  A block of mass $m$ moving with a velocity $v$ strikes squarely the free end $A$ of the nonuniform rod $ABC$, Fig. 12.13. Determine (1) the displacement of point $A$, and (2) the maximum normal stress in the rod. Knowing that $E$=100 GPa, $d_1$=10 mm, and $d_2$=20 mm.

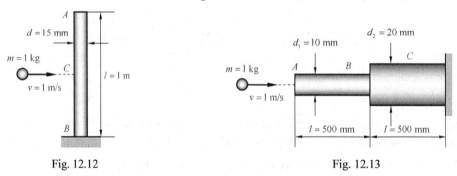

Fig. 12.12  Fig. 12.13

# Chapter 13  Statically Indeterminate Structures

Structures can be classified into two general categories: statically determinate and statically indeterminate. A structure in which all reactions and internal forces can be completely determined by using the equations of equilibrium alone is called statically determinate. It then follows that a statically indeterminate structure is one in which all reactions, and/or internal forces cannot be obtained from the equations of equilibrium only. The primary advantage of a statically indeterminate structure is that its strength is higher than a corresponding statically determinate structure. Another advantage of a statically indeterminate structure is that its rigidity is also greater than a corresponding statically determinate structure. To solve a statically indeterminate structure, we must combine the equilibrium conditions of force with the compatibility conditions of deformation.

## 13.1  Static Indeterminacy

For a statically indeterminate structure, when the reaction is required to be taken as the redundant force, the structure is externally statically indeterminate. If the internal force is needed to be regarded as the redundant force, the structure is internally statically indeterminate. It is also possible that a statically indeterminate structure will have a combination of external and internal static indeterminacy.

**1. External Static Indeterminacy**

The structure shown in Fig. 13.1 is a typical example of external static indeterminacy. The structure has four reactions, but only three independent equations of equilibrium are available, therefore there is one reaction that cannot be determined from the equation of equilibrium. The number of unknown reactions in excess of the equilibrium equations is called the degree of indeterminacy, thus the structure shown in Fig. 13.1 is externally statically indeterminate to the first degree.

**2. Internal Static Indeterminacy**

A structure is internally statically indeterminate when it is not possible to determine all internal forces by using the equations of equilibrium. The structure shown in Fig. 13.2 is internally statically indeterminate to the third degree.

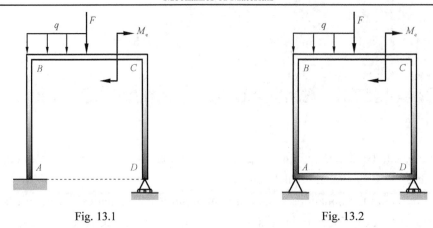

Fig. 13.1                                  Fig. 13.2

## 13.2 Force Method for Analysis of Statically Indeterminate Structures

The force method has been developed to solve a statically indeterminate structure. The force method takes the redundant force as unknown and utilizes the compatibility condition of displacement to determine the redundant force. The basic idea of the force method can be illustrated by considering a statically indeterminate beam, Fig. 13.3(a). This beam has four unknown reactions and three independent equilibrium equations, and is therefore statically indeterminate to the first degree. Taking one unknown reaction, say the reaction $X_1$ at $B$, as redundant and eliminating the corresponding support at $B$, then the original statically indeterminate beam can be reduced to a statically determinate beam in stability, Fig. 13.3(b).

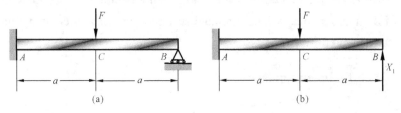

Fig. 13.3

Denoting by $\varDelta_1$ the displacement at $B$ in the direction of $X_1$, then this displacement can be considered to be the superposition of two independent displacements, i.e.,

$$\varDelta_1 = \varDelta_{1X_1} + \varDelta_{1F} \tag{13.1}$$

where $\varDelta_{1X_1}$ is the displacement at $B$ in the direction of $X_1$, caused by the redundant reaction $X_1$, and $\varDelta_{1F}$ is the displacement at $B$ in the direction of $X_1$, caused by the external load $F$ applied to the beam. It should be noted that it is not possible to determine the displacement $\varDelta_{1X_1}$ prior to the determination of the redundant force $X_1$, however, by applying the principle of superposition, $\varDelta_{1X_1}$ can be expressed as

Chapter 13　Statically Indeterminate Structures　　　　· 133 ·

$$\Delta_{X_1} = \delta_{11} X_1 \tag{13.2}$$

where $\delta_{11}$ is the displacement at $B$ in the direction of $X_1$, caused by a unit load acting at $B$ in the direction of $X_1$. From the compatibility condition of displacement, the displacement $\Delta_1$ at $B$ in the direction of $X_1$ must be equal to zero, i.e.,

$$\delta_{11} X_1 + \Delta_{1F} = 0 \tag{13.3}$$

Eq. (13.3) is called the canonical equation of force method. The coefficients, $\delta_{11}$ and $\Delta_{1F}$, can be computed as

$$\delta_{11} = \int \frac{\overline{M}_1^2 \mathrm{d}x}{EI}, \quad \Delta_{1F} = \int \frac{M_F \overline{M}_1 \mathrm{d}x}{EI} \tag{13.4}$$

where $\overline{M}_1$ is the bending moment caused by a unit load acting at $B$ in the direction of $X_1$, $M_F$ is the bending moment caused by the external load applied to the beam, and $EI$ is the flexural rigidity of the beam.

By substitution of Eq. (13.4) into Eq. (13.3), we can obtain

$$X_1 = -\frac{\Delta_{1F}}{\delta_{11}} = -\int \frac{M_F \overline{M}_1 \mathrm{d}x}{EI} \bigg/ \int \frac{\overline{M}_1^2 \mathrm{d}x}{EI} \tag{13.5}$$

**Example 13.1**　For the beam and loading shown in Fig. 13.4(a) and assuming that $EI$ is constant, determine the reaction at the roller support.

**Solution**　The original beam is statically indeterminate to the first degree. Taking the constraint at $B$ as redundant and releasing this constraint from $B$, Fig. 13.4(b), then the canonic equation of force method can, using the compatibility condition of displacement at $B$ where the deflection of the beam must be zero, be given as

$$\delta_{11} X_1 + \Delta_{1F} = 0$$

where $X_1$ is the redundant reaction. From Appendix III, the deflection at $B$ caused by a unit load acting at $B$ in the direction of $X_1$ can be given as

$$\delta_{11} = \frac{l^3}{3EI}$$

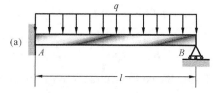

The deflection at $B$ produced by the uniformly distributed load $q$ can be expressed as

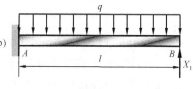

$$\Delta_{1F} = -\frac{ql^4}{8EI}$$

Fig. 13.4

By substitution of $\delta_{11}$ and $\Delta_{1F}$ into $\delta_{11} X_1 + \Delta_{1F} = 0$, then we have

$$\frac{l^3}{3EI} X_1 - \frac{ql^4}{8EI} = 0$$

Solving for $X_1$, we obtain the redundant reaction at the roller support $B$

$$X_1 = \frac{3}{8}ql$$

**Example 13.2**  Two rods $AB$ and $BC$ of the same flexural rigidity $EI$ are welded together at $B$ to form a frame, Fig. 13.5(a), which has a fixed end at $A$ and is supported by a roller at $C$. Using the force method, determine the reaction at $C$.

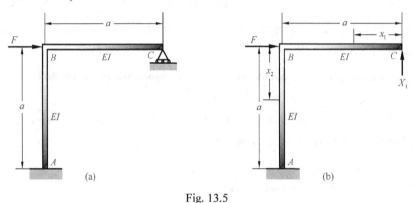

Fig. 13.5

**Solution**  The structure shown in Fig. 13.5(a) is statically indeterminate to the first degree. We designate the reaction $X_1$ at $C$ as redundant and eliminate the corresponding support, Fig. 13.5(b). The bending moment caused by the external load $F$ can be expressed as

$$M_F(x_1) = 0 \quad (0 \leq x_1 \leq a), \qquad M_F(x_2) = -Fx_2 \quad (0 \leq x_2 < a)$$

And the bending moment caused by a unit load acting at $C$ in the direction of $X_1$ can be expressed as

$$\bar{M}_1(x_1) = x_1 \quad (0 \leq x_1 \leq a), \qquad \bar{M}_1(x_2) = a \quad (0 \leq x_2 < a)$$

From Eq. (13.7), we have

$$\delta_{11} = \int_0^a \frac{[\bar{M}_1(x_1)]^2}{EI} dx_1 + \int_0^a \frac{[\bar{M}_1(x_2)]^2}{EI} dx_2 = \frac{4a^3}{3EI}, \qquad \Delta_{1F} = \int_0^a \frac{M_F(x_2)\bar{M}_1(x_2)}{EI} dx_2 = -\frac{Fa^3}{2EI}$$

Using the canonic equation of force method, $\delta_{11}X_1 + \Delta_{1F} = 0$, then we obtain

$$X_1 = \frac{3}{8}F$$

**Example 13.3**  A beam $AB$ of flexural rigidity $EI$ and a bar $BC$ of tensile rigidity $EA$ are hinged together at $B$, Fig. 13.6(a). The beam has a fixed support at $A$ and the bar has a pin support at $C$. Knowing that $EA=3EI/(2a^2)$, determine the reaction at $C$ using the force method.

**Solution**  The structure in Fig. 13.6(a) is statically indeterminate to the first degree. We take the reaction $X_1$ at $C$ as redundant and eliminate the corresponding support, Fig. 13.6(b).

The bending moment in beam $AB$ and the axial force in bar $BC$ caused by the external load $F$ can be expressed as

$$M_F = 0 \quad (0 \leq x \leq a), \qquad M_F = -F(x-a) \quad (a \leq x < 2a), \qquad N_F = 0$$

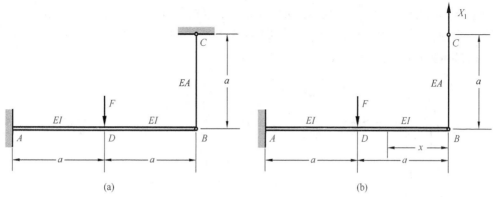

Fig. 13.6

And the bending moment in beam $AB$ and the axial force in bar $BC$ caused by a unit load acting at $C$ in the direction of $X_1$ can be expressed as

$$\overline{M}_1(x) = x \ (0 \leqslant x < 2a), \qquad \overline{N}_1 = 1$$

From Eq. (13.7), we have

$$\delta_{11} = \int_0^{2a} \frac{\overline{M}_1^2}{EI} dx + \int_0^a \frac{\overline{N}_1^2}{EA} dx = \frac{8a^3}{3EI} + \frac{a}{EA} = \frac{10a^3}{3EI}, \qquad \Delta_{1F} = \int_a^{2a} \frac{M_F \overline{M}_1}{EI} dx = -\frac{5Fa^3}{6EI}$$

Using the canonic equation of force method $\delta_{11} X_1 + \Delta_{1F} = 0$, then we obtain

$$X_1 = \frac{1}{4} F$$

Considering a statically indeterminate structure having the $n$th degree of indeterminacy, and denoting by $X_1, X_2, \cdots, X_n$ the redundant forces caused by the redundant constraints, then the canonical equations of force method for this statically indeterminate structure can be expressed as

$$\begin{aligned} \delta_{11} X_1 + \delta_{12} X_2 + \cdots + \delta_{1n} X_n + \Delta_{1F} &= 0 \\ \delta_{21} X_1 + \delta_{22} X_2 + \cdots + \delta_{2n} X_n + \Delta_{2F} &= 0 \\ &\vdots \\ \delta_{n1} X_1 + \delta_{n2} X_2 + \cdots + \delta_{nn} X_n + \Delta_{nF} &= 0 \end{aligned} \qquad (13.6)$$

where $\delta_{ij}$ is the generalized displacement at the application point of $X_i$ in the direction of $X_i$, caused by a unit load acting at the application point of $X_j$ in the direction of $X_j$, and $\Delta_{iF}$ is the generalized displacement at the application point of $X_i$ in the direction of $X_i$, caused by the external loads applied to the structure. $\delta_{ij}$ and $\Delta_{iF}$ can be given as

$$\delta_{ij} = \int_l \frac{\overline{M}_i \overline{M}_j dx}{EI} \ (i, j = 1, 2, \cdots, n), \qquad \Delta_{iF} = \int_l \frac{M_F \overline{M}_i dx}{EI} \ (i = 1, 2, \cdots, n) \qquad (13.7)$$

where $\overline{M}_i$ is the bending moment caused by a unit load acting at the application point of $X_i$ in the direction of $X_i$, $M_F$ is the bending moment caused by the external loads applied to the statically indeterminate structure.

**Example 13.4** Two rods *AB* and *BC* of the same flexural rigidity *EI* are welded together at *B* to form a frame, Fig. 13.7(a), which is fixed at *A* and *C* and subjected to a uniform load $q$. Using the force method, determine the reactions at *C*.

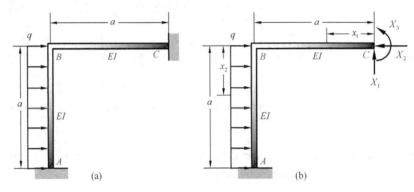

Fig. 13.7

**Solution** The structure shown in Fig. 13.7(a) is statically indeterminate to the third degree. We designate the reactions $X_1$, $X_2$ and $X_3$ at *C* as redundant and eliminate the corresponding support at *C*, Fig. 13.7(b).

The bending moment caused by the uniform load $q$ can be expressed as

$$M_q(x_1) = 0 \quad (0 < x_1 \leq a), \qquad M_q(x_2) = -\frac{1}{2}qx_2^2 \quad (0 \leq x_2 < a)$$

The bending moments caused by the unit loads acting at *C*, which are respectively in the directions of $X_1$, $X_2$ and $X_3$, can be expressed as

$$\overline{M}_1(x_1) = x_1 \quad (0 < x_1 \leq a), \qquad \overline{M}_1(x_2) = a \quad (0 \leq x_2 < a)$$

$$\overline{M}_2(x_1) = 0 \quad (0 < x_1 \leq a), \qquad \overline{M}_2(x_2) = x_2 \quad (0 \leq x_2 < a)$$

$$\overline{M}_3(x_1) = 1 \quad (0 < x_1 \leq a), \qquad \overline{M}_3(x_2) = 1 \quad (0 \leq x_2 < a)$$

From Eq. (13.7), we have

$$\delta_{11} = \frac{4a^3}{3EI}, \quad \delta_{12} = \frac{a^3}{2EI}, \quad \delta_{13} = \frac{3a^2}{2EI}, \quad \delta_{22} = \frac{a^3}{3EI}, \quad \delta_{23} = \frac{a^2}{2EI}, \quad \delta_{33} = \frac{2a}{EI}$$

$$\Delta_{1F} = -\frac{qa^4}{6EI}, \quad \Delta_{2F} = -\frac{qa^4}{8EI}, \quad \Delta_{3F} = -\frac{qa^3}{6EI}$$

Solving the canonic equations of force method:

$$8aX_1 + 3aX_2 + 9X_3 = qa^2$$

$$12aX_1 + 8aX_2 + 12X_3 = 3qa^2$$

$$9aX_1 + 3aX_2 + 12X_3 = qa^2$$

we have
$$\begin{bmatrix} X_1 \\ X_2 \\ X_3 \end{bmatrix} = \text{inv}\left(\begin{bmatrix} 8a & 3a & 9 \\ 12a & 8a & 12 \\ 9a & 3a & 12 \end{bmatrix}\right)\begin{bmatrix} qa^2 \\ 3qa^2 \\ qa^2 \end{bmatrix} = \frac{1}{48}\begin{bmatrix} -3qa \\ 21qa \\ qa^2 \end{bmatrix}$$

where inv represents the matrix inverse.

The procedure of analysis for a statically indeterminate structure by using the force method can be outlined as follows: ① Identify the degree of indeterminacy of a statically indeterminate structure, and release the redundant constraint to obtain a statically determinate structure in stability; ② Calculate the generalized displacements, $\delta_{ij}$ and $\Delta_{iF}$, using the energy method or the method of superposition; ③ Using the displacement compatibility, determine the redundant force $X_i$ from the given canonic equations of force method.

## 13.3 Force Method for Analysis of Symmetrical Statically-Indeterminate Structures

**1. Symmetrical Structures Subjected to Symmetrical Loads**

For a symmetrical indeterminate structure subjected to symmetrical loads, the antisymmetrical internal forces (torsional moment and shearing force) on the plane of symmetry of the structure must be identically equal to zero. For example, the structure and the loads are symmetrical about the midsection $E$ of the structure, Fig. 13.8, thus the shearing force on the section $E$ is equal to zero.

**2. Symmetrical Structures Subjected to Antisymmetrical Loads**

For a symmetrical indeterminate structure subjected to antisymmetrical loads, the symmetrical internal forces (axial force and bending moment) on the plane of symmetry of the structure must be identically equal to zero. For example, the structure is symmetrical and the loads are antisymmetrical about the midsection $E$ of the structure, Fig. 13.9, thus the axial force and bending moment on the section $E$ are equal to zero.

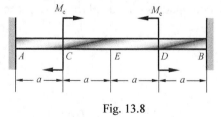

Fig. 13.8          Fig. 13.9

**3. Symmetrical Structures Subjected to General Loads**

A symmetrical indeterminate structure subjected to general loads, Fig. 13.10(a), can be considered to be the superposition of two symmetrical indeterminate structures, one subjected to symmetrical loads, Fig. 13.10(b), and the other subjected to antisymmetrical loads, Fig. 13.10(c).

**Example 13.5** A beam $AB$ of constant flexural rigidity $EI$ is fixed at $A$ and $B$, Fig. 13.11(a). Neglecting the axial force in the beam, determine the internal forces at the midsection $E$ using the force method.

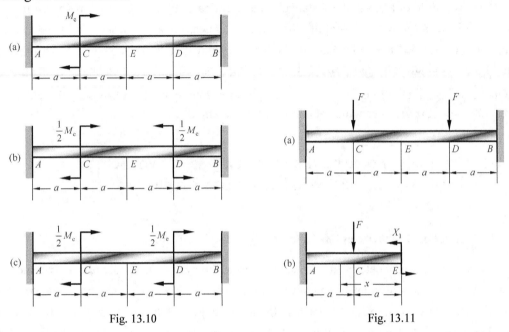

Fig. 13.10          Fig. 13.11

**Solution** Using the symmetry of the structure and of the external loads, we know that the shearing force on the midsection $E$ is equal to zero. When the axial force in the beam is also neglected, then the bending moment will be the only internal force on the midsection $E$. If the structure is sectioned from the midsection $E$, the original structure is statically indeterminate to the first degree. Taking the bending moment $X_1$ at $E$ as redundant and using the fact that the angle of rotation at $E$ is equal to zero, Fig. 13.11(b), then we have $\delta_{11}X_1 + \Delta_{1F} = 0$.

The bending moment in beam $AE$ caused by the external load $F$ can be expressed as
$$M_F = 0 \quad (0 \leqslant x \leqslant a), \qquad M_F = -F(x-a) \quad (a \leqslant x < 2a)$$

And the bending moment in beam $AE$ caused by a unit load acting in place of $X_1$ can be expressed as
$$\overline{M}_1(x) = 1 \quad (0 \leqslant x < 2a)$$

From Eq. (13.7), we have
$$\delta_{11} = \int_0^{2a} \frac{\overline{M}_1^2}{EI} dx = \frac{2a}{EI}, \qquad \Delta_{1F} = \int_a^{2a} \frac{M_F \overline{M}_1}{EI} dx = -\frac{Fa^2}{2EI}$$

Using the canonic equation $\delta_{11}X_1 + \Delta_{1F} = 0$, then we obtain
$$X_1 = \frac{1}{4}F$$

**Example 13.6** A statically indeterminate frame $ABC$ is fixed at $A$ and $C$, Fig. 13.12(a). Knowing that the flexural rigidity $EI$ is constant, determine the internal forces at the section $B$ using the force method.

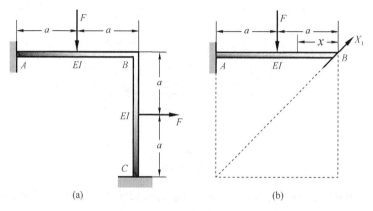

Fig. 13.12

**Solution** Using the symmetry of the structure and the antisymmetry of the external loads, we obtain that the axial force and bending moment on the section $B$ is equal to zero. The shearing force will be the only internal force on the section $E$. If the structure is sectioned from the section $B$, the original structure is statically indeterminate to the first degree. Taking the shearing force $X_1$ at $B$ as redundant, Fig. 13.12(b), then we have $\delta_{11}X_1 + \Delta_{1F} = 0$.

The bending moment in beam $AB$ caused by the external load $F$ can be expressed as

$$M_F = 0 \ (0 \leqslant x \leqslant a), \qquad M_F = -F(x-a) \ (a \leqslant x < 2a)$$

And the bending moment in beam $AB$ caused by a unit load acting in place of $X_1$ can be expressed as

$$\overline{M}_1(x) = \frac{1}{\sqrt{2}} x \ (0 \leqslant x < 2a)$$

From Eq. (13.7), we have

$$\delta_{11} = \int_0^{2a} \frac{\overline{M}_1^{\,2}}{EI} dx = \frac{4a^3}{3EI}, \qquad \Delta_{1F} = \int_a^{2a} \frac{M_F \overline{M}_1}{EI} dx = -\frac{5\sqrt{2}Fa^3}{12EI}$$

Using the canonic equation $\delta_{11}X_1 + \Delta_{1F} = 0$, then we obtain

$$X_1 = \frac{5\sqrt{2}}{16} F$$

# Problems

13.1 For the beam and loading shown in Fig. 13.13 and assuming that $EI$ is constant, determine the reaction at the roller support.

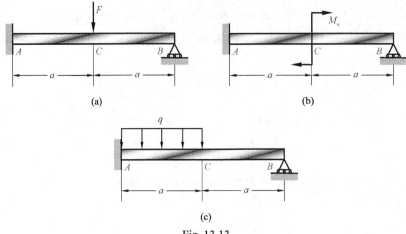

Fig. 13.13

13.2 Two rods *AB* and *BC* of the same flexural rigidity *EI* are welded together at *B* to form a frame, which has a fixed end at *A* and is supported by a roller at *C*, Fig. 13.14. Using the force method, determine the reaction at point *C*.

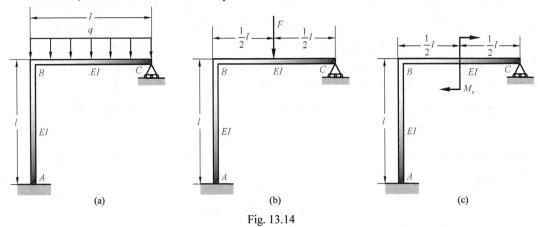

Fig. 13.14

13.3 A beam *AB* of flexural rigidity *EI* and a bar *BC* of tensile rigidity *EA* are hinged together at *B*. The beam has a fixed support at *A* and the bar pin support at *C*, Fig. 13.15. Knowing that $EA=3EI/(10a^2)$, determine the reaction at *C* using the force method.

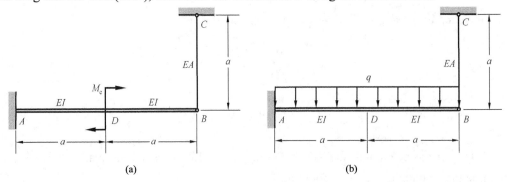

Fig. 13.15

13.4  A beam *AB* of constant flexural rigidity *EI* is fixed at *A* and *B*, Fig. 13.16. Neglecting the axial force in the beam, determine the internal forces at the midsection *E* using the force method.

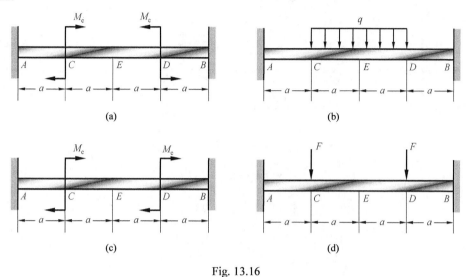

Fig. 13.16

13.15  A statically indeterminate frame *ABC* is fixed at *A* and *C*, Fig. 13.17. Knowing that the flexural rigidity *EI* is constant, determine the internal forces at the section *B* using the force method.

13.16  A statically indeterminate frame, fixed at *A* and *B*, is subjected to a uniform load *q*, Fig. 13.18. Knowing that the flexural rigidity *EI* is constant, determine the internal force at section *C* using the force method. (Hint: Know that the axial and shearing forces at section *C* are both equal to zero.)

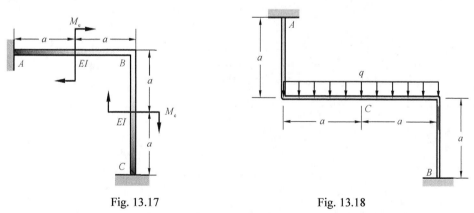

Fig. 13.17  Fig. 13.18

13.17  A statically indeterminate frame *ABC* is fixed at *A* and supported by a roller at *C*, Fig. 13.19. Knowing that the flexural rigidity *EI* is constant, determine the reaction at *C* using the force method.

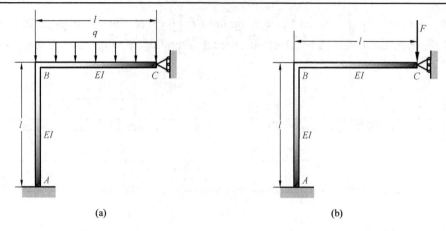

Fig. 13.19

# Appendix I  Properties of Area

## I.1  First Moment (Static Moment)

Considering a shaded area $A$ shown in Fig. I.1, the first moments of the area $A$ with respect to the $z$ and $y$ axes can be defined, respectively, as

$$Q_z = \int y\,dA, \qquad Q_y = \int z\,dA \tag{I.1}$$

In SI units, the first moments, $Q_z$ and $Q_y$, are expressed in m$^3$.

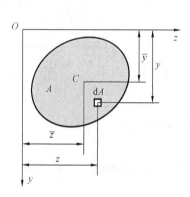
Fig. I.1

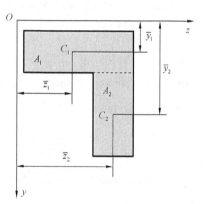
Fig. I.2

The centroid of the area $A$ can be defined as

$$\bar{y} = \frac{\int y\,dA}{A}, \qquad \bar{z} = \frac{\int z\,dA}{A} \tag{I.2}$$

Comparing Eq. (I.1) with Eq. (I.2), we have

$$Q_z = A\bar{y}, \qquad Q_y = A\bar{z} \tag{I.3}$$

For a composite area, consisting of areas $A_1$ and $A_2$, shown in Fig. I.2, the first moments of the composite area with respect to the $z$ and $y$ axes can, respectively, be expressed as

$$Q_z = \sum (Q_z)_i = \sum A_i \bar{y}_i, \qquad Q_y = \sum (Q_y)_i = \sum A_i \bar{z}_i \tag{I.4}$$

The centroid of the composite area can be written as

$$\bar{y} = \frac{\sum A_i \bar{y}_i}{\sum A_i}, \qquad \bar{z} = \frac{\sum A_i \bar{z}_i}{\sum A_i} \tag{I.5}$$

## I.2  Moment of Inertia and Polar Moment of Inertia

### 1. Moment of Inertia

Considering a shaded area $A$ shown in Fig. I.3, the moments of inertia of the area $A$ with respect to the $z$ and $y$ axes can be defined as

$$I_z = \int y^2 dA, \qquad I_y = \int z^2 dA \tag{I.6}$$

In SI units, the moments of inertia, $I_z$ and $I_y$, are expressed in m$^4$.

### 2. Polar Moment of Inertia

Considering a shaded area $A$ shown in Fig. I.4, the polar moment of inertia of the area $A$ with respect to the origin $O$ can be defined as

$$I_p = \int \rho^2 dA \tag{I.7}$$

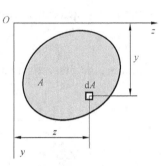

Fig. I.3

In SI units, the polar moment of inertia, $I_p$, is expressed in m$^4$.

Using $\rho^2 = y^2 + z^2$, we have

$$I_p = \int \rho^2 dA = \int y^2 dA + \int z^2 dA = I_z + I_y \tag{I.8}$$

## I.3  Radius of Gyration and Polar Radius of Gyration

### 1. Radius of Gyration

Considering the shaded area $A$ shown in Fig. I.3, the radius of gyration of the area $A$ with respect to the $z$ and $y$ axes can be defined, respectively, as

$$i_z = \sqrt{\frac{I_z}{A}}, \qquad i_y = \sqrt{\frac{I_y}{A}} \tag{I.9}$$

In SI units, the radius of gyration, $i_z$ and $i_y$, are expressed in m.

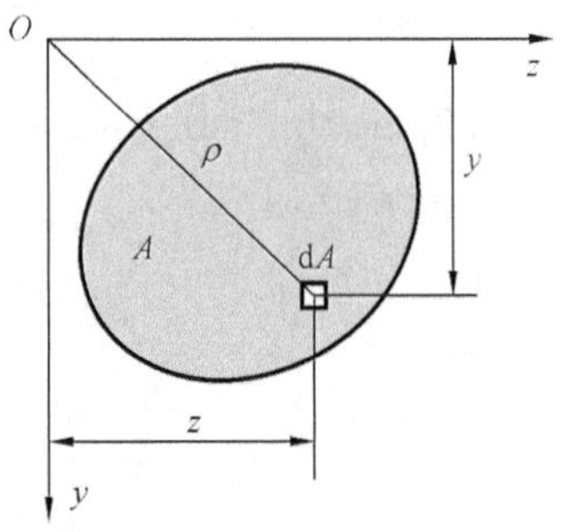

### 2. Polar Radius of Gyration

Considering the shaded area $A$ shown in Fig. I.4, the polar radius of gyration of the area $A$ with respect to the origin $O$ can be defined as

$$i_p = \sqrt{\frac{I_p}{A}} \tag{I.10}$$

Fig. I.4

In SI units, the polar radius of gyration, $i_p$, is expressed in m.

## I.4 Product of Inertia

Considering a shaded area $A$ shown in Fig. I.3, the product of inertia of the area $A$ can be defined as

$$I_{zy} = \int zy \, dA \tag{I.11}$$

In SI units, the product of inertia, $I_{zy}$, is expressed in $m^4$.

## I.5 Parallel-Axis Theorem

Considering a shaded area $A$ shown in Fig. I.5, the moments of inertia of the area $A$ with respect to the $z$ and $z_C$ axes can be defined, respectively, as

$$I_z = \int y^2 \, dA, \qquad I_{z_C} = \int y_C^2 \, dA \tag{I.12}$$

Using $y = y_C + b$, we have

$$I_z = \int y^2 \, dA = \int (y_C + b)^2 \, dA = \int y_C^2 \, dA + 2b \int y_C \, dA + b^2 \int dA \tag{I.13}$$

Using $\int y_C \, dA = 0$, Eq. (I.13) can be simplified as

$$I_z = I_{z_C} + Ab^2 \tag{I.14}$$

Similarly, we can obtain

$$I_y = I_{y_C} + Aa^2, \qquad I_{zy} = I_{z_C y_C} + Aab, \qquad I_p = (I_p)_C + A(a^2 + b^2) \tag{I.15}$$

The relations expressed by Eq. (I.14) and Eq. (I.15) are called the parallel-axis theorem. These relations are often used to determine the moment of inertia of an area with respect to an arbitrary axis and the polar moment of inertia of an area with respect to an arbitrary point if the moment of inertia with respect to the centroidal axis and the polar moment of inertia with respect to the centroid are known.

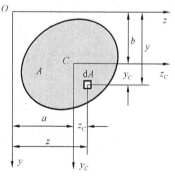

Fig. I.5

## I.6 Properties of Commonly-Used Areas

| Type of area | | Moment of inertia and modulus of section |
|---|---|---|
| Rectangle | | $A = bh$<br>$I_z = \dfrac{1}{12}bh^3, \quad S_z = \dfrac{1}{6}bh^2$<br>$I_{z'} = I_z + Aa^2$ |
| Solid circle | | $A = \dfrac{1}{4}\pi d^2$<br>$I_z = \dfrac{1}{64}\pi d^4, \quad S_z = \dfrac{1}{32}\pi d^3$<br>$(I_p)_C = \dfrac{1}{32}\pi d^4, \quad (S_p)_C = \dfrac{1}{16}\pi d^3$<br>$I_{z'} = I_z + Aa^2$ |
| Hollow circle | | $A = \dfrac{1}{4}\pi D^2(1-\alpha^2) \quad (\alpha = d/D)$<br>$I_z = \dfrac{1}{64}\pi D^4(1-\alpha^4), \quad S_z = \dfrac{1}{32}\pi D^3(1-\alpha^4)$<br>$(I_p)_C = \dfrac{1}{32}\pi D^4(1-\alpha^4), \quad (S_p)_C = \dfrac{1}{16}\pi D^3(1-\alpha^4)$<br>$I_{z'} = I_z + Aa^2$ |

# Appendix II  Shape Steels

## II.1  I-Steel

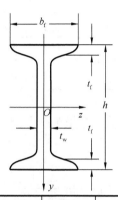

| Design-ation | Mass per unit length/(kg/m) | Depth $h$/mm | Flange | | Web | Area $A$/cm$^2$ | Moment of inertia | | Section modulus | |
|---|---|---|---|---|---|---|---|---|---|---|
| | | | Width $b_f$/mm | Thickness $t_f$/mm | Thickness $t_w$/mm | | $I_z$/cm$^4$ | $I_y$/cm$^4$ | $S_z$/cm$^3$ | $S_y$/cm$^3$ |
| 10 | 11.261 | 100 | 68 | 7.6 | 4.5 | 14.345 | 245 | 33.0 | 49.0 | 9.72 |
| 12.6 | 14.223 | 126 | 74 | 8.4 | 5.0 | 18.118 | 488 | 46.9 | 77.5 | 12.7 |
| 14 | 16.890 | 140 | 80 | 9.1 | 5.5 | 21.516 | 712 | 64.4 | 102 | 16.1 |
| 16 | 20.513 | 160 | 88 | 9.9 | 6.0 | 26.131 | 1130 | 93.1 | 141 | 21.2 |
| 18 | 24.143 | 180 | 94 | 10.7 | 6.5 | 30.756 | 1660 | 122 | 185 | 26.0 |
| 20a | 27.929 | 200 | 100 | 11.4 | 7.0 | 35.578 | 2370 | 158 | 237 | 31.5 |
| 20b | 31.069 | 200 | 102 | 11.4 | 9.0 | 39.578 | 2500 | 169 | 250 | 33.1 |
| 22a | 33.070 | 220 | 110 | 12.3 | 7.5 | 42.128 | 3400 | 225 | 309 | 40.9 |
| 22b | 36.524 | 220 | 112 | 12.3 | 9.5 | 46.528 | 3570 | 239 | 325 | 42.7 |
| 25a | 38.105 | 250 | 116 | 13.0 | 8.0 | 48.541 | 5020 | 280 | 402 | 48.3 |
| 25b | 42.030 | 250 | 118 | 13.0 | 10.0 | 53.541 | 52800 | 309 | 423 | 52.4 |
| 28a | 43.492 | 280 | 122 | 13.7 | 8.5 | 55.404 | 7110 | 345 | 508 | 56.6 |
| 28b | 47.888 | 280 | 124 | 13.7 | 10.5 | 61.004 | 7480 | 379 | 534 | 61.2 |
| 32a | 52.717 | 320 | 130 | 15.0 | 9.5 | 67.156 | 11100 | 460 | 692 | 70.8 |
| 32b | 57.741 | 320 | 132 | 15.0 | 11.5 | 73.556 | 11600 | 502 | 723 | 76.0 |
| 32c | 62.765 | 320 | 134 | 15.0 | 13.5 | 79.956 | 12200 | 544 | 760 | 81.2 |
| 36a | 60.037 | 360 | 136 | 15.8 | 10.0 | 76.480 | 15800 | 552 | 875 | 81.2 |
| 36b | 65.689 | 360 | 138 | 15.8 | 12.0 | 83.680 | 16500 | 582 | 919 | 84.3 |
| 36c | 71.341 | 360 | 140 | 15.8 | 14.0 | 90.880 | 17300 | 612 | 962 | 87.4 |
| 40a | 67.598 | 400 | 142 | 16.5 | 10.5 | 86.112 | 21700 | 660 | 1090 | 93.2 |
| 40b | 73.878 | 400 | 144 | 16.5 | 12.5 | 94.112 | 22800 | 692 | 1140 | 96.2 |
| 40c | 80.158 | 400 | 146 | 16.5 | 14.5 | 102.112 | 23900 | 727 | 1190 | 99.6 |
| 45a | 80.420 | 450 | 150 | 18.0 | 11.5 | 102.446 | 32200 | 855 | 1430 | 114 |
| 45b | 87.485 | 450 | 152 | 18.0 | 13.5 | 111.446 | 33800 | 894 | 1500 | 118 |
| 45c | 94.550 | 450 | 154 | 18.0 | 15.5 | 120.116 | 35300 | 938 | 1570 | 122 |

Continued

| Design-ation | Mass per unit length/(kg/m) | Depth $h$/mm | Flange | | Web Thickness $t_w$/mm | Area $A$/cm² | Moment of inertia | | Section modulus | |
|---|---|---|---|---|---|---|---|---|---|---|
| | | | Width $b_f$/mm | Thickness $t_f$/mm | | | $I_z$/cm⁴ | $I_y$/cm⁴ | $S_z$/cm³ | $S_y$/cm³ |
| 50a | 93.654 | 500 | 158 | 20.0 | 12.0 | 119.304 | 46500 | 1120 | 1860 | 142 |
| 50b | 101.504 | 500 | 160 | 20.0 | 14.0 | 129.304 | 48600 | 1170 | 1940 | 146 |
| 50c | 109.354 | 500 | 162 | 20.0 | 16.0 | 139.304 | 50600 | 1220 | 2080 | 151 |
| 56a | 106.316 | 560 | 166 | 21.0 | 12.5 | 135.435 | 65600 | 1370 | 2340 | 165 |
| 56b | 115.108 | 560 | 168 | 21.0 | 14.5 | 146.635 | 68500 | 1490 | 2450 | 174 |
| 56c | 123.900 | 560 | 170 | 21.0 | 16.5 | 157.835 | 71400 | 1560 | 2550 | 183 |
| 63a | 121.407 | 630 | 176 | 22.0 | 13.0 | 154.658 | 93900 | 1700 | 2980 | 193 |
| 63b | 131.298 | 630 | 178 | 22.0 | 15.0 | 167.258 | 98100 | 1810 | 3160 | 204 |
| 63c | 141.189 | 630 | 180 | 22.0 | 17.0 | 179.858 | 102000 | 1920 | 3300 | 214 |

## II.2  Channel Steel

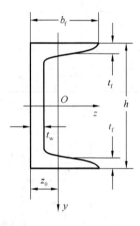

| Desig-nation | Mass per unit length/(kg/m) | Depth $h$/mm | Flange | | Web Thickness $t_w$/mm | Area $A$/cm² | Moment of inertia | | Section modulus | | $z_0$/cm |
|---|---|---|---|---|---|---|---|---|---|---|---|
| | | | Width $b_f$/mm | Thickness $t_f$/mm | | | $I_z$/cm⁴ | $I_y$/cm⁴ | $S_z$/cm³ | $S_y$/cm³ | |
| 5 | 5.438 | 50 | 37 | 7 | 4.5 | 6.928 | 26.0 | 8.3 | 10.4 | 3.55 | 1.35 |
| 6.3 | 6.634 | 63 | 40 | 7.5 | 4.8 | 8.451 | 50.8 | 11.9 | 16.1 | 4.50 | 1.36 |
| 8 | 8.045 | 80 | 43 | 8 | 5.0 | 10.248 | 101 | 16.6 | 25.3 | 5.79 | 1.43 |
| 10 | 10.007 | 100 | 48 | 8.5 | 5.3 | 42.748 | 198 | 25.6 | 39.7 | 7.8 | 1.52 |
| 12.6 | 12.318 | 126 | 53 | 9 | 5.5 | 15.692 | 391 | 38.0 | 62.1 | 10.2 | 1.59 |
| 14a | 14.535 | 140 | 58 | 9.5 | 6.0 | 18.516 | 564 | 53.2 | 80.5 | 13.0 | 1.71 |
| 14b | 16.733 | 140 | 60 | 9.5 | 8.0 | 21.316 | 609 | 61.1 | 87.1 | 14.1 | 1.67 |
| 16a | 17.240 | 160 | 63 | 10 | 6.5 | 21.962 | 866 | 73.3 | 108 | 16.3 | 1.80 |
| 16 | 19.752 | 160 | 65 | 10 | 8.5 | 25.162 | 935 | 83.4 | 117 | 17.6 | 1.75 |
| 18a | 20.174 | 180 | 68 | 10.5 | 7.0 | 25.699 | 1270 | 98.6 | 141 | 20.0 | 1.88 |
| 18 | 23.000 | 180 | 70 | 10.5 | 9.0 | 29.299 | 1370 | 111 | 152 | 21.5 | 1.84 |
| 20a | 22.637 | 200 | 73 | 11 | 7.0 | 28.837 | 1780 | 128 | 178 | 24.2 | 2.01 |
| 20 | 25.777 | 200 | 75 | 11 | 9.0 | 32.837 | 1910 | 144 | 191 | 25.9 | 1.95 |
| 22a | 24.999 | 220 | 77 | 11.5 | 7.0 | 31.846 | 2390 | 158 | 218 | 28.2 | 2.10 |
| 22 | 28.453 | 220 | 79 | 11.5 | 9.0 | 36.246 | 2570 | 176 | 234 | 30.1 | 2.03 |

Continued

| Desig-nation | Mass per unit length/(kg/m) | Depth $h$/mm | Flange | | Web | Area $A$/cm$^2$ | Moment of inertia | | Section modulus | | $z_0$/cm |
|---|---|---|---|---|---|---|---|---|---|---|---|
| | | | Width $b_f$/mm | Thickness $t_f$/mm | Thickness $t_w$/mm | | $I_z$/cm$^4$ | $I_y$/cm$^4$ | $S_z$/cm$^3$ | $S_y$/cm$^3$ | |
| 25a | 27.410 | 250 | 78 | 12 | 7.0 | 34.917 | 3370 | 176 | 270 | 30.6 | 2.07 |
| 25b | 31.335 | 250 | 80 | 12 | 9.0 | 39.917 | 3530 | 196 | 282 | 32.7 | 1.98 |
| 25c | 35.260 | 250 | 82 | 12 | 11.0 | 44.917 | 3690 | 218 | 295 | 35.9 | 1.92 |
| 28a | 31.427 | 280 | 82 | 12.5 | 7.5 | 40.034 | 4760 | 218 | 340 | 35.7 | 2.10 |
| 28b | 35.823 | 280 | 84 | 12.5 | 9.5 | 45.634 | 5130 | 242 | 366 | 37.9 | 2.02 |
| 28c | 40.219 | 280 | 86 | 12.5 | 11.5 | 51.234 | 5500 | 268 | 393 | 40.3 | 1.95 |
| 32a | 38.083 | 320 | 88 | 14 | 8.0 | 48.513 | 7600 | 305 | 475 | 46.5 | 2.24 |
| 32b | 43.107 | 320 | 90 | 14 | 10.0 | 54.913 | 8140 | 336 | 509 | 49.2 | 2.16 |
| 32c | 48.131 | 320 | 92 | 14 | 12.0 | 61.313 | 8690 | 374 | 543 | 52.6 | 2.09 |
| 36a | 47.814 | 360 | 96 | 16 | 9.0 | 60.910 | 11900 | 455 | 660 | 63.5 | 2.44 |
| 36b | 53.466 | 360 | 98 | 16 | 11.0 | 68.110 | 12700 | 497 | 703 | 66.9 | 2.37 |
| 36c | 59.118 | 360 | 100 | 16 | 13.0 | 75.310 | 13400 | 536 | 746 | 70.0 | 2.34 |
| 40a | 58.928 | 400 | 100 | 18 | 10.5 | 75.068 | 17600 | 592 | 879 | 78.8 | 2.49 |
| 40b | 65.208 | 400 | 102 | 18 | 12.5 | 93.068 | 18600 | 640 | 932 | 82.5 | 2.44 |
| 40c | 71.488 | 400 | 104 | 18 | 14.5 | 91.068 | 19700 | 688 | 986 | 86.2 | 2.42 |

## II.3 Equal Angle Steel

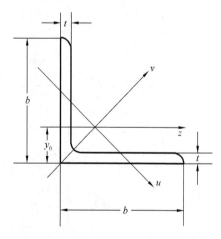

| Designation | Mass per unit length/(kg/m) | Width $b$/mm | Thickness $t$/mm | Area $A$/cm$^2$ | Moment of inertia | | | $y_0$/cm |
|---|---|---|---|---|---|---|---|---|
| | | | | | $I_z$/cm$^4$ | $I_v$/cm$^4$ | $I_u$/cm$^4$ | |
| 2 | 0.889 | 20 | 3 | 1.132 | 0.40 | 0.63 | 0.17 | 0.60 |
| | 1.145 | | 4 | 1.459 | 0.50 | 0.78 | 0.22 | 0.64 |
| 2.5 | 1.124 | 25 | 3 | 1.432 | 0.82 | 1.29 | 0.34 | 0.73 |
| | 1.459 | | 4 | 1.859 | 1.03 | 1.62 | 0.43 | 0.76 |
| 3.0 | 1.373 | 30 | 3 | 1.749 | 1.46 | 2.31 | 0.61 | 0.85 |
| | 1.786 | | 4 | 2.276 | 1.84 | 2.92 | 0.77 | 0.89 |
| 3.6 | 1.656 | 36 | 3 | 2.109 | 2.58 | 4.09 | 1.07 | 1.00 |
| | 2.163 | | 4 | 2.756 | 3.29 | 5.22 | 1.37 | 1.04 |
| | 2.654 | | 5 | 3.382 | 3.95 | 6.24 | 1.65 | 1.07 |

Continued

| Designation | Mass per unit length/(kg/m) | Width $b$/mm | Thickness $t$/mm | Area $A$/cm$^2$ | Moment of inertia | | | $y_0$/cm |
|---|---|---|---|---|---|---|---|---|
| | | | | | $I_z$/cm$^4$ | $I_y$/cm$^4$ | $I_u$/cm$^4$ | |
| 4.0 | 1.852 | 40 | 3 | 2.359 | 3.59 | 5.69 | 1.49 | 1.09 |
| | 2.422 | | 4 | 3.086 | 4.60 | 7.29 | 1.91 | 1.13 |
| | 2.976 | | 5 | 3.791 | 5.53 | 8.76 | 2.30 | 1.17 |
| 4.5 | 2.088 | 45 | 3 | 2.659 | 5.17 | 8.20 | 2.14 | 1.22 |
| | 2.736 | | 4 | 3.486 | 6.65 | 10.56 | 2.75 | 1.26 |
| | 3.369 | | 5 | 4.292 | 8.04 | 12.74 | 3.33 | 1.30 |
| | 3.985 | | 6 | 5.076 | 9.33 | 14.76 | 3.89 | 1.33 |
| 5 | 2.332 | 50 | 3 | 2.971 | 7.18 | 11.37 | 2.98 | 1.34 |
| | 3.059 | | 4 | 3.897 | 9.26 | 14.70 | 3.82 | 1.38 |
| | 3.770 | | 5 | 4.803 | 11.21 | 17.79 | 4.64 | 1.42 |
| | 4.465 | | 6 | 5.688 | 13.05 | 20.68 | 5.42 | 1.46 |
| 5.6 | 2.624 | 56 | 3 | 3.343 | 10.19 | 16.14 | 4.24 | 1.48 |
| | 3.446 | | 4 | 4.390 | 13.18 | 20.92 | 5.46 | 1.53 |
| | 4.251 | | 5 | 5.415 | 16.02 | 25.42 | 6.61 | 1.57 |
| | 6.568 | | 8 | 8.367 | 23.63 | 37.37 | 9.89 | 1.68 |
| 6.3 | 3.907 | 63 | 4 | 4.978 | 19.03 | 30.17 | 7.89 | 1.70 |
| | 4.882 | | 5 | 6.143 | 23.17 | 36.77 | 9.57 | 1.74 |
| | 5.721 | | 6 | 7.288 | 27.12 | 43.03 | 11.20 | 1.78 |
| | 7.469 | | 8 | 9.515 | 34.46 | 54.56 | 14.33 | 1.85 |
| | 9.151 | | 10 | 11.657 | 41.09 | 64.85 | 17.33 | 1.93 |
| 7 | 4.372 | 70 | 4 | 5.570 | 26.39 | 41.80 | 10.99 | 1.86 |
| | 5.397 | | 5 | 6.875 | 32.21 | 51.08 | 13.34 | 1.91 |
| | 6.406 | | 6 | 8.160 | 37.77 | 59.93 | 15.61 | 1.95 |
| | 7.398 | | 7 | 9.424 | 43.09 | 68.35 | 17.82 | 1.99 |
| | 8.373 | | 8 | 10.667 | 48.17 | 76.37 | 19.98 | 2.03 |
| 7.5 | 5.818 | 75 | 5 | 7.412 | 39.97 | 63.30 | 16.63 | 2.04 |
| | 6.905 | | 6 | 8.797 | 46.95 | 74.38 | 19.51 | 2.07 |
| | 7.976 | | 7 | 10.160 | 53.57 | 84.96 | 22.18 | 2.11 |
| | 9.030 | | 8 | 11.503 | 59.96 | 95.07 | 24.86 | 2.15 |
| | 11.089 | | 10 | 14.126 | 71.98 | 113.92 | 30.05 | 2.22 |
| 8 | 6.211 | 80 | 5 | 7.912 | 48.79 | 77.33 | 20.25 | 2.15 |
| | 7.376 | | 6 | 9.397 | 57.35 | 90.98 | 23.72 | 2.19 |
| | 8.525 | | 7 | 10.860 | 65.58 | 104.07 | 27.09 | 2.23 |
| | 9.658 | | 8 | 12.303 | 73.49 | 116.60 | 30.39 | 2.27 |
| | 11.874 | | 10 | 15.126 | 88.43 | 140.09 | 36.77 | 2.35 |
| 9 | 8.350 | 90 | 6 | 10.637 | 82.77 | 131.26 | 34.28 | 2.44 |
| | 9.656 | | 7 | 12.301 | 94.83 | 150.47 | 39.18 | 2.48 |
| | 10.946 | | 8 | 13.944 | 106.47 | 168.97 | 43.97 | 2.52 |
| | 13.476 | | 10 | 17.167 | 128.58 | 203.90 | 53.26 | 2.59 |
| | 15.940 | | 12 | 20.306 | 149.22 | 236.21 | 62.22 | 2.67 |
| 10 | 9.366 | 100 | 6 | 11.932 | 114.95 | 181.98 | 47.92 | 2.67 |
| | 10.830 | | 7 | 13.796 | 131.86 | 208.97 | 54.74 | 2.71 |
| | 12.276 | | 8 | 15.638 | 148.24 | 235.07 | 61.41 | 2.76 |
| | 15.120 | | 10 | 19.161 | 179.51 | 284.68 | 74.35 | 2.84 |
| | 17.898 | | 12 | 22.800 | 208.90 | 330.95 | 86.84 | 2.91 |
| | 20.611 | | 14 | 26.256 | 236.53 | 374.06 | 99.00 | 2.99 |
| | 23.257 | | 16 | 29.627 | 262.53 | 414.16 | 110.89 | 3.06 |

Continued

| Designation | Mass per unit length/(kg/m) | Width $b$/mm | Thickness $t$/mm | Area $A$/cm² | Moment of inertia | | | $y_0$/cm |
|---|---|---|---|---|---|---|---|---|
| | | | | | $I_z$/cm⁴ | $I_y$/cm⁴ | $I_u$/cm⁴ | |
| 11 | 11.928 | 110 | 7 | 15.196 | 177.16 | 280.94 | 73.38 | 2.96 |
| | 13.532 | | 8 | 17.238 | 199.46 | 316.49 | 82.42 | 3.01 |
| | 16.690 | | 10 | 21.261 | 242.19 | 384.39 | 99.98 | 3.09 |
| | 19.782 | | 12 | 25.200 | 282.55 | 448.17 | 116.93 | 3.16 |
| | 22.809 | | 14 | 29.056 | 320.71 | 508.01 | 133.40 | 3.24 |
| 12.5 | 15.504 | 125 | 8 | 19.750 | 297.03 | 470.89 | 123.16 | 3.37 |
| | 19.133 | | 10 | 24.373 | 361.67 | 573.89 | 149.46 | 3.45 |
| | 22.696 | | 12 | 28.912 | 423.16 | 671.44 | 174.88 | 3.53 |
| | 26.193 | | 14 | 33.367 | 481.65 | 763.73 | 199.57 | 3.61 |
| 14 | 21.488 | 140 | 10 | 27.372 | 514.65 | 817.27 | 212.04 | 3.82 |
| | 25.522 | | 12 | 32.512 | 603.68 | 958.79 | 248.57 | 3.90 |
| | 29.490 | | 14 | 37.567 | 688.81 | 1093.56 | 284.06 | 3.98 |
| | 33.393 | | 16 | 42.539 | 770.24 | 1221.81 | 318.67 | 4.06 |
| 16 | 24.729 | 160 | 10 | 31.502 | 779.63 | 1237.30 | 321.76 | 4.31 |
| | 29.391 | | 12 | 37.441 | 916.58 | 1455.68 | 377.49 | 4.39 |
| | 33.987 | | 14 | 43.296 | 1048.36 | 1665.02 | 431.70 | 4.47 |
| | 38.518 | | 16 | 49.067 | 1175.08 | 1865.57 | 484.59 | 4.55 |
| 18 | 33.159 | 180 | 12 | 42.241 | 1321.35 | 2100.10 | 542.61 | 4.89 |
| | 38.383 | | 14 | 48.896 | 1514.48 | 2407.42 | 621.53 | 4.97 |
| | 43.542 | | 16 | 55.467 | 1700.99 | 2703.37 | 698.60 | 5.05 |
| | 48.634 | | 18 | 61.955 | 1875.12 | 2988.24 | 762.01 | 5.13 |
| 20 | 42.894 | 200 | 14 | 54.642 | 2103.55 | 3343.26 | 863.83 | 5.46 |
| | 48.680 | | 16 | 62.013 | 2366.15 | 3760.89 | 971.41 | 5.54 |
| | 54.401 | | 18 | 69.301 | 2620.64 | 4164.54 | 1076.74 | 5.62 |
| | 60.056 | | 20 | 76.505 | 2867.30 | 4554.55 | 1180.04 | 5.69 |
| | 71.168 | | 24 | 90.661 | 3338.25 | 5294.97 | 1381.53 | 5.87 |

## II.4 Unequal Angle Steel

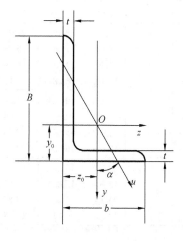

| Design-ation | Mass per unit length/(kg/m) | Width | | Thickness t/mm | Area A/cm² | Moment of inertia | | | z₀/cm | y₀/cm | tanα |
|---|---|---|---|---|---|---|---|---|---|---|---|
| | | B/mm | b/mm | | | $I_z$/cm⁴ | $I_y$/cm⁴ | $I_u$/cm⁴ | | | |
| 2.5/1.6 | 0.192 | 25 | 16 | 3 | 1.162 | 0.70 | 0.22 | 0.14 | 0.42 | 0.86 | 0.392 |
| | 1.176 | | | 4 | 1.499 | 0.88 | 0.27 | 0.17 | 0.46 | 0.90 | 0.381 |
| 3.2/2 | 1.171 | 32 | 20 | 3 | 1.492 | 1.53 | 0.46 | 0.28 | 0.49 | 1.08 | 0.382 |
| | 1.522 | | | 4 | 1.939 | 1.93 | 0.57 | 0.35 | 0.53 | 1.12 | 0.374 |
| 4/2.5 | 1.484 | 40 | 25 | 3 | 1.890 | 3.08 | 0.93 | 0.56 | 0.59 | 1.32 | 0.385 |
| | 1.936 | | | 4 | 2.467 | 3.93 | 1.18 | 0.71 | 0.63 | 1.37 | 0.381 |
| 4.5/2.8 | 1.687 | 45 | 28 | 3 | 2.149 | 4.45 | 1.34 | 0.80 | 0.64 | 1.47 | 0.383 |
| | 2.203 | | | 4 | 2.806 | 5.69 | 1.70 | 1.02 | 0.68 | 1.51 | 0.380 |
| 5/3.2 | 1.908 | 50 | 32 | 3 | 2.431 | 6.24 | 2.02 | 1.20 | 0.73 | 1.60 | 0.404 |
| | 2.494 | | | 4 | 3.177 | 8.02 | 2.58 | 1.53 | 0.77 | 1.65 | 0.402 |
| 5.6/3.6 | 2.153 | 56 | 36 | 3 | 2.743 | 8.88 | 2.92 | 1.73 | 0.80 | 1.78 | 0.408 |
| | 2.818 | | | 4 | 3.590 | 11.45 | 3.76 | 2.23 | 0.85 | 1.82 | 0.408 |
| | 3.466 | | | 5 | 4.415 | 13.86 | 4.49 | 2.67 | 0.88 | 1.87 | 0.404 |
| 6.3/4 | 3.185 | 63 | 40 | 4 | 4.058 | 16.49 | 5.23 | 3.12 | 0.92 | 2.04 | 0.398 |
| | 3.920 | | | 5 | 4.993 | 20.02 | 6.31 | 3.76 | 0.95 | 2.08 | 0.396 |
| | 4.638 | | | 6 | 5.908 | 23.36 | 7.29 | 4.34 | 0.99 | 2.12 | 0.393 |
| | 5.339 | | | 7 | 6.802 | 26.53 | 8.24 | 4.97 | 1.03 | 2.15 | 0.389 |
| 7/4.5 | 3.570 | 70 | 45 | 4 | 4.547 | 23.17 | 7.55 | 4.40 | 1.02 | 2.24 | 0.410 |
| | 4.403 | | | 5 | 5.609 | 27.95 | 9.13 | 5.40 | 1.06 | 2.28 | 0.407 |
| | 5.218 | | | 8 | 6.647 | 32.54 | 10.62 | 6.35 | 1.09 | 2.32 | 0.404 |
| | 6.011 | | | 7 | 7.657 | 37.22 | 12.01 | 7.16 | 1.13 | 2.36 | 0.402 |
| (7.5/5) | 4.808 | 75 | 50 | 5 | 6.125 | 34.86 | 12.61 | 7.41 | 1.17.201 | 2.40 | 0.435 |
| | 5.699 | | | 6 | 7.260 | 41.12 | 14.70 | 8.54 | 1 1.21 | 2.44 | 0.435 |
| | 7.431 | | | 8 | 9.467 | 52.39 | 18.53 | 10.87 | 1.29 | 2.52 | 0.429 |
| | 9.098 | | | 10 | 11.590 | 62.71 | 21.96 | 13.10 | 1.36 | 2.60 | 0.423 |
| 8/5 | 5.005 | 80 | 50 | 5 | 6.375 | 41.96 | 12.82 | 7.66 | 1.14 | 2.60 | 0.388 |
| | 5.935 | | | 6 | 7.560 | 49.49 | 14.95 | 8.85 | 1.18 | 2.65 | 0.387 |
| | 6.848 | | | 7 | 8.724 | 56.16 | 16.96 | 10.18 | 1.21 | 2.69 | 0.384 |
| | 7.745 | | | 8 | 9.867 | 62.83 | 18.85 | 11.38 | 1.25 | 2.73 | 0.381 |
| 9/5.6 | 5.661 | 90 | 56 | 5 | 7.212 | 60.45 | 18.32 | 10.98 | 1.25 | 2.91 | 0.385 |
| | 6.717 | | | 6 | 8.557 | 71.03 | 21.42 | 12.90 | 1.29 | 2.95 | 0.384 |
| | 7.756 | | | 7 | 9.880 | 81.01 | 24.36 | 14.67 | 1.33 | 3.00 | 0.382 |
| | 8.779 | | | 8 | 11.183 | 91.03 | 27.15 | 16.34 | 1.36 | 3.04 | 0.380 |
| 10/6.3 | 7.550 | 100 | 63 | 6 | 9.617 | 99.06 | 30.94 | 18.42 | 1.43 | 3.24 | 0.394 |
| | 8.722 | | | 7 | 11.111 | 113.45 | 35.26 | 21.00 | 1.47 | 3.28 | 0.394 |
| | 9.878 | | | 8 | 12.584 | 127.37 | 39.39 | 23.50 | 1.50 | 3.32 | 0.391 |
| | 12.142 | | | 10 | 15.467 | 153.81 | 47.12 | 28.33 | 1.58 | 3.40 | 0.387 |
| 10/8 | 8.350 | 100 | 80 | 6 | 10.637 | 107.04 | 61.24 | 31.65 | 1.97 | 2.95 | 0.627 |
| | 9.656 | | | 7 | 12.301 | 122.73 | 70.08 | 36.17 | 2.01 | 3.00 | 0.626 |
| | 10.946 | | | 8 | 13.944 | 137.92 | 78.58 | 40.58 | 2.05 | 3.04 | 0.625 |
| | 13.476 | | | 10 | 17.167 | 166.87 | 94.65 | 49.10 | 2.13 | 3.12 | 0.622 |
| 11/7 | 8.350 | 110 | 70 | 6 | 10.637 | 133.37 | 42.92 | 25.36 | 1.57 | 3.53 | 0.403 |
| | 9.656 | | | 7 | 12.301 | 153.00 | 49.01 | 28.95 | 1.61 | 3.57 | 0.402 |
| | 10.946 | | | 8 | 13.944 | 172.04 | 54.87 | 32.45 | 1.65 | 3.62 | 0.401 |
| | 13.467 | | | 10 | 17.167 | 208.39 | 65.88 | 39.20 | 1.72 | 3.07 | 0.397 |

Continued

| Design-ation | Mass per unit length/(kg/m) | Width | | Thickness $t$/mm | Area $A$/cm² | Moment of inertia | | | $z_0$/cm | $y_0$/cm | $\tan\alpha$ |
|---|---|---|---|---|---|---|---|---|---|---|---|
| | | $B$/mm | $b$/mm | | | $I_z$/cm⁴ | $I_y$/cm⁴ | $I_u$/cm⁴ | | | |
| 12.5/8 | 11.066 | 125 | 80 | 7 | 14.096 | 227.98 | 74.42 | 43.81 | 1.80 | 4.01 | 0.408 |
| | 12.551 | | | 8 | 15.989 | 256.77 | 83.49 | 49.15 | 1.84 | 4.06 | 0.407 |
| | 15.474 | | | 10 | 19.712 | 312.04 | 100.67 | 59.45 | 1.92 | 4.14 | 0.404 |
| | 18.330 | | | 12 | 23.351 | 364.41 | 116.67 | 69.35 | 2.00 | 4.22 | 0.400 |
| 14/9 | 14.160 | 140 | 90 | 8 | 18.038 | 365.64 | 120.69 | 70.83 | 2.04 | 4.50 | 0.411 |
| | 17.475 | | | 10 | 22.261 | 445.50 | 140.03 | 85.82 | 2.12 | 4.58 | 0.409 |
| | 20.724 | | | 12 | 26.400 | 521.59 | 169.79 | 100.21 | 2.19 | 4.66 | 0.406 |
| | 23.908 | | | 14 | 30.456 | 594.10 | 192.10 | 114.13 | 2.27 | 4.74 | 0.403 |
| 16/10 | 19.872 | 160 | 100 | 10 | 25.315 | 668.69 | 205.03 | 121.74 | 2.28 | 5.24 | 0.390 |
| | 23.592 | | | 12 | 30.054 | 784.91 | 239.06 | 142.33 | 2.36 | 5.32 | 0.388 |
| | 27.247 | | | 14 | 34.709 | 896.30 | 271.20 | 162.23 | 2.43 | 5.40 | 0.385 |
| | 30.835 | | | 16 | 39.281 | 1003.04 | 301.60 | 182.57 | 2.51 | 5.48 | 0.382 |
| 18/11 | 22.273 | 180 | 110 | 10 | 28.373 | 956.25 | 278.11 | 166.50 | 2.44 | 5.89 | 0.376 |
| | 26.464 | | | 12 | 33.712 | 1124.72 | 325.03 | 194.87 | 2.52 | 5.98 | 0.374 |
| | 30.589 | | | 14 | 38.967 | 1286.91 | 369.55 | 222.30 | 2.59 | 60.6 | 0.372 |
| | 34.649 | | | 16 | 44.139 | 1443.06 | 411.85 | 248.94 | 2.67 | 6.14 | 0.369 |
| 20/12.5 | 29.761 | 200 | 125 | 12 | 37.912 | 1570.90 | 483.16 | 285.79 | 2.83 | 6.54 | 0.392 |
| | 34.436 | | | 14 | 43.867 | 1800.97 | 550.83 | 326.58 | 2.91 | 6.62 | 0.390 |
| | 39.045 | | | 16 | 49.739 | 2023.35 | 615.44 | 366.21 | 2.99 | 6.70 | 0.388 |
| | 43.588 | | | 18 | 55.526 | 2238.30 | 677.19 | 404.83 | 3.06 | 6.78 | 0.385 |

# Appendix III  Deflection Curves

| Beam and load | Deflection Dequation | Critical deflection | Critical slope |
|---|---|---|---|
| Cantilever with point load $F$ at $C$ (distance $a$ from $A$) | $w = \dfrac{Fx^2}{6EI}(3a-x)$  $(0 \leqslant x \leqslant a)$<br>$w = \dfrac{Fa^2}{6EI}(3x-a)$  $(a \leqslant x \leqslant l)$ | $w_{max} = w_B = \dfrac{Fa^2}{6EI}(3l-a)$<br>$w_C = \dfrac{Fa^3}{3EI}$ | $\theta_{max} = \theta_B = \theta_C = \dfrac{Fa^2}{2EI}$ |
| Cantilever with moment $M_e$ at $C$ | $w = \dfrac{M_e x^2}{2EI}$  $(0 \leqslant x \leqslant a)$<br>$w = \dfrac{M_e a}{2EI}(2x-a)$  $(a \leqslant x \leqslant l)$ | $w_{max} = w_B = \dfrac{M_e a}{2EI}(2l-a)$<br>$w_C = \dfrac{M_e a^2}{2EI}$ | $\theta_{max} = \theta_B = \theta_C = \dfrac{M_e a}{EI}$ |
| Cantilever with uniform load $q$ | $w = \dfrac{qx^2}{24EI}(6l^2 - 4lx + x^2)$ | $w_{max} = w_B = \dfrac{ql^4}{8EI}$ | $\theta_{max} = \theta_B = \dfrac{ql^3}{6EI}$ |
| Simply supported beam with point load $F$ at $C$ | $w = \dfrac{Fbx}{6EIl}(l^2 - b^2 - x^2)$<br>$(0 \leqslant x \leqslant a)$<br>$w = \dfrac{F}{6EIl}[l(x-a)^3$<br>$+ b(l^2 - b^2)x - bx^3]$<br>$(a \leqslant x \leqslant l)$ | For $a \geqslant b$:<br>$w_{max} = w\left[\sqrt{\dfrac{1}{3}(l^2 - b^2)}\right]$<br>$= \dfrac{Fb\sqrt{(l^2 - b^2)^3}}{9\sqrt{3}EIl}$<br>$w\left(\dfrac{1}{2}l\right) = \dfrac{Fb(3l^2 - 4b^2)}{48EI}$ | For $a \geqslant b$:<br>$\theta_A = \dfrac{Fab(l+b)}{6EIl}$<br>$\theta_{max} = -\theta_B = \dfrac{Fab(l+a)}{6EIl}$ |
| Simply supported beam with moment $M_e$ at $C$ | $w = \dfrac{M_e x}{6EIl}(x^2 + 3b^2 - l^2)$<br>$(0 \leqslant x \leqslant a)$<br>$w = \dfrac{M_e}{6EIl}[x^3 - 3l(x-a)^2$<br>$- (l^2 - 3b^2)x]$<br>$(a \leqslant x \leqslant l)$ | For $a \geqslant b$:<br>$w_{max} = -w\left[\sqrt{\dfrac{1}{3}(l^2 - 3b^2)}\right]$<br>$= \dfrac{M_e \sqrt{(l^2 - 3b^2)^3}}{9\sqrt{3}EIl}$<br>$w\left(\dfrac{1}{2}l\right) = -\dfrac{M_e(l^2 - 4b^2)}{16EI}$ | $\theta_A = -\dfrac{M_e}{6EIl}(l^2 - 3b^2)$<br>$\theta_B = -\dfrac{M_e}{6EIl}(l^2 - 3a^2)$<br>$\theta_{max} = \theta_C = \dfrac{M_e}{6EIl}[3(a^2 + b^2) - l^2]$ |
| Simply supported beam with uniform load $q$ | $w = \dfrac{qx}{24EI}(l^3 - 2lx^2 + x^3)$ | $w_{max} = w\left(\dfrac{1}{2}l\right) = \dfrac{5ql^4}{384EI}$ | $\theta_{max} = \theta_A = -\theta_B = \dfrac{ql^3}{24EI}$ |

# References

BEER F P, 2015. Mechanics of materials [M]. 7th ed. New York: McGraw-Hill Education.
BARBER J R, 2011. Intermediate mechanics of materials [M]. 2nd ed. New York: Springer.
HIBBELER R C, 2017. Mechanics of materials [M]. 10th ed. New York: Pearson Prentice Hall.

# 第1章 材料力学基础

在理论力学中，物体被假设为刚体。然而，就结构或机械抵抗失效而言，物体变形就显得极其重要。因此，在材料力学中不再假设物体具有完全刚性。

材料力学研究结构或机械抵抗失效的能力，其主要研究任务包括：①强度，即构件支撑规定载荷而不经历过高应力的能力；②刚度，即构件支撑规定载荷而不经历过大变形的能力；③稳定性，即构件支撑规定轴向压缩载荷而不引起突然横向挠曲的能力。

在材料力学中，所涉及的材料都假设为：①连续，即材料由连续分布的物质构成；②均匀，即材料在各点处都具有相同的力学性能；③各向同性，即材料在所有方向上都具有相同的力学性能。

材料的强度和刚度依赖于材料支撑规定载荷而不出现过高应力和过大变形的能力。这些能力是材料本身的固有属性，必须通过试验方法进行测定。测定材料力学性能的最重要试验之一就是拉伸或压缩试验。该试验经常用于测定所用材料的应力应变关系。

## 1.1 外　　力

作用于物体上的外力分为面力和体力。

**1. 面力**

作用于物体表面的外力称为表面力，简称面力。

如果面力分布在物体表面的有限区域，则称为面分布载荷，如图1.1(a)所示。如果面力作用在物体表面的狭窄区域，则称为线分布载荷，如图1.1(b)所示。如果面力作用区域非常小，则称为集中载荷，如图1.1(c)所示。

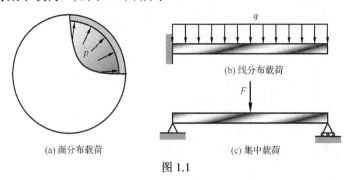

图 1.1

**2. 体力**

作用于物体内每一点的外力称为体积力，简称体力。重力是典型的体力。

## 1.2 内　力

当外载荷作用于构件时,在构件内将产生相应的分布内力。构件截面上的分布内力可通过截面法进行求解。

如图1.2(a)所示,用平面$\Pi$截开构件上需要确定分布内力的截面。为了求截开平面上的分布内力,移去截开平面右侧构件,用分布内力代替其对左侧构件的作用,如图1.2(b)所示。

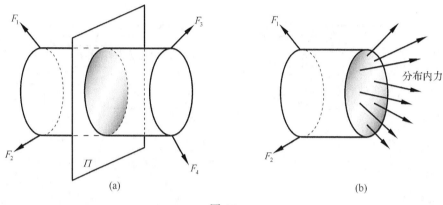

图 1.2

考虑构件保留部分的平衡,则分布内力可通过静平衡方程确定。虽然内力的精确分布未知,但是利用静平衡方程可以确定截面上分布内力引起的合力 $R$ 和对 $O$ 点的合力偶 $M_O$,如图1.3(a)所示。

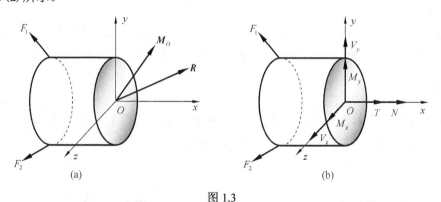

图 1.3

一般来说,合力 $R$ 和合力偶 $M_O$ 具有任意方向,即既不垂直截面,也不平行截面。然而,可以把合力和合力偶分解为六个分量,如图1.3(b)所示。

(1)轴力。合力的法向分量称为轴力,由 $N$ 表示。
(2)剪力。合力的切向分量称为剪力,由 $V_y$ 和 $V_z$ 表示。
(3)扭矩。合力偶的法向分量称为扭矩,由 $T$ 表示。
(4)弯矩。合力偶的切向分量称为弯矩,由 $M_y$ 和 $M_z$ 表示。

## 1.3 应力

当各种外载荷作用于构件时，在构件内任何一点将产生分布内力。图1.4(a)为定义截面上点 $P$ 处的应力，围绕点 $P$ 取面元 $\Delta A$，并假设面元上的合力为 $\Delta F$。通常 $\Delta F$ 在截面上特定点有唯一确定的方向，因而可分解为三个分量 $\Delta N$、$\Delta V_y$ 和 $\Delta V_z$，如图1.4(b)所示。$\Delta N$ 为垂直面元 $\Delta A$ 的法向分量，$\Delta V_y$ 和 $\Delta V_z$ 为面元 $\Delta A$ 内的切向分量。

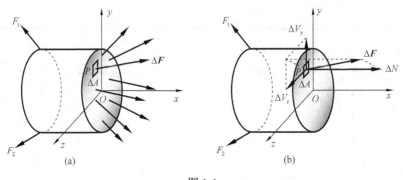

图1.4

### 1. 正应力

垂直截面的法向力的集度(即单位面积上的法向力)定义为正应力，用 $\sigma$ 表示。如图1.5所示，构件截面上点 $P$ 处的正应力定义为

$$\sigma_x = \lim_{\Delta A \to 0} \frac{\Delta N}{\Delta A} \tag{1.1}$$

式中，$\sigma_x$ 垂直于截面，下标表示截面的外法向和正应力方向。正号表示拉应力，负号表示压应力。在国际单位制中，$\Delta N$ 的单位为 N，$\Delta A$ 的单位为 $m^2$，正应力 $\sigma_x$ 的单位为 Pa。

### 2. 剪应力

平行截面的切向力的集度(即单位面积上的切向力)定义为剪应力，用 $\tau$ 表示。如图1.6所示，构件截面上点 $P$ 处的两个剪应力分量分别定义为

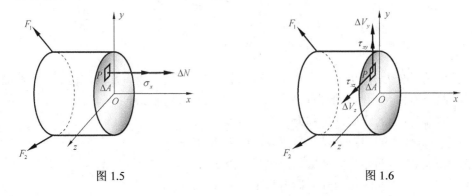

图1.5　　　　　　　　　　　　图1.6

$$\tau_{xy} = \lim_{\Delta A \to 0} \frac{\Delta V_y}{\Delta A}, \qquad \tau_{xz} = \lim_{\Delta A \to 0} \frac{\Delta V_z}{\Delta A} \tag{1.2}$$

式中，$\tau_{xy}$ 和 $\tau_{xz}$ 位于截面内。两个下标用于表示剪应力分量：第一个表示截面的外法向，第二个表示剪应力的方向。在国际单位制中，剪应力 $\tau_{xy}$ 和 $\tau_{xz}$ 的单位为 Pa。

为了显示剪应力的产生，考虑如图 1.7(a) 所示构件，受大小为 $F$ 的两个横向力作用。

把两力作用点之间的截面 $CC'$ 截开，考虑如图 1.7(b) 所示截面左侧部分的平衡，则在横截面上一定存在分布内力。分布内力的合力称为直接剪力，由 $V$ 表示。直接剪力 $V$ 除以横截面面积 $A$ 即可得到直剪应力。用 $\tau$ 表示直剪应力，则有

$$\tau = \frac{V}{A} \tag{1.3}$$

应该注意，式(1.3)所得值为整个截面上的剪应力的平均值。

用于连接各种构件的螺钉、销钉和铆钉通常会出现直剪应力。如图 1.8(a) 所示，考虑通过螺钉连接的两块薄板。如果薄板受大小为 $F$ 的拉力作用，在与薄板接触面对应的螺钉横截面上将出现直剪应力。画接触面下侧螺钉简图，如图 1.8(b) 所示，则截面上直剪应力 $\tau$ 等于 $V/A$。

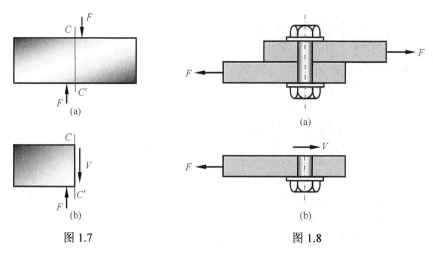

图 1.7  图 1.8

**例 1.1** 载荷 $F$ 作用于如图 1.9(a) 所示钢杆，钢杆由开有直径为 15 mm 孔的薄板支撑。已知钢的剪应力不能超过 120 MPa，求能够作用于钢杆的最大载荷 $F_{\max}$。

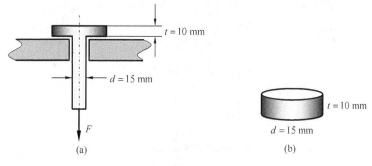

图 1.9

**解** 剪切面为圆柱面，如图 1.9(b) 所示，剪切面积 $A = \pi d t$。因最大剪应力 $\tau_{max} = 120$ MPa，则利用式(1.3)，得最大载荷 $F_{max}$ 为

$$F_{max} = \tau_{max} A = 56.5 \text{ kN}$$

#### 3. 挤压应力

由螺钉、销钉和铆钉连接的构件的挤压面上将产生应力。如图 1.10(a) 所示，考虑通过螺钉连接的两块薄板。螺钉作用于上板的力 $F_{bs}$（图 1.10(b)）与上板反作用于螺钉的力 $F'_{bs}$（图 1.10(c)）大小相等，方向相反。螺钉施加的力 $F_{bs}$ 表示半圆柱内表面上分布力的合力。

因为构件接触面上的力的分布十分复杂，所以通过合力 $F_{bs}$ 除以螺钉在薄板横截面上的投影面积 $A_{bs}$ 而得到的应力平均值是挤压应力 $\sigma_{bs}$。因投影面积 $A_{bs}$ 等于 $td$，其中 $t$ 为薄板厚度，$d$ 为螺钉直径，故有

$$\sigma_{bs} = \frac{F_{bs}}{A_{bs}} = \frac{F_{bs}}{td} \tag{1.4}$$

**例 1.2** 载荷 $F$ 作用于如图 1.11(a) 所示钢杆，钢杆由开有直径为 15 mm 孔的薄板支撑。已知钢的挤压应力不能超过 150 MPa，求能够作用于钢杆的最大载荷 $F_{max}$。

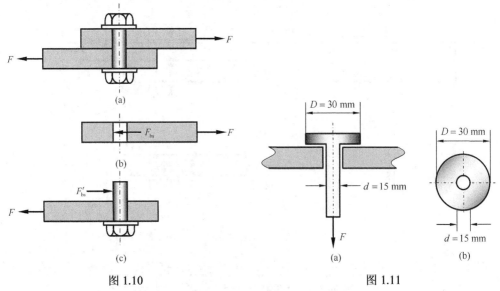

图 1.10 　　　　　　　　　　图 1.11

**解** 如图 1.11(b) 所示，挤压面为内径 $d = 15$ mm 和外径 $D = 30$ mm 的环形表面，因此挤压面积等于 $A_{bs} = \frac{1}{4}\pi(D^2 - d^2)$。因最大挤压应力 $\sigma_{bs} = 150$ MPa，则利用式(1.4)，得最大载荷 $F_{max}$ 为

$$F_{max} = \sigma_{bs} A_{bs} = 79.5 \text{ kN}$$

## 1.4　应　　变

当外载荷作用于构件时，载荷将倾向于改变构件的尺寸和形状。这些变化称为构件的变形。

1. 线应变

单位长度线段的伸长或缩短称为线应变，用 $\varepsilon$ 表示。如图 1.12(a)所示，考虑构件上过点 $P$ 的线段 $\Delta x$。在外载荷作用下构件发生变形，线段 $\Delta x$ 的长度变为 $\Delta x'$，如图 1.12(b)所示。线段 $\Delta x$ 的变形等于 $\Delta \delta = \Delta x' - \Delta x$。

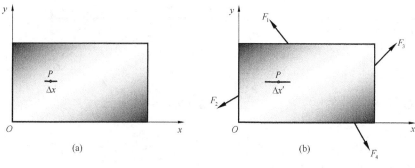

图 1.12

当 $\Delta x$ 趋于零时，$\Delta \delta$ 与 $\Delta x$ 的商将趋于极限值。该极限值称为点 $P$ 处沿线段方向的线应变，可表示为

$$\varepsilon_x = \lim_{\Delta x \to 0} \frac{\Delta \delta}{\Delta x} \tag{1.5}$$

式中，线应变 $\varepsilon_x$ 为无量纲量。如果线段伸长，线应变为正；如果线段缩短，线应变为负。

2. 剪应变

两垂直线段之间的角度变化称为剪应变，用 $\gamma$ 表示。如图 1.13(a)所示，考虑两垂直线段 $\Delta x$ 和 $\Delta y$，在点 $P$ 处相交。在外载荷作用下构件发生变形，两线段的夹角由 $\frac{1}{2}\pi$ 变为 $\theta'$，如图 1.13(b)所示。

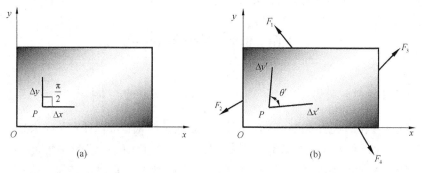

图 1.13

点 $P$ 处在包含两线段的平面内的剪应变可写为

$$\gamma_{xy} = \frac{1}{2}\pi - \lim_{\substack{\Delta x \to 0 \\ \Delta y \to 0}} \theta' \tag{1.6}$$

如果 $\theta'$ 小于 $\frac{1}{2}\pi$，则剪应变 $\gamma_{xy}$ 为正；否则，为负。剪应变 $\gamma_{xy}$ 的单位为 rad。

**例 1.3** 如图 1.14(a)所示，考虑矩形薄板 $OACB$。薄板发生均匀平面变形后的尺寸和形状如图 1.14(b)所示。求薄板点 $O$ 处的线应变 $\varepsilon_x$ 和 $\varepsilon_y$ 以及剪应变 $\gamma_{xy}$。

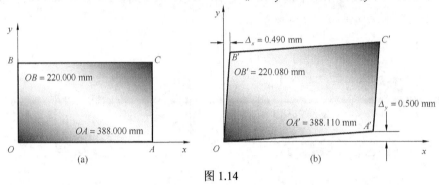

图 1.14

**解** 根据线应变和剪应变的定义，有

$$\varepsilon_x = \frac{OA' - OA}{OA} = 2.84 \times 10^{-4}, \quad \varepsilon_y = \frac{OB' - OB}{OB} = 3.64 \times 10^{-4}$$

$$\gamma_{xy} = \frac{1}{2}\pi - \angle A'OB' \approx \frac{\Delta_y}{OA} + \frac{\Delta_x}{OB} = 3.51 \times 10^{-3} \text{ rad}$$

## 1.5 胡 克 定 律

大部分工程结构都设计为承受较小的变形，即仅包含应力应变曲线的直线部分。对应力应变曲线的直线部分，应力 $\sigma$ 与应变 $\varepsilon$ 成正比，则有

$$\sigma = E\varepsilon \tag{1.7}$$

上述关系称为胡克定律。系数 $E$ 称为材料的弹性模量或杨氏模量。应变 $\varepsilon$ 为无量纲量，因此弹性模量 $E$ 具有与应力 $\sigma$ 相同的单位。

## 1.6 低碳钢拉伸性能

1. 标准试样

具有确定尺寸和形状的标准试样必须用于低碳钢拉伸试验，以便实验结果具有可比性。

图 1.15(a)为一种常用的低碳钢标准拉伸试样。试样中间部位的横截面面积 $A_0$ 已经精确测量，间距为 $l_0$ 的两根标线也已经刻划在试样中间部位。距离 $l_0$ 称为试样标距。

2. 拉伸图

把试样放入万能试验机，施加中心载荷 $F$。当载荷 $F$ 增加时，两标线之间的距离 $l$ 也随之增加，如图 1.15(b)所示。对每个 $F$ 值，记录相应的伸长量 $\delta = l - l_0$。

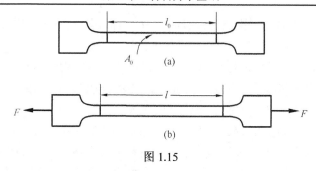

图 1.15

画载荷大小 $F$ 与变形 $\delta$ 的曲线，得载荷变形图，如图 1.16 所示。该图称为低碳钢的拉伸图。

3. 应力应变曲线

根据每对读数 $F$ 和 $\delta$，载荷 $F$ 除以原始横截面面积 $A_0$ 计算应力 $\sigma$，伸长量 $\delta$ 除以两标线之间的原始距离 $l_0$ 计算应变 $\varepsilon$。

以 $\varepsilon$ 为横坐标、$\sigma$ 为纵坐标绘图而得到的曲线，既能够表征低碳钢力学性能，又与所用特定试样的尺寸无关。该曲线称为低碳钢的应力应变曲线，如图 1.17 所示。

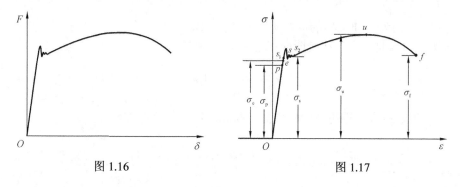

图 1.16　　　　　　　　　　图 1.17

根据如图 1.17 所示低碳钢的应力应变曲线，可看出材料所表现出的四个不同范围。

(1) 弹性阶段($Oe$)。应力应变曲线的 $Oe$ 区域称为弹性范围，如图 1.17 所示。该区域材料呈现弹性特征。对应弹性范围的最大应力称为弹性极限，用 $\sigma_e$ 表示。

弹性范围 $Oe$ 的直线区域 $Op$ 称为线弹性范围。在该范围材料呈现线弹性特征，即应力与应变成正比。满足线性关系的最大应力称为比例极限，用 $\sigma_p$ 表示。如果应力稍微超过比例极限，则材料仍然呈现弹性特征，直到应力达到弹性极限。

在弹性范围，如果卸载，则试样将恢复到其原始尺寸和形状。通常对低碳钢而言，弹性极限非常接近比例极限，因而很难确定弹性极限。

(2) 屈服阶段($s_1s_2$)。稍微高于弹性极限的应力将会引起材料失效，引起材料发生永久变形，如图 1.17 所示。这种现象称为屈服，屈服由应力应变曲线的 $s_1s_2$ 区域表示。屈服开始产生所对应的应力称为屈服强度、屈服应力或屈服点，用 $\sigma_s$ 表示。屈服范围产生的变形称为塑性变形。塑性变形由材料斜截面上的剪应力引起材料沿斜截面的滑移而形成。达到屈服点后，试样在不增大载荷的情况下还会继续伸长。

对于低碳钢和热轧钢，屈服点常常有两个值。上屈服点首先出现，紧接着承载能力突然下降到下屈服点。

(3) 强化阶段($s_2u$)。当屈服结束时，更大的载荷能够作用于试样，引起曲线继续上升但变得更为平坦，直到达到最大应力。该最大应力称为极限强度或极限应力，用$\sigma_u$表示。以这种方式引起的曲线上升称为应变强化，应变强化由图 1.17 的 $s_2u$ 区域表示。

(4) 颈缩阶段($uf$)。达到极限强度后，横截面面积在试样局部区域开始急剧减小，如图 1.17 所示。这种现象称为颈缩，颈缩现象由材料内部形成的滑移面引起，实际应变由剪应力引起。因此，当试样进一步伸长时，横向收缩逐渐形成了颈缩区域。在颈缩区域横截面面积不断减小，因此越来越小面积仅能承受不断减小的载荷。应力应变曲线向下弯曲直到试样达到断裂应力$\sigma_f$时而发生断裂。

应该注意，断裂沿着与试样原始表面大约成 45° 的锥形面发生。

**4. 伸长率和断面收缩率**

材料塑性的标准度量方法是伸长率$\delta_1$，定义为

$$\delta_1 = \frac{l_1 - l_0}{l_0} \times 100\% \tag{1.8}$$

式中，$l_0$ 和 $l_1$ 分别为拉伸试样的原始标距和断后标距。

材料塑性的其他度量方法是断面收缩率$\psi_1$，定义为

$$\psi_1 = \frac{A_0 - A_1}{A_0} \times 100\% \tag{1.9}$$

式中，$A_0$ 和 $A_1$ 分别为试样的原始横截面面积和断后最小横截面面积。

## 1.7 无明显屈服点塑性材料的应力应变曲线

许多塑性材料的屈服并不是由应力应变曲线的波动或水平部分进行表征的，因为应力一直不断增大，直到达到极限强度，然后颈缩开始，最终断裂。对于这类材料，屈服强度可以通过偏移方法进行定义。过横坐标$\varepsilon = 0.2\%$处，画平行于应力应变曲线直线部分的倾斜直线，交应力应变曲线于$s$点，如图 1.18 所示。对应于$s$点的应力$\sigma_{0.2}$定义为材料的条件屈服强度。

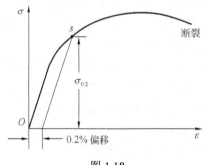

图 1.18

## 1.8 塑性材料和脆性材料

各种材料的应力应变曲线相差很大。然而，从各种材料的应力应变曲线中可以区分一些共同特征，并根据这些特征把材料分为两大类，即塑性材料和脆性材料。

由结构钢和许多其他金属合金组成的塑性材料由其在常温时的屈服能力进行表征，如图 1.17 所示。

由铸铁、玻璃和石块组成的脆性材料由其伸长率在没有明显变化的情况下而发生的断裂进行表征，如图 1.19 所示。脆性材料试样没有颈缩阶段，断裂沿垂直于载荷的截面发生。

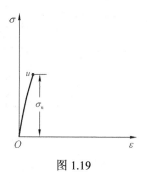

图 1.19

## 1.9 材料压缩性能

当塑性材料试样压缩时，应力应变曲线的直线部分和对应屈服及强化的开头部分与塑性材料拉伸时的应力应变曲线相同。尤其值得注意的是，低碳钢的屈服强度在拉伸和压缩中相同。对于较大应变，拉伸和压缩应力应变曲线开始分叉，应该注意压缩没有颈缩现象，如图 1.20(a) 所示。

对于大部分脆性材料，如铸铁，压缩极限强度远远高于拉伸极限强度，如图 1.20(b) 所示。

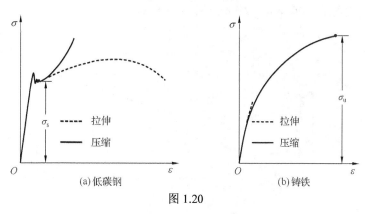

图 1.20

## 习 题

**1.1** 厚度 $t = 20$ mm、宽度 $b = 80$ mm 的两块木板通过胶合榫眼接头进行连接（图 1.21）。已知胶的平均剪应力达到 $\tau = 1.2$ MPa 时接头失效，求切口长度 $a = 30$ mm 时的最大轴向载荷 $F_{max}$。

**1.2** 载荷 $F$ 通过方板均匀作用于混凝土地基（图 1.22）。已知混凝土地基的挤压应力不能超过 12 MPa，求满足最经济安全设计的方板边长 $a$。

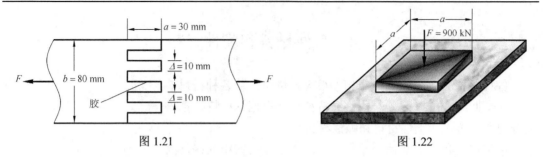

图 1.21                    图 1.22

1.3  铝合金应力应变曲线如图 1.23 所示，如果应力施加到 550 MPa，求卸载后的永久应变。

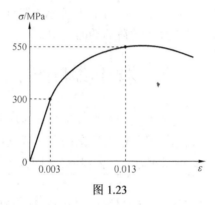

图 1.23

1.4  如图 1.24(a)所示圆截面杆受轴力 $F$ 作用，应力应变曲线如图 1.24(b)所示，求加载后杆件伸长。已知 $a = 50$ mm，$d = 4$ mm，$F = 1131$ N。

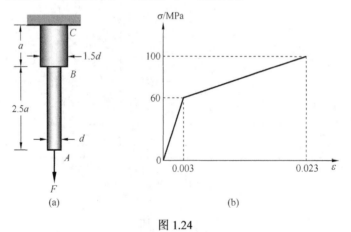

图 1.24

# 第 2 章　轴向拉伸与压缩

考虑等截面杆，两端受等值反向并与纵向形心轴重合的一对外力作用。如果外力指向杆外，则杆为拉伸，如图 2.1(a) 所示；如果外力指向杆内，则杆为压缩，如图 2.1(b) 所示。

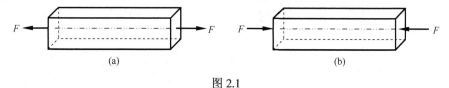

图 2.1

## 2.1　轴　　力

如图 2.1(a) 所示，在一对外力作用下，杆的横截面上将产生分布内力。这些分布内力的合力称为轴力或法向力，用 $N$ 表示，如图 2.2 所示。

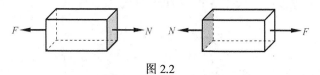

图 2.2

假想用一个平面在需要求轴力的位置沿杆的横截面把杆截开，为了得到横截面上的轴力，可移去截开面右侧（或左侧）的杆，用轴力代替对左侧（或右侧）部分的作用。通过截开横截面，相对杆的保留部分而言，原来的轴力就变成了外力。考虑杆的保留部分的平衡，轴力就可通过沿杆方向的静平衡方程求解。

## 2.2　横截面正应力

如图 2.3(a) 所示，关于拉伸或压缩杆横截面上的轴力分布，需要提出一些假设。外力通过截面形心，因此通常假设轴力在横截面上均匀分布，如图 2.3(b) 所示。因此横截面上的正应力可表示为

$$\sigma = \frac{N}{A} \quad (2.1)$$

式中，$A$ 为横截面面积；$N$ 为横截面上的轴力。正号表示拉应力（杆受拉），而负号表示压应力（杆受压）。

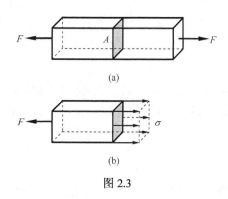

图 2.3

**例 2.1** 如图 2.4(a) 所示，矩形截面连杆 BD 和 CE 的横截面尺寸均为 16 mm×26 mm，销钉的直径为 6 mm。求连杆 BD 和 CE 的最大平均正应力。

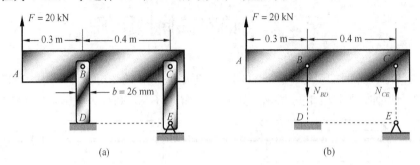

图 2.4

**解** 如图 2.4(b) 所示，考虑水平梁的平衡，有

$$N_{BD} = F\frac{l_{AC}}{l_{BC}} = 35 \text{ kN}, \quad N_{CE} = F - N_{BD} = -15 \text{ kN}$$

连杆 BD 受拉，因此其最小横截面面积等于

$$A_{BD} = t(b-d) = 16 \times (26-6) = 320 (\text{mm}^2)$$

根据式(2.1)，得连杆 BD 的最大平均正应力为

$$\sigma_{BD} = \frac{N_{BD}}{A_{BD}} = 109.4 \text{ MPa}$$

同理，连杆 CE 的最大平均正应力为

$$\sigma_{CE} = \frac{N_{CE}}{A_{CE}} = \frac{-15 \times 10^3}{16 \times 26 \times 10^{-6}} = -36.1 (\text{MPa})$$

## 2.3 斜截面正应力和剪应力

考虑图 2.5(a) 所示二力杆，受一对拉力 F 作用。把杆沿斜截面（与横截面成 θ 角）截开，画左侧部分的受力图。根据平衡条件，作用于斜截面上的轴力 N 必定等于拉力 F。如图 2.5(b) 所示，把 N 分解为分别垂直和平行斜截面的分量 $N_n$ 和 $N_t$，则有

$$\begin{aligned} N_n &= N\cos\theta \\ N_t &= N\sin\theta \end{aligned} \quad (2.2)$$

式中，$N_n$ 为斜截面上法向分布力的合力；$N_t$ 为切向分布力的合力。$N_n$ 和 $N_t$ 分别除以斜截面面积 $A_\theta$，可得斜截面上的正应力和剪应力为

$$\sigma_\theta = \frac{N_n}{A_\theta}, \quad \tau_\theta = \frac{N_t}{A_\theta} \quad (2.3)$$

观察图 2.5(a)，有 $A_\theta = \dfrac{A}{\cos\theta}$，其中 A 为横截面面积，因此得

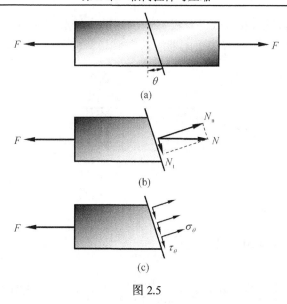

图 2.5

$$\sigma_\theta = \frac{N}{A}\cos^2\theta = \sigma\cos^2\theta, \quad \tau_\theta = \frac{N}{2A}\sin(2\theta) = \frac{1}{2}\sigma\sin(2\theta) \tag{2.4}$$

式中，$\sigma$为横截面上的正应力。

由式(2.4)得，若$\theta = 0$，则$\sigma_\theta = \sigma_{\max} = \sigma$和$\tau_\theta = 0$；若$\theta = 45°$，则$\sigma_\theta = \tau_\theta = \tau_{\max} = \frac{1}{2}\sigma$；若$\theta = 90°$，则$\sigma_\theta = \tau_\theta = 0$。

与抗拉能力相比，塑性材料的抗剪能力较弱。因此，塑性材料试样在受拉时，将会沿与横截面成45°的截面发生失效，如图2.6(a)所示。然而，脆性材料的抗拉能力比抗剪能力弱。因此，脆性材料试样在受拉时，将会沿横截面发生失效，如图2.6(b)所示。

塑性材料的抗剪能力比抗压能力弱，因此，塑性材料试样在受压时，将会沿与横截面成45°的截面发生失效，如图2.7(a)所示。脆性材料的抗剪能力也比抗压能力弱，因此，脆性材料试样在受压时，将会沿与横截面成45°~55°的截面发生破坏，如图2.7(b)所示。

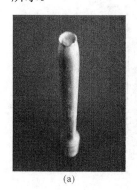

图 2.6　　　　　　　　　　　图 2.7

**例 2.2** 具有 90 mm×120 mm 矩形横截面的两根木杆通过如图 2.8 所示胶合接头进行连接。已知 $F=100$ kN，求胶合接头面上的正应力和剪应力。

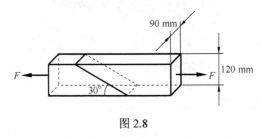

图 2.8

**解** 利用式(2.4)，有

$$\sigma_{60°} = \sigma\cos^2 60° = \frac{100\times 10^3}{90\times 120\times 10^{-6}}\times \cos^2 60° = 2.31(\text{MPa})$$

$$\tau_{60°} = \frac{\sigma}{2}\sin(2\times 60°) = \frac{100\times 10^3}{2\times 90\times 120\times 10^{-6}}\times \sin 120° = 4.01(\text{MPa})$$

## 2.4 线 应 变

**1. 纵向线应变**

长度为 $l$、横截面面积为 $A$ 的等截面杆，上端悬挂，如图 2.9(a)所示。如果在下端施加外力 $F$，则杆将伸长，如图 2.9(b)所示。

在轴向载荷作用下杆的纵向线应变可表示为

$$\varepsilon = \frac{\delta}{l} \tag{2.5}$$

式中，$\delta$ 为杆的纵向变形；$l$ 为杆长。

对于变截面杆，横截面上的正应力沿杆将发生变化，因此有必要定义在一点处的纵向线应变。考虑未变形长度为 $\Delta x$ 的微段，如图 2.10(a)所示。用 $\Delta\delta$ 表示微段在载荷作用下的

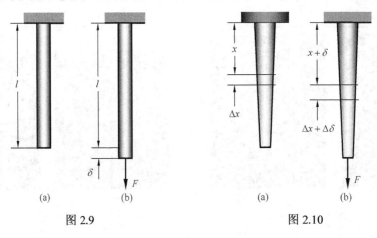

图 2.9　　　　　　　图 2.10

变形，如图 2.10(b)所示。那么在 $x$ 截面上一点处的纵向线应变可表示为

$$\varepsilon = \lim_{\Delta x \to 0} \frac{\Delta \delta}{\Delta x} = \frac{\mathrm{d}\delta}{\mathrm{d}x} \tag{2.6}$$

**2. 横向线应变**

当杆轴向受载时，只要不超过材料的比例极限，横截面上正应力与纵向线应变将满足胡克定律。所有材料都假设为均匀和各向同性，即力学性能与位置和方向无关。因此，对如图 2.11 所示载荷，有 $\varepsilon_y = \varepsilon_z$，其中 $\varepsilon_y$ 和 $\varepsilon_z$ 分别为沿 $y$ 和 $z$ 轴的横向线应变。

材料的重要常数之一是泊松比，用 $\mu$ 表示。泊松比定义为

$$\mu = -\frac{\varepsilon_y}{\varepsilon_x} = -\frac{\varepsilon_z}{\varepsilon_x} \tag{2.7}$$

因此，横向线应变可表示为

$$\varepsilon_y = \varepsilon_z = -\mu \varepsilon_x \tag{2.8}$$

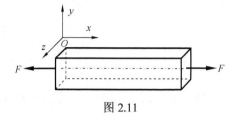

图 2.11

## 2.5 拉压变形

考虑长度为 $l$、横截面面积为 $A$ 的等截面杆，受中心轴向载荷作用，如图 2.12 所示。如果横截面上正应力不超过材料的比例极限，则胡克定律成立，即 $\sigma = E\varepsilon$。利用 $\sigma = N/A$ 和 $\varepsilon = \delta/l$，有

$$\delta = \frac{Nl}{EA} \tag{2.9}$$

式中，$EA$ 为杆的轴向(抗拉或抗压)刚度。式(2.9)仅适用于满足均匀和各向同性假设、具有不变横截面和受中心轴向载荷作用的杆。

对变截面杆，如图 2.13 所示，则长度为 $\mathrm{d}x$ 的微段变形可表示为

$$\mathrm{d}\delta = \frac{N\mathrm{d}x}{EA} \tag{2.10}$$

因此，通过沿杆长积分，得杆的总变形为

$$\delta = \int \frac{N\mathrm{d}x}{EA} \tag{2.11}$$

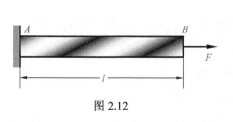

图 2.12

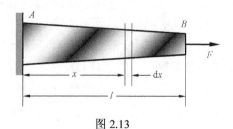

图 2.13

**例 2.3** 直径为 4 mm 的绳子 $BC$ 由弹性模量 $E=200$ GPa 的钢制成。已知绳中最大应力不超过 190 MPa，绳的伸长不超过 6 mm，求可施加到图 2.14(a) 所示结构的最大载荷 $F_{max}$。

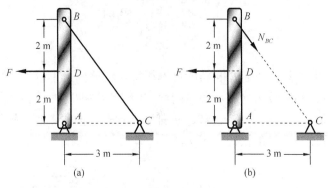

图 2.14

**解** 如图 2.14(b) 所示，取 $AB$ 为自由体，考虑其平衡，有

$$\sum M_A = 0: Fl_{AD} - \left(N_{BC} \frac{l_{AC}}{l_{BC}}\right) l_{AB} = 0, \quad \text{即 } F = 1.2 N_{BC}$$

根据绳中应力求最大载荷，有

$$F_{max}^{\sigma} = 1.2 N_{BC} = 1.2 \sigma_{max} A_{BC} = 1.2 \sigma_{max} \left(\frac{1}{4} \pi d_{BC}^2\right) = 2.87 \text{ kN}$$

根据绳的变形求最大载荷，有

$$F_{max}^{\delta} = 1.2 N_{BC} = 1.2 \frac{\delta E A_{BC}}{l_{BC}} = 1.2 \frac{\delta E \left(\frac{1}{4} \pi d_{BC}^2\right)}{l_{BC}} = 3.62 \text{ kN}$$

应该取两个最大载荷中的较小者，即可施加到结构的最大载荷为

$$F_{max} = \min[F_{max}^{\sigma}, F_{max}^{\delta}] = 2.87 \text{ kN}$$

## 2.6 静不定拉压杆

前面各节所考虑的问题中，利用静平衡方程即可求解载荷作用下杆各个部分产生的内力。然而，工程中有些问题的内力仅根据静平衡方程并不能进行求解。对这类问题，静平衡方程必须用满足变形协调关系的额外方程进行补充。静平衡方程不足以求解全部反力或内力，因此这类问题称为静不定问题。

考虑长度为 $l$、横截面面积为 $A$ 和弹性模量为 $E$ 的等截面杆 $AB$，受载前两端 $A$ 和 $B$ 固定于刚性支座，如图 2.15(a) 所示。

如图 2.15(b) 所示，画出杆的受力图，得平衡方程为

$$\sum F_y = 0: R_A - R_B - F = 0 \tag{2.12}$$

1 个方程不足以求两个未知反力 $R_A$ 和 $R_B$，因此本问题为 1 次静不定。

观察几何关系，得杆的总变形 $\delta$ 必须等于零。用 $\delta_1$ 和 $\delta_2$ 分别表示 $AC$ 和 $BC$ 部分的变形，有

$$\delta = \delta_1 + \delta_2 = 0 \tag{2.13}$$

用相应的内力 $N_1$ 和 $N_2$ 表达 $\delta_1$ 和 $\delta_2$，得

$$\delta_1 = \frac{N_1 a}{EA}, \quad \delta_2 = \frac{N_2 b}{EA} \tag{2.14}$$

根据图 2.15(c) 和图 2.15(d)，有 $N_1 = R_A$ 和 $N_2 = R_B$。把这些关系代入式 (2.14)，并利用式 (2.13)，有

$$R_A a + R_B b = 0 \tag{2.15}$$

式 (2.12) 和式 (2.15) 联立求解，得

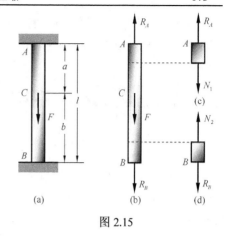

图 2.15

$$R_A = \frac{Fb}{l}, \quad R_B = -\frac{Fa}{l} \tag{2.16}$$

## 2.7 拉压杆设计

在工程应用中，应力用于结构设计，以保证所设计的结构能够既安全又经济地履行规定的功能。

能够作用于杆的最大力称为极限载荷，用 $F_u$ 表示。中心受载，因此极限载荷除以原始横截面面积即可得到极限强度，可表示为

$$\sigma_u = \frac{F_u}{A} \tag{2.17}$$

在正常应用条件下，杆允许承受的工作载荷要小于极限载荷。能够施加到杆的最大工作载荷称为许可载荷，用 $F_{\text{allow}}$ 表示。因此，施加许可载荷仅发挥杆的部分极限承载能力，剩余的承载能力予以保留以确保杆的安全。极限载荷与许可载荷的比值定义为安全因数，可表示为

$$n = \frac{F_u}{F_{\text{allow}}} \tag{2.18}$$

通过应力，安全因数还可表示为

$$n = \frac{\sigma_u}{\sigma_{\text{allow}}} \tag{2.19}$$

安全因数的选择是最重要的工程任务之一。一方面，如果安全因数太小，杆失效的可能性就会很大；另一方面，如果安全因数太大，杆的设计就会变得不经济。选择合适的安全因数需要具备基于多种考虑的工程判断能力。

## 习 题

2.1 实心圆柱杆 $AB$ 和 $BC$ 在截面 $B$ 处焊接为整体，受载如图 2.16 所示。已知 $d_1$=50 mm 和 $d_2$=30 mm，求杆 $AB$ 和杆 $BC$ 中间截面上的正应力。

2.2 直径为 $d$ 的实心圆柱杆 $AB$ 受两轴向载荷作用，$F_1$ 作用于 $C$ 处，$F_2$ 作用于 $B$ 处（图 2.17）。已知 $d=50$ mm、$F_1=200$ kN 和 $F_2=50$ kN，求 $AC$ 段和 $BC$ 段的正应力。

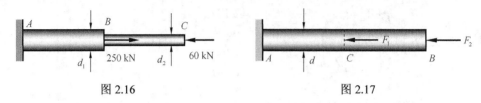

图 2.16    图 2.17

2.3 宽度 $b=50$ mm、厚度 $t=6$ mm 的连杆 $AB$ 用于支撑水平梁的末端（图 2.18）。已知连杆的平均正应力为 138 MPa，销钉的平均剪应力为 82 MPa，求销钉直径 $d$ 和连杆的平均挤压应力。

2.4 矩形截面连杆 $BD$ 和 $CE$ 的横截面面积均为 16 mm×26 mm，销钉的直径为 6 mm（图 2.19）。求：(1) $B$ 处销钉的平均剪应力；(2) 连杆 $BD$ 在 $B$ 处的平均挤压应力；(3) 梁 $ABC$ 在 $B$ 处的平均挤压应力，已知矩形截面梁的横截面尺寸为 10 mm×50 mm。

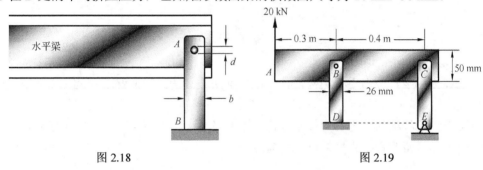

图 2.18    图 2.19

2.5 轴向载荷作用于钢杆 $ABC$ 的 $C$ 端（图 2.20）。已知 $E=200$ GPa，求当 $C$ 点位移为 3 mm 时 $BC$ 段的直径 $d_2$。

2.6 考虑如图 2.21 所示钢杆 $AB$（$E=200$ GPa），受轴向载荷 $F$ 作用。如果杆的应力不超过 120 MPa，杆的最大长度变化不超过杆长的 0.001，求杆的最小直径。

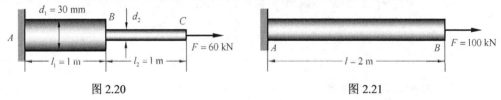

图 2.20    图 2.21

2.7 杆 $ABC$ 的两段均由黄铜（$E=105$ GPa）制成（图 2.22）。已知 $d_2=10$ mm 和 $\sigma_{\text{allow}}=100$ MPa，如果 $C$ 点位移不超过 4 mm，求所能施加的最大载荷 $F_{\max}$。

2.8 连杆 $AB$ 和 $CD$ 由铝（$E=75$ GPa）制成，横截面面积为 125 mm$^2$（图 2.23）。已知连杆用于支撑刚性梁 $AC$，求点 $E$ 处的位移。

2.9 刚性梁 $AD$ 由两根直径为 1.5 mm 的钢丝（$E=200$ GPa）和 $A$ 处的固定铰支座支撑（图 2.24）。已知受载前钢丝正好拉紧，求当 1.0 kN 的载荷 $F$ 作用于点 $D$ 处时：(1) 每根钢丝的附加拉力；(2) 点 $D$ 处的位移。

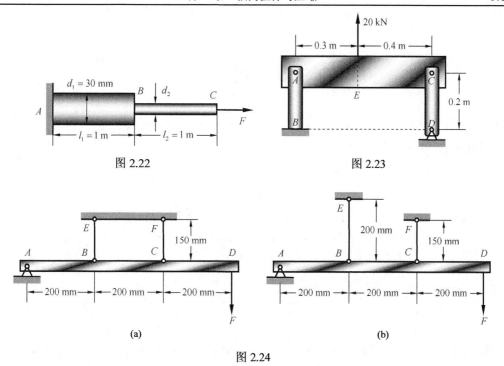

图 2.22

图 2.23

(a)　　　　　　　　　　　　(b)

图 2.24

# 第 3 章 扭 转

如图 3.1 所示，圆截面杆两端受等值反向并位于纵轴垂直面内的一对外力偶 $M_e$ 作用，则该杆称为扭转轴。

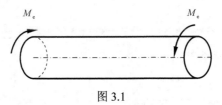

图 3.1

## 3.1 扭 矩

如图 3.1 所示，在一对外力偶作用下，轴的横截面上将产生分布内力。分布内力对纵轴的合力矩称为扭矩，用 $T$ 表示，如图 3.2 所示。

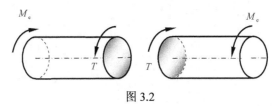

图 3.2

## 3.2 剪切胡克定律

考虑如图 3.3(a)所示纯剪切单元体，受剪应力 $\tau_{xy}$ 和 $\tau_{yx}$ 作用，$\tau_{xy}$ 和 $\tau_{yx}$ 分别垂直于 $x$ 轴和 $y$ 轴。根据单元体平衡，得

$$\tau_{xy} = \tau_{yx} \tag{3.1}$$

上述关系称为剪应力互等定理。

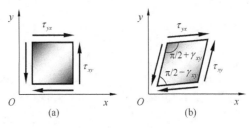

图 3.3

在纯剪应力作用下，单元体变形为菱形，如图 3.3(b)所示。其中，两个角从 $\frac{1}{2}\pi$ 减小为 $\frac{1}{2}\pi - \gamma_{xy}$，另外两个角则从 $\frac{1}{2}\pi$ 增加为 $\frac{1}{2}\pi + \gamma_{xy}$，其中 $\gamma_{xy}$ 为剪应变。如果剪应力不超过比例极限，对均匀、各向同性材料，则有

$$\tau_{xy} = G\gamma_{xy} \tag{3.2}$$

上述关系称为剪切胡克定律，常数 $G$ 称为剪切弹性模量，其单位与 $\tau_{xy}$ 相同。

均匀、各向同性材料的拉压和剪切弹性模量满足如下方程：

$$G = \frac{E}{2(1+\mu)} \tag{3.3}$$

式中，$\mu$ 为泊松比。

## 3.3 横截面剪应力

### 1. 平衡条件

考虑受扭圆轴 $AB$，在任意位置 $C$ 处沿横截面把圆轴截开，如图 3.4(a)所示。截面 $C$ 上各点必存在垂直于半径的微剪力 $dF$，如图 3.4(b)所示。这些微剪力是轴发生扭转时由 $BC$ 段施加于 $AC$ 段的。

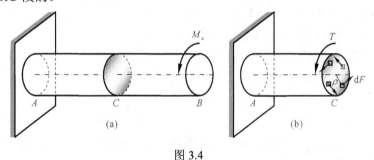

图 3.4

根据静平衡条件，微剪力对纵向形心轴的合力矩等于横截面上的扭矩 $T$。用 $\rho$ 表示微剪力 $dF$ 到纵向形心轴的距离，则有

$$T = \int \rho dF = \int \rho \tau dA \tag{3.4}$$

式中，$\tau$ 为面元 $dA$ 上的剪应力。

### 2. 变形条件

如图 3.5(a)所示，圆轴 $AB$ 受外力偶 $M_e$ 作用时将发生扭转，每个横截面保持平面。考虑长度为 $dx$ 的微段 $CD$，具有相对扭转角 $d\varphi$，如图 3.5(b)所示。

从微段 $CD$ 中取出半径为 $\rho$ 的圆柱，如图 3.5(c)所示。对于小变形，圆弧 $ab$ 的长度可表示为 $l_{ab} = \gamma dx$，也可表示为 $l_{ab} = \rho d\varphi$。由此得 $\gamma dx = \rho d\varphi$，即

$$\gamma = \frac{d\varphi}{dx}\rho \tag{3.5}$$

式中，$\gamma$ 和 $d\varphi$ 的单位为 rad。式(3.5)表明，圆轴的剪应变与到纵向形心轴的距离 $\rho$ 成正比。

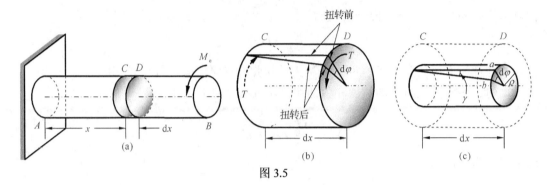

图 3.5

**3. 应力应变关系**

假设圆轴的剪应力不超过比例极限，根据剪切胡克定律 $\tau = G\gamma$，有

$$\tau = G\frac{d\varphi}{dx}\rho \tag{3.6}$$

式中，$G$ 为剪切弹性模量。式(3.6)表明，只要不超过比例极限，圆轴的剪应力与到纵向形心轴的距离 $\rho$ 成正比。图 3.6(a) 和图 3.6(b) 分别表示实心圆轴和空心圆轴的剪应力分布。

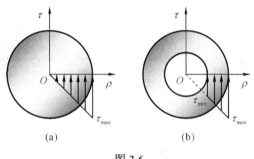

图 3.6

将式(3.6)代入式(3.4)，得

$$T = \int \rho \tau dA = G\frac{d\varphi}{dx}\int \rho^2 dA = GI_p\frac{d\varphi}{dx} \tag{3.7}$$

式中，$I_p = \int \rho^2 dA$ 为横截面对形心 $O$ 的极惯性矩。对于直径为 $d$ 的实心圆轴，$I_p = \frac{1}{32}\pi d^4$；对于内外径分别为 $d$ 和 $D$ 的空心圆轴，$I_p = \frac{1}{32}\pi(D^4 - d^4) = \frac{1}{32}\pi D^4(1 - \alpha^4)$，其中 $\alpha = d/D$。

利用式(3.6)和式(3.7)，有

$$\tau = \frac{T\rho}{I_p} \tag{3.8}$$

由式(3.8)可以看出，最大剪应力可表示为

$$\tau_{max} = \frac{T_{max}\rho_{max}}{I_p} \tag{3.9}$$

注意，比值 $I_p/\rho_{max}$ 仅与截面几何尺寸有关。该比值称为极截面模量或抗扭截面系数，用 $S_p$ 表示。因此，式(3.9)可写为

$$\tau_{max} = \frac{T_{max}}{S_p} \tag{3.10}$$

对于直径为 $d$ 的实心圆轴，$S_p = \dfrac{1}{16}\pi d^3$；对于内外径分别为 $d$ 和 $D$ 的空心圆轴，$S_p = \dfrac{1}{16}\pi D^3(1-\alpha^4)$，其中 $\alpha = d/D$。

**例 3.1** 如图 3.7 所示，外力偶 $M_B$ 和 $M_C$ 分别作用于变截面实心轴 $ABC$ 的截面 $B$ 和 $C$ 处。已知 $D=46$ mm、$d=30$ mm、$M_B=400$ N·m 和 $M_C=300$ N·m，求轴 $AB$ 和轴 $BC$ 的最大剪应力。

**解** 利用截面法，得扭矩为
$$T_{AB} = M_B + M_C = 700\ \text{N·m}, \quad T_{BC} = M_C = 300\ \text{N·m}$$

对于轴 $AB$，由式(3.10)，得
$$(\tau_{\max})_{AB} = \dfrac{T_{AB}}{(S_p)_{AB}} = \dfrac{T_{AB}}{\dfrac{1}{16}\pi D^3} = 36.6\ \text{MPa}$$

图 3.7

同理，对于轴 $BC$，有
$$(\tau_{\max})_{BC} = \dfrac{T_{BC}}{(S_p)_{BC}} = \dfrac{T_{BC}}{\dfrac{1}{16}\pi d^3} = 56.6\ \text{MPa}$$

## 3.4 斜截面正应力和剪应力

前面考虑了横截面上的剪应力，下面考虑斜截面上的应力。如图 3.8 所示，在受扭圆轴的前表面取三个单元体 $a$、$b$ 和 $c$。单元体 $a$ 的面分别平行和垂直纵轴，因此单元体上只有剪应力，并且剪应力达到最大。单元体 $b$ 的面与纵轴成任意角，因此单元体上同时有正应力和剪应力。单元体 $c$ 的面与纵轴成 $45°$，因此单元体上只有正应力。

塑性材料的抗剪能力较差，因此塑性材料受扭时将沿横截面破坏，如图 3.9(a)所示。脆性材料抗拉能力较差，因此脆性材料受扭时将沿垂直最大拉应力的截面破坏，即沿与横截面成 $45°$ 的截面破坏，如图 3.9(b)所示。

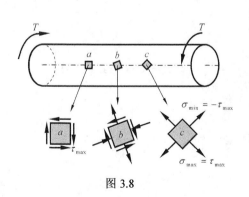

图 3.8

(a)      (b)

图 3.9

## 3.5 扭 转 角

如图 3.10 所示，考虑圆轴 $AB$ 自由端受外力偶 $M_e$ 作用，假设扭矩用 $T$ 表示，并且圆轴保持线弹性，利用式(3.7)，则微段 $CD$ 的扭转角 $d\varphi$ 可写为

$$d\varphi = \frac{Tdx}{GI_p} \tag{3.11}$$

式中，$GI_p$ 为抗扭刚度。式(3.11)表明，在线弹性范围，扭转角 $d\varphi$ 与扭矩 $T$ 成正比。积分得总扭转角为

$$\varphi = \int \frac{Tdx}{GI_p} \tag{3.12}$$

对等截面圆轴受不变扭矩作用，则式(3.12)简化为

$$\varphi = \frac{Tl}{GI_p} \tag{3.13}$$

**例 3.2** 如图 3.11 所示，外力偶 $M_B$ 和 $M_C$ 分别作用于变截面实心铝($G=77\ \text{GPa}$)轴 $ABC$ 的截面 $B$ 和 $C$ 处。已知 $D=46$ mm、$d=30$ mm、$a=750$ mm、$b=900$ mm、$M_B=400$ N·m 和 $M_C=300$ N·m，求截面 $B$、$C$ 之间的扭转角和截面 $C$ 的扭转角。

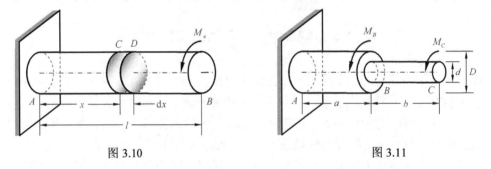

图 3.10　　　　　　　　　　图 3.11

**解** 利用式(3.13)，得截面 $C$ 相对截面 $B$ 的扭转角为

$$\varphi_{C/B} = \frac{T_{BC}l_{BC}}{G(I_p)_{BC}} = \frac{T_{BC}l_{BC}}{G\left(\dfrac{1}{32}\pi d^4\right)} = 4.41\times 10^{-2}\ \text{rad} = 2.53°$$

截面 $C$ 的绝对扭转角为

$$\varphi_C = \varphi_{C/A} = \varphi_{C/B} + \varphi_{B/A} = \frac{T_{BC}l_{BC}}{G\left(\dfrac{1}{32}\pi d^4\right)} + \frac{T_{AB}l_{AB}}{G\left(\dfrac{1}{32}\pi D^4\right)} = 5.96\times 10^{-2}\ \text{rad} = 3.41°$$

## 3.6 静 不 定 轴

如图 3.12(a)所示，考虑圆轴 $AB$，两端固支，在截面 $C$ 处受外力偶 $M_e$ 作用。反力包含两个未知量，而平衡条件只有 1 个，因此圆轴为 1 次静不定。

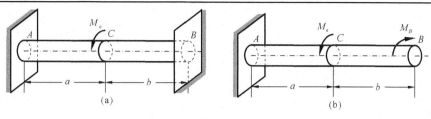

图 3.12

取支座 $B$ 处的反力为多余反力，消除对应的支座 $B$，如图 3.12(b)所示。多余反力可作为未知载荷 $M_B$，连同其余载荷 $M_e$，共同作用所产生的变形必须与原始支座 $B$ 协调。利用圆轴在截面 $B$ 处的总扭转角必须等于零即可求得多余反力 $M_B$。

分别考虑 $B$ 处多余反力 $M_B$ 所产生的扭转角 $(\varphi_B)_{M_B}$ 和外力偶 $M_e$ 在同一截面所产生的扭转角 $(\varphi_B)_{M_e}$ 即可得到解。根据式(3.13)，得 $(\varphi_B)_{M_B} = \dfrac{M_B(a+b)}{GI_p}$，$(\varphi_B)_{M_e} = (\varphi_C)_{M_e} = \dfrac{M_e a}{GI_p}$。$B$ 处总扭转角必须为零，即 $\varphi_B = (\varphi_B)_{M_B} + (\varphi_B)_{M_e} = 0$ 或 $\dfrac{M_B(a+b)}{GI_p} + \dfrac{M_e a}{GI_p} = 0$。求解得 $M_B = -\dfrac{M_e a}{a+b}$。

## 3.7 扭转轴设计

扭转轴设计的主要规范是传输的功率以及轴的转速。选择合适的材料和截面尺寸以保证轴在以规定的转速传输所要求的功率时所产生的应力不超过许用应力。

旋转轴以转速 $\omega(\text{rad/s})$ 或 $n(\text{r/min})$ 传输功率 $P(\text{kW})$ 时所产生的扭矩 $T$ 可表示为

$$T(\text{N}\cdot\text{m}) = \dfrac{P(\text{W})}{\omega(\text{rad/s})} \quad \text{或} \quad T(\text{N}\cdot\text{m}) = 9549\dfrac{P(\text{kW})}{n(\text{r/min})} \tag{3.14}$$

得到扭矩 $T$ 和选择材料后，将许用剪应力代入式(3.9)并将许可扭转角代入式(3.12)，得

$$\tau_{\max} = \dfrac{T_{\max}\rho_{\max}}{I_p} \leqslant \tau_{\text{allow}}, \quad \varphi = \int\dfrac{T\mathrm{d}x}{GI_p} \leqslant \varphi_{\text{allow}} \tag{3.15}$$

式中，$\tau_{\text{allow}}$ 为许用剪应力；$\varphi_{\text{allow}}$ 为许可扭转角。

**例 3.3** 长度为 1.8 m、外径为 45 mm 的空心圆截面钢轴的许用剪应力 $\tau_{\text{allow}} = 75$ MPa，剪切弹性模量 $G = 77$ GPa（图 3.13）。已知钢轴受 $M_e = 900$ N·m 的外力偶作用时的扭转角不能超过 $5°$，求最大内径 $d$。

**解** 根据 $\tau_{\max} = \dfrac{T\rho_{\max}}{I_p} \leqslant \tau_{\text{allow}}$，有

$$(I_p)_\tau \geqslant \dfrac{T\rho_{\max}}{\tau_{\text{allow}}} = 2.70 \times 10^{-7} \text{ m}^4$$

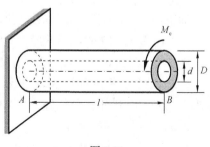

图 3.13

根据 $\varphi = \dfrac{Tl}{GI_p} \le \varphi_{\text{allow}}$，有

$$(I_p)_\varphi \ge \frac{Tl}{G\varphi_{\text{allow}}} = 2.41\times 10^{-7} \text{ m}^4$$

应该选用较大的 $I_p$ 值，因此

$$I_p = \max[(I_p)_\tau, (I_p)_\varphi] \ge 2.70\times 10^{-7} \text{ m}^4$$

对于内径为 $d$、外径为 $D$ 的空心圆轴，利用 $I_p = \dfrac{1}{32}\pi(D^4 - d^4)$，有

$$d \le \sqrt[4]{D^4 - \frac{32I_p}{\pi}} = 34 \text{ mm}$$

因此，能够用于设计的最大内径 $d$ 等于 34 mm。

# 习　题

3.1　如图 3.14 所示，已知 $d$=30 mm 和 $D$=40 mm，求空心轴产生 52 MPa 最大剪应力时所施加的外力偶 $M_e$。

3.2　外力偶 $M_B$ 和 $M_C$ 分别作用于变截面实心轴 $ABC$ 的截面 $B$ 和 $C$ 处（图 3.15）。已知 $D$=46 mm、$d$=30 mm、$M_B$=400 N·m 和 $M_C$=300 N·m，为了降低轴的总质量，在轴的最大剪应力不增大的情况下，求轴 $AB$ 的最小直径。

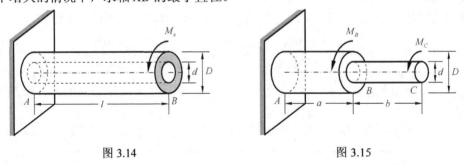

图 3.14　　　　　　　　　　图 3.15

3.3　空心管 $AB$ 的外径为 90 mm、壁厚为 6 mm，实心杆 $BC$ 的直径为 60 mm（图 3.16）。已知管和杆均由钢制成，许用剪应力为 95 MPa，求在截面 $C$ 处所能施加的最大外力偶 $M_e$。

3.4　铝杆 $AB$ 的许用剪应力为 25 MPa，铜杆 $BC$ 的许用剪应力为 50 MPa（图 3.17）。已知 $M_e$=1.5 kN·m，求杆 $AB$ 和杆 $BC$ 的所需直径。

3.5　空心铝杆如图 3.18 所示。已知 $G$=27 GPa、$l$=1.1 m、$d$=10 mm 和 $D$=20 mm，求：(1)产生 5° 扭转角的外力偶 $M_e$；(2)相同长度和相同横截面面积的实心圆轴在相同外力偶 $M_e$ 作用下的扭转角。

3.6　当如图 3.19 所示横截面的钢杆以 120 r/min 的角速度转动时，频闪测量显示 4 m 长度的扭转角为 2°。已知 $G$=77 GPa，求传输的功率。

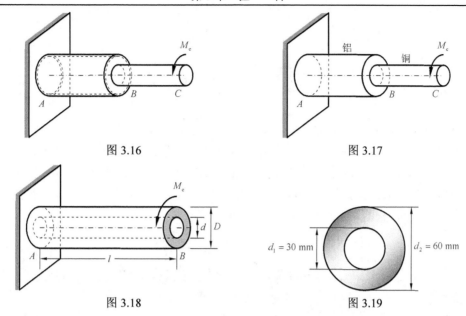

图 3.16　　　　　　　　　　　　　图 3.17

图 3.18　　　　　　　　　　　　　图 3.19

3.7　长度为 1.8 m、内径为 45 mm 的空心圆截面钢轴的许用剪应力 $\tau_{\text{allow}}$=75 MPa，剪切弹性模量 $G$=77 GPa（图 3.20）。已知钢轴受 $M_e$=900 N·m 的外力偶作用时的扭转角不能超过 5°，求最小外径 $D$。

3.8　机器构件的设计要求为外径 $D$=38 mm 的轴传输 45 kW 功率。(1) 如果转速为 800 r/min，求如图 3.21(a) 所示实心轴的最大剪应力；(2) 如果转速增加 50% 达到 1200 r/min，但保持最大剪应力不变，求如图 3.21(b) 所示空心轴的最大内径。

3.9　长度为 1.5 m、外径 $D$ 为 40 mm、内径 $d$ 为 30 mm 的空心圆截面钢轴（$G$=77 GPa）传输 120 kW 的功率（图 3.22）。已知许用剪应力为 65 MPa，扭转角不超过 3°，求钢轴的最小转速。

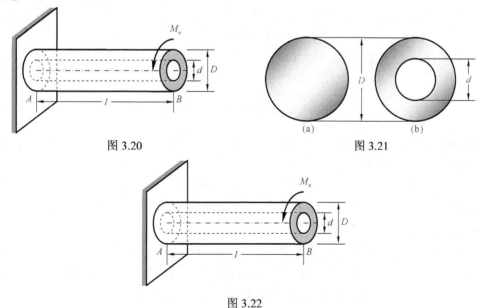

图 3.20　　　　　　　　　　　　　图 3.21

图 3.22

# 第4章 弯曲内力

受垂直杆纵轴的外力或杆纵轴所在平面内的外力偶作用的杆称为弯曲梁。

梁通过支撑方式进行分类。图 4.1 为几种类型的常用梁。简支梁为一端固定铰支，另一端可动铰支，如图 4.1(a)所示。外伸梁为两点支撑，梁的一端或两端越过支座，如图 4.1(b)所示。悬臂梁为一端固支，另一端自由，如图 4.1(c)所示。作用于梁的外载荷包括集中力(图 4.1(a))、分布力(图 4.1(b))和集中力偶(图 4.1(c))。

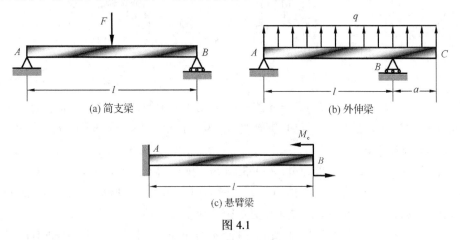

图 4.1

上述梁的支座反力共有 3 个未知量，因此通过静平衡方程即可确定全部未知量。这类梁称为静定梁。如果梁的支座反力有 3 个以上的未知量，就不能仅仅通过静平衡方程确定全部未知量，这类梁称为静不定梁。图 4.2 为 1 次静不定梁。

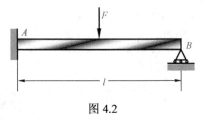

图 4.2

## 4.1 剪力图和弯矩图

若外载荷作用于梁，则梁将产生剪力和弯矩，如图 4.3 所示。一般而言，剪力和弯矩是截面位置的函数，可表示为

$$V = V(x), \quad M = M(x) \tag{4.1}$$

上述关系分别称为剪力方程和弯矩方程。根据上述方程，梁的剪力和弯矩可以用图形表示，该图形分别称为剪力图和弯矩图。

为了绘制剪力图和弯矩图，有必要首先确立剪力和弯矩的符号规则。虽然符号规则的选择具有任意性，但是为了方便，将采用常用的符号规则。引起梁段顺时针转动的剪力为正，如图 4.4(a)所示；引起梁段上侧受压的弯矩为正，如图 4.4(b)所示。

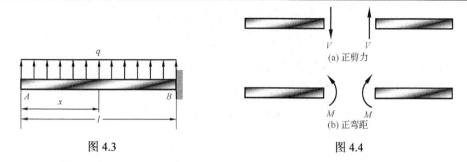

图 4.3　　　　　　　　　图 4.4

梁上任何截面的剪力和弯矩可通过截面法确定，即在需求剪力和弯矩的位置把梁截开（图 4.5(a)），并考虑截面左侧梁段的平衡（图 4.5(b)）或截面右侧梁段的平衡（图 4.5(c)）。

如图 4.5(b) 所示，考虑左段梁 AC 的平衡，并利用平衡方程，得

$$\sum F_y = 0: \quad qx - V = 0 \quad (0 \leq x < l)$$
$$\sum M_C = 0: \quad M - \frac{1}{2}qx^2 = 0 \quad (0 \leq x < l) \tag{4.2}$$

解方程，得

$$V = qx \quad (0 \leq x < l)$$
$$M = \frac{1}{2}qx^2 \quad (0 \leq x < l) \tag{4.3}$$

得到剪力方程和弯矩方程后，接着可以根据方程绘制剪力图和弯矩图。图 4.6(a) 中梁的剪力图和弯矩图分别如图 4.6(b) 和图 4.6(c) 所示。

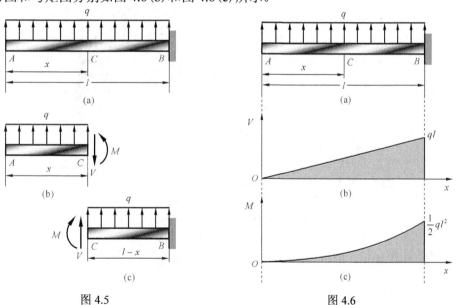

图 4.5　　　　　　　　　图 4.6

由图 4.6(b) 和图 4.6(c) 可以看出，截面 B 是危险截面，因为该截面有最大剪力和最大弯矩。

**例 4.1**　绘制如图 4.7(a) 所示结构的弯矩图，并求最大弯矩。

**解**　利用截面法和平衡条件，可得弯矩方程为

$$M_1 = F_1 x_1 \qquad (0 \leqslant x_1 \leqslant a)$$
$$M_2 = F_1 a + F_2 x_2 \qquad (0 \leqslant x_2 < b)$$

由弯矩方程可以看出,结构每段的弯矩图均为倾斜直线。整个结构的弯矩图如图4.7(b)所示,最大弯矩位于截面 $A$ 处,其最大值等于

$$M_{\max} = F_1 a + F_2 b$$

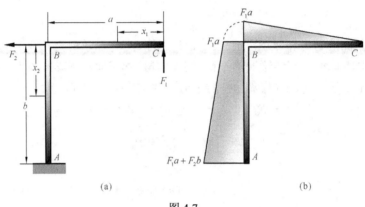

图 4.7

**例 4.2** 绘制如图 4.8(a)所示结构的弯矩图,并求最大弯矩。

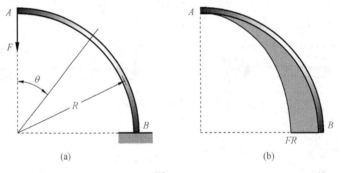

图 4.8

**解** 利用截面法和平衡条件,可得弯矩方程为

$$M = FR\sin\theta \quad \left(0 \leqslant \theta < \frac{1}{2}\pi\right)$$

结构的弯矩图如图 4.8(b)所示,固定端的最大弯矩为

$$M_{\max} = FR$$

## 4.2 分布载荷、剪力和弯矩之间的关系

当梁承受多个外载荷作用时,4.1 节讨论的绘制剪力图和弯矩图的方法非常麻烦。如果考虑载荷集度、剪力和弯矩之间的关系,那么剪力图和弯矩图绘制将非常方便。

如图 4.9(a)所示，考虑简支梁 AB，受分布载荷 q 作用(q 为分布载荷集度，即沿梁轴单位长度上的作用力)，设 C、D 为梁上相距 dx 的两点。截取梁段 CD 并画受力图，如图 4.9(b)所示。

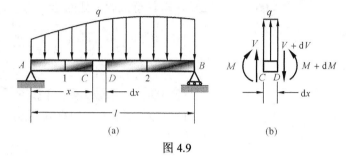

图 4.9

1. 分布载荷和剪力的关系

根据垂直方向力平衡条件，有
$$\sum F_y = 0: \ V - (V + dV) + q dx = 0$$

即
$$dV = q dx \tag{4.4}$$

除以 $dx$，得
$$\frac{dV}{dx} = q \tag{4.5}$$

在点 1 和点 2 之间进行积分，有
$$V_2 = V_1 + \int_{x_1}^{x_2} q dx \tag{4.6}$$

式中，$\int_{x_1}^{x_2} q dx$ 为点 1 和点 2 之间分布载荷的合力。

2. 分布载荷和弯矩的关系

根据对 $D$ 的矩平衡条件，得
$$\sum M_D = 0: \ (M + dM) - M - V dx - \frac{1}{2} q (dx)^2 = 0 \tag{4.7}$$

即
$$dM = V dx + \frac{1}{2} q (dx)^2$$

忽略二阶无穷小量 $(dx)^2$，并除以 $dx$，得
$$\frac{dM}{dx} = V \tag{4.8}$$

在点 1 和点 2 之间进行积分，得
$$M_2 = M_1 + \int_{x_1}^{x_2} V dx \tag{4.9}$$

式中，$\int_{x_1}^{x_2} V dx$ 为点 1 和点 2 之间剪力图的面积。

利用上述关系可以快速绘制在分布载荷作用下梁的剪力图和弯矩图。

**例 4.3**  绘制如图 4.10(a)所示简支梁的剪力图和弯矩图，并求剪力和弯矩的最大值。

**解**  (1) A 和 B 处反力分别为 $R_A = R_B = \frac{1}{2} q l$。

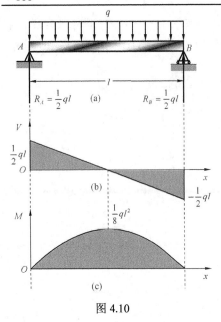

(2) 截面 A 右侧的剪力为 $V_A^R = R_A = \frac{1}{2}ql$。利用式 (4.6)，得任意截面 $x$ 处的剪力为

$$V = V_A^R + \int_0^x (-q)\mathrm{d}x = \frac{1}{2}ql - qx \quad (0 < x < l)$$

显然，剪力图是倾斜直线，如图 4.10(b) 所示。在 A 或 B 处的剪力达到最大绝对值。

(3) 利用 $M_A = 0$，根据式 (4.9)，任意截面 $x$ 处的弯矩为

$$M = M_A + \int_0^x V\mathrm{d}x = \frac{1}{2}qlx - \frac{1}{2}qx^2 \quad (0 \leqslant x \leqslant l)$$

显然，弯矩图是抛物线，如图 4.10(c) 所示。在中间截面处有最大正弯矩。

图 4.10

## 4.3 集中载荷、剪力和弯矩之间的关系

如图 4.11(a) 所示，考虑简支梁 AB，受集中力 F 和集中力偶 $M_e$ 作用，设 C、D 为梁上相距 $\mathrm{d}x$ 的两点。假设集中力和集中力偶作用于梁段 CD 中点，截取梁段 CD 并画受力图，如图 4.11(b) 所示。

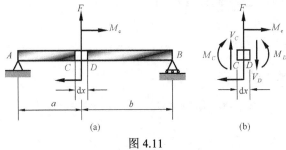

图 4.11

**1. 集中载荷和剪力的关系**

根据垂直方向力平衡条件，有

$$\sum F_y = 0: \ V_C - V_D + F = 0 \tag{4.10}$$

即

$$V_D = V_C + F \tag{4.11}$$

显然，剪力图在集中力作用的截面将发生突变。

**2. 集中载荷和弯矩的关系**

根据对 D 的矩平衡条件，得

$$\sum M_D = 0: \ M_D - M_C - V_C\mathrm{d}x - M_e - F\left(\frac{1}{2}\mathrm{d}x\right) = 0 \tag{4.12}$$

忽略无穷小量 $\mathrm{d}x$，得
$$M_D = M_C + M_e \tag{4.13}$$

显然，弯矩图在集中力偶作用的截面将发生突变。

利用上述关系可以快速绘制在集中载荷作用下梁的剪力图和弯矩图。

**例 4.4** 绘制如图 4.12(a)所示简支梁的剪力图和弯矩图，并求剪力和弯矩的最大值。

**解** (1)根据梁的平衡条件，有
$$R_A = \frac{3}{4}F, \quad R_B = \frac{1}{4}F$$

(2)截面 $A$ 右侧的剪力等于 $V_A^R = R_A = \frac{3}{4}F$。因没有外载荷作用于 $AC$ 段，故 $AC$ 段上剪力为常量，该段剪力图为水平直线。利用 $V_C^L = V_A^R = \frac{3}{4}F$ 和式(4.11)，得截面 $C$ 右侧的剪力为
$$V_C^R = V_C^L + (-F) = -\frac{1}{4}F$$

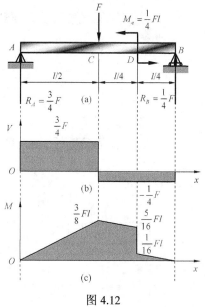

图 4.12

同理，$BC$ 段上剪力为常量，因此该段剪力图也为水平直线。

整段梁的剪力图如图 4.12(b)所示，最大剪力位于 $AC$ 段，其最大值等于 $V_{\max} = \frac{3}{4}F$。

(3)利用 $M_A = 0$，截面 $C$ 处弯矩为
$$M_C = M_A + \int_{x_A}^{x_C} V \mathrm{d}x = \frac{3}{4}F \cdot \frac{1}{2}l = \frac{3}{8}Fl$$

因 $AC$ 段上剪力为常量，故相应的弯矩图为倾斜直线。同理，$CD$ 段上的弯矩图也为倾斜直线，截面 $D$ 左侧的弯矩为
$$M_D^L = M_C + \int_{x_C}^{x_D} V \mathrm{d}x = \frac{5}{16}Fl$$

在截面 $D$ 有集中力偶，因此弯矩在该截面将有突变。利用式(4.13)，有
$$M_D^R = M_D^L + (-M_e) = \frac{1}{16}Fl$$

同理，可绘制 $DB$ 段的弯矩图，并且截面 $B$ 的弯矩为
$$M_B = M_D^R + \int_{x_D}^{x_B} V \mathrm{d}x = 0$$

整段梁的弯矩图如图 4.12(c)所示，最大弯矩位于截面 $C$，其最大值等于 $M_{\max} = \frac{3}{8}Fl$。

## 习 题

**4.1** 绘制如图 4.13 所示梁在载荷作用下的剪力图和弯矩图。

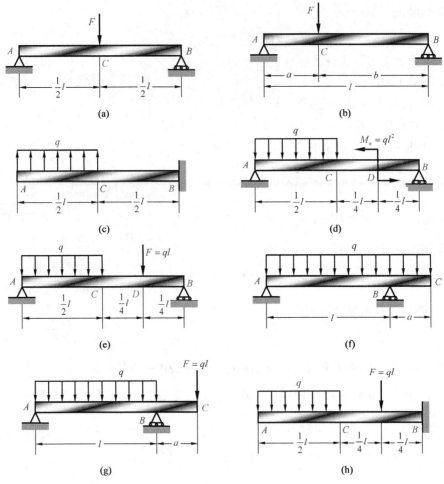

图 4.13

# 第5章 弯曲应力

作用于梁的外载荷包括垂直梁轴的外力和梁轴所在平面内的外力偶。这些作用于梁上的外力和外力偶产生的作用效果之一就是在垂直于梁轴的横截面上产生正应力和剪应力。

如果仅外力偶作用于梁，则称为纯弯曲，如图5.1(a)所示。承受纯弯曲的梁在横截面上仅存在正应力，而没有剪应力。

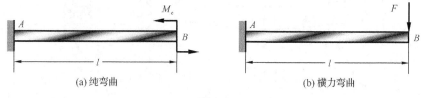

(a) 纯弯曲　　　　　　　　　　(b) 横力弯曲

图 5.1

由不能构成力偶的外力产生的弯曲称为横力弯曲，如图5.1(b)所示。承受横力弯曲的梁在横截面上既有正应力，又有剪应力。

## 5.1 纯弯曲横截面正应力

等截面梁具有纵向对称面，承受纵向对称面内等值反向外力偶作用，如图5.2所示。

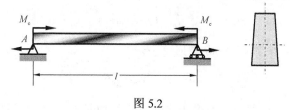

图 5.2

把梁分成许多表面平行于三个坐标轴的单元体，如图5.3(a)所示，则当梁发生纯弯曲时这些单元体将产生如图5.3(b)所示的变形。

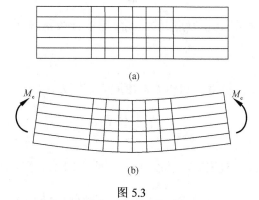

图 5.3

单元体上唯一不为零的应力分量为垂直于纵轴的单元体面上的正应力。因此，纯弯曲梁的每一点都处于单向应力状态。若弯矩为正，则梁的上、下表面分别缩短和伸长，即梁的上部压缩，下部拉伸。

上述分析表明，在梁的内部必定存在与上、下表面平行的线应变和正应力为零的面。该面称为中性层，如图 5.4(a) 所示。中性层与横截面的交线称为中性轴，如图 5.4(b) 所示。坐标原点选在中性层上，以便任何一点到中性层的距离可通过坐标 $y$ 进行表示。

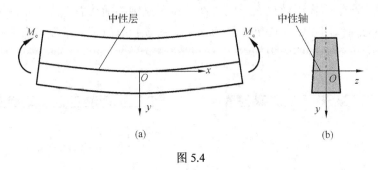

图 5.4

1. 变形条件

用 $\rho$ 表示中性层 $aa$ 的曲率半径，$\theta$ 表示中性层 $aa$ 所对应的中心角，如图 5.5 所示。观察表明，变形后中性层 $aa$ 的长度等于变形前中性层的长度 $l$，故有

$$l = \rho\theta \tag{5.1}$$

现考虑中性层下方 $y$ 处的弧线 $bb$，变形后其长度为

$$l'_{bb} = (\rho + y)\theta \tag{5.2}$$

因弧线 $bb$ 原长为 $l_{bb} = l$，则弧线 $bb$ 变形可表示为

$$\delta = l'_{bb} - l_{bb} = y\theta \tag{5.3}$$

弧线 $bb$ 的纵向线应变 $\varepsilon$ 可通过弧线变形 $\delta$ 除以弧线原长 $l_{bb}$ 而得到，即

$$\varepsilon = \frac{\delta}{l_{bb}} = \frac{y}{\rho} \tag{5.4}$$

式(5.4)表明，纵向线应变 $\varepsilon$ 与到中性层的距离 $y$ 呈线性关系。

2. 应力应变关系

假设正应力不超过比例极限，则胡克定律成立，即 $\sigma = E\varepsilon$。由式(5.4)得

$$\sigma = \frac{E}{\rho} y \tag{5.5}$$

式中，$E$ 为弹性模量。式(5.5)表明，在线弹性范围，正应力随到中性层的距离线性变化，如图 5.6 所示。

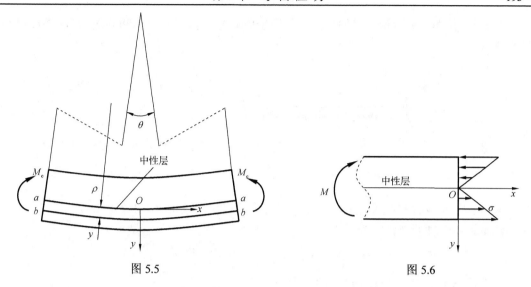

图 5.5　　　　　　　　　　　图 5.6

### 3. 平衡条件

根据 $x$ 方向平衡条件，得

$$\sum F_x = 0: \quad N = \int \sigma dA = \frac{E}{\rho} \int y dA = 0 \tag{5.6}$$

即

$$\int y dA = 0 \tag{5.7}$$

式(5.7)表明，对于纯弯曲梁，只要应力在线弹性范围，则中性轴通过截面形心。

根据绕中性轴的平衡条件，得

$$\sum M_z = 0: \quad M = \int y\sigma dA = \frac{E}{\rho} \int y^2 dA \tag{5.8}$$

即

$$\frac{1}{\rho} = \frac{M}{E \int y^2 dA} = \frac{M}{EI} \tag{5.9}$$

式中，$I = \int y^2 dA$ 为横截面对形心轴的惯性矩。把式(5.9)代入式(5.5)，得

$$\sigma = \frac{My}{I} \tag{5.10}$$

当弯矩为正时，中性轴上方的正应力为压应力；当弯矩为负时，中性轴上方的正应力为拉应力。

根据式(5.10)，最大正应力等于 $\sigma_{max} = \dfrac{My_{max}}{I}$。比值 $I/y_{max}$ 仅与截面几何尺寸有关。该比值称为截面模量或抗弯截面系数，用 $S$ 表示。根据式(5.10)，得

$$\sigma_{max} = \frac{My_{max}}{I} = \frac{M}{S} \tag{5.11}$$

对宽为 $b$、高为 $h$ 的矩形截面梁，有 $S = \dfrac{1}{6}bh^2$。

**例 5.1**  如图 5.7 所示，已知外力偶作用于竖直平面内，求梁中间截面 $C$ 上 $P$ 点和 $Q$ 点的应力。

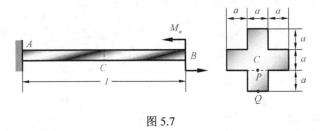

图 5.7

**解**  从中间截面 $C$ 把梁截开，考虑右侧梁的平衡，得截面 $C$ 的弯矩 $M=M_e$。截面对中性轴的惯性矩为

$$I = \frac{1}{12}a^4 + \frac{1}{12}a(3a)^3 + \frac{1}{12}a^4 = \frac{29}{12}a^4$$

利用式(5.10)，得 $P$ 点和 $Q$ 点的应力为

$$\sigma_P = \frac{My_P}{I} = \frac{M_e\left(\frac{1}{2}a\right)}{\frac{29}{12}a^4} = \frac{6M_e}{29a^3}, \quad \sigma_Q = \frac{My_Q}{I} = \frac{M_e\left(\frac{3}{2}a\right)}{\frac{29}{12}a^4} = \frac{18M_e}{29a^3}$$

## 5.2 横力弯曲横截面正应力和剪应力

作用于梁的横向载荷将在横截面上产生正应力和剪应力。正应力由弯矩引起，而剪应力则由剪力引起。

**1. 正应力**

纯弯曲梁横截面正应力公式对横向载荷作用的细长梁仍然有效，如图 5.8 所示，因此横力弯曲梁的正应力为

$$\sigma = \frac{My}{I} \tag{5.12}$$

式中，弯矩 $M$ 随截面不同而变化，即 $M = M(x)$ 是位置 $x$ 的函数。

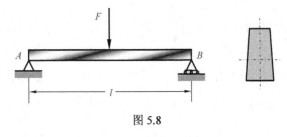

图 5.8

横力弯曲梁的最大正应力为

$$\sigma_{\max} = \frac{M_{\max} y_{\max}}{I} = \frac{M_{\max}}{S} \tag{5.13}$$

## 2. 剪应力

因梁强度设计的主要准则是梁的最大正应力，故正应力计算在梁的设计中显得十分重要。然而，在粗短梁的设计中，剪应力也显得很重要。

考虑具有竖直对称面的等截面梁，该梁受外载荷(集中力 $F$、集中力偶 $M_e$ 和分布载荷 $q$)作用。在距 $A$ 端为 $x$ 处取长度为 $\mathrm{d}x$ 的单元体 $abcd$，设单元体的上表面到中性轴的距离为 $y$，如图 5.9 所示。

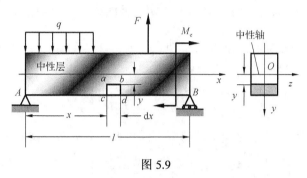

图 5.9

当 $\mathrm{d}x$ 趋于零时，单元体左、右表面上的剪应力大小相等。只要梁的横截面的宽度比高度小，那么可认为剪应力沿截面宽度均匀分布。假设单元体左表面(或右表面)上到中性轴的距离为 $y$ 处的剪应力用 $\tau$ 表示，则根据剪应力互等定理，单元体上表面的剪应力也等于 $\tau$，如图 5.10(a) 所示。

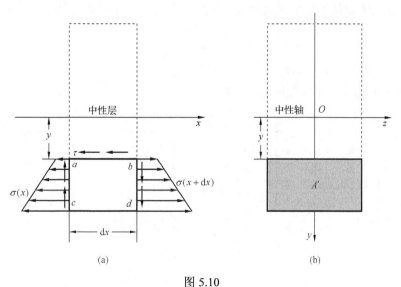

图 5.10

考虑单元体 $abcd$ 在水平方向的平衡，则有

$$\sum F_x = 0: \quad \int [\sigma(x+\mathrm{d}x) - \sigma(x)] \mathrm{d}A' - \tau(b\mathrm{d}x) = 0 \tag{5.14}$$

式中，$b$ 为梁的横截面宽度；$A'$ 为距中性轴为 $y$ 的直线下方截面的面积，如图 5.10(b) 所示。

利用 $\sigma(x+\mathrm{d}x)-\sigma(x)=\dfrac{M(x+\mathrm{d}x)y'}{I}-\dfrac{M(x)y'}{I}=\dfrac{\mathrm{d}My'}{I}=\dfrac{(V\mathrm{d}x)y'}{I}$, 解式(5.14), 有

$$\tau=\dfrac{V}{Ib}\int y'\mathrm{d}A'=\dfrac{VQ'}{Ib} \tag{5.15}$$

式中, $Q'=\int y'\mathrm{d}A'$ 为距中性轴为 $y$ 的直线下方(或上方)截面对中性轴的静矩; $I$ 为整个截面的形心惯性矩; $V$ 为截面上的剪力。

如图 5.11(a)所示, 对宽为 $b$、高为 $h$ 的矩形截面梁, $I=\dfrac{1}{12}bh^3$, 以及

$$Q'=A'\overline{y}'=\left[b\left(\dfrac{1}{2}h-y\right)\right]\left[\dfrac{1}{2}\left(\dfrac{1}{2}h+y\right)\right]=\dfrac{1}{8}b(h^2-4y^2) \tag{5.16}$$

则有
$$\tau=\dfrac{3}{2}\dfrac{V}{bh}\left[1-\left(\dfrac{2y}{h}\right)^2\right] \tag{5.17}$$

式(5.17)表明, 矩形截面梁横截面上的剪应力满足抛物线分布, 如图 5.11(b)所示。在上、下表面($y=\pm\dfrac{1}{2}h$), 剪应力为零。设 $y=0$, 横截面上剪应力的最大值为

$$\tau_{\max}=\dfrac{3V}{2bh}=\dfrac{3V}{2A} \tag{5.18}$$

式中, $A$ 为矩形截面梁的横截面面积。

如图 5.12 所示, 对工字截面梁(简称工字梁), 腹板上的剪应力在截面竖直方向仅发生微小变化, 并且剪力几乎全由腹板承担。因此, 剪力除以腹板横截面面积即可得到剪应力的最大值:

$$\tau_{\max}=\dfrac{V}{A_{\mathrm{w}}}=\dfrac{V}{b_0 h_0} \tag{5.19}$$

式中, $b_0$ 和 $h_0$ 分别为工字梁腹板的宽度和高度。

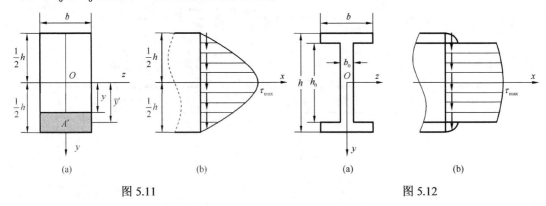

图 5.11    图 5.12

如图 5.13 所示, 对实心圆截面梁, 最大剪应力位于中性轴上, 最大值可表示为

$$\tau_{\max}=\dfrac{4V}{3A} \tag{5.20}$$

式中, $A$ 为实心圆截面梁的横截面面积。

如图 5.14 所示，对于空心圆截面梁，中性轴上有最大剪应力，最大值可表示为

$$\tau_{\max} = \frac{2V}{A} \tag{5.21}$$

式中，$A$ 为空心圆截面梁的横截面面积。

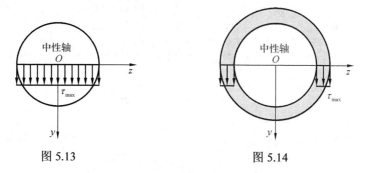

图 5.13　　　　　　　　图 5.14

**例 5.2**　如图 5.15 所示，已知外力作用于竖直面内，求中间截面 $C$ 上点 $P$ 处的剪应力。

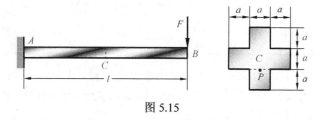

图 5.15

**解**　根据截面法，剪力等于截面 $B$ 处的外力，即 $V=F$。截面对中性轴的惯性矩为 $I = \frac{29}{12}a^4$，距离中性轴 $\frac{1}{2}a$ 的水平线下方截面对中性轴的静矩为 $Q' = A'\overline{y}' = a^3$。因此点 $P$ 处的剪应力为

$$\tau_P = \frac{VQ'}{Ib} = \frac{Fa^3}{\left(\frac{29}{12}a^4\right)a} = \frac{12F}{29a^2}$$

## 5.3　弯曲梁设计

弯曲梁设计通常由弯矩最大绝对值 $|M|_{\max}$ 控制。危险截面（即 $|M|_{\max}$）处梁在上、下表面正应力达到最大值 $\sigma_{\max}$，最大值可表示为

$$\sigma_{\max} = \frac{|M|_{\max} y_{\max}}{I} = \frac{|M|_{\max}}{S} \tag{5.22}$$

安全设计要求满足 $\sigma_{\max} \leqslant \sigma_{\text{allow}}$，其中 $\sigma_{\text{allow}}$ 为许用应力。把 $\sigma_{\text{allow}}$ 代入式(5.22)，得

$$S \geqslant \frac{|M|_{\max}}{\sigma_{\text{allow}}} \tag{5.23}$$

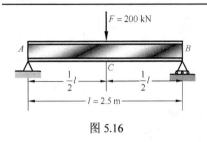

图 5.16

在梁的设计中，合理设计应该满足最经济原则，即在类型和材料相同的梁中，当其他因素相同时，应该选择单位长度质量最小或横截面面积最小的梁，以降低梁的费用。

**例 5.3** 已知钢的许用正应力为 165 MPa，选择支撑如图 5.16 所示外载荷的工字钢梁。

**解** 弯矩在截面 $C$ 处达到最大，$M_{\max} = \dfrac{1}{4}Fl = 125 \text{ kN} \cdot \text{m}$。因此，最小截面模量为

$$S_{\min} = \frac{|M|_{\max}}{\sigma_{\text{allow}}} = 757.58 \text{ cm}^3$$

查阅附录 II 中型钢表，选择截面模量不小于 $S_{\min}$ 的一组工字钢梁，如表 5.1 所示。

表 5.1

| 型号 | 单位长度质量/(kg/m) | 截面模量/cm³ |
| --- | --- | --- |
| 32c | 62.765 | 760 |
| 36a | 60.037 | 875 |
| 36b | 65.689 | 919 |

根据表 5.1，可选 32c 或 36a 工字钢梁用于安全设计。然而，最经济的设计应该选择 36a 工字钢梁，因为与 32c 工字钢梁相比，36a 工字钢梁的单位长度质量虽然只有 60.037 kg/m，但是却具有较大的截面模量。

所选梁的总重量为 1.47 kN。该重量与 200 kN 外载荷相比显得很小，因而可忽略不计。

# 习 题

5.1 如图 5.17 所示外力偶作用于竖直面内，求中间截面 $C$ 上点 $P_1$ 和点 $P_2$ 处的应力。

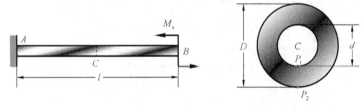

图 5.17

5.2 已知如图 5.18 所示工字钢梁的许用应力 $\sigma_{\text{allow}} = 150$ MPa，求所能施加的最大力偶。忽略倒角效应。

5.3 外力偶作用于如图 5.19 所示 T 形截面梁。求梁的最大拉应力和最大压应力。忽略倒角效应。

5.4 已知如图 5.20 所示外力作用于竖直面内，求中间截面 $C$ 上点 $P$ 处的剪应力。

5.5 已知如图 5.21 所示外力作用于竖直面内，求中间截面 $C$ 上点 $P$ 处的剪应力。

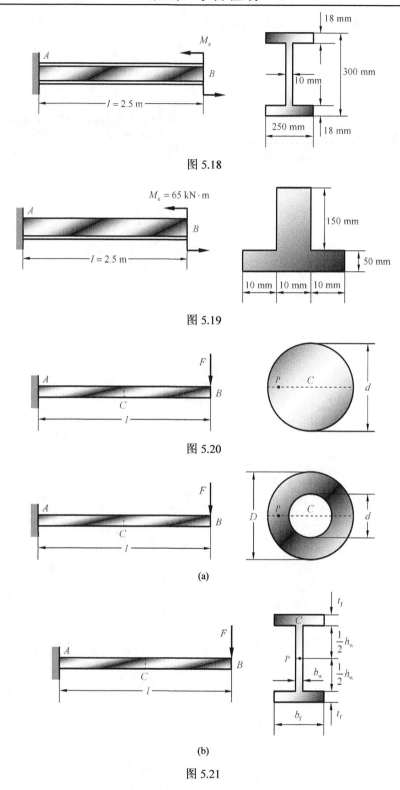

图 5.18

图 5.19

图 5.20

(a)

(b)

图 5.21

5.6 已知钢的许用正应力为 165 MPa，选择工字钢梁以支撑如图 5.22 所示外载荷。

5.7 已知钢的许用正应力为 165 MPa，选择最经济的工字钢梁以支撑如图 5.23 所示外载荷。

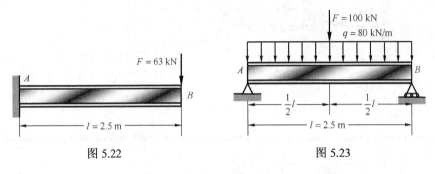

图 5.22　　　　　　　　图 5.23

# 第 6 章 弯 曲 变 形

如图 6.1 所示,纯弯曲等截面梁在变形后将弯成圆弧。在线弹性范围,中性层的曲率可表示为

$$\frac{1}{\rho} = \frac{M}{EI} \tag{6.1}$$

式中,弯矩 $M$ 为常数；$EI$ 为梁的抗弯刚度。

当如图 6.2 所示细长梁承受横向载荷时,式(6.1)仍然成立。不过弯矩和中性层曲率将会随着截面的不同而发生变化。设 $x$ 表示截面到左端的距离,则有

$$\frac{1}{\rho} = \frac{M(x)}{EI} \tag{6.2}$$

式中,弯矩 $M(x)$ 为 $x$ 的函数。

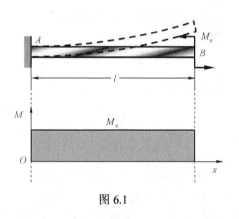

图 6.1

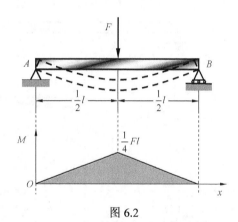

图 6.2

梁在变形后中性层的曲率可表示为

$$\frac{1}{\rho} = \frac{w''}{[1+(w')^2]^{3/2}} \tag{6.3}$$

式中,$w = w(x)$ 为挠度函数,见图 6.3,$w'$ 和 $w''$ 分别为函数 $w = w(x)$ 的一阶和二阶导数。对于小变形梁,与 1 相比,$w'$ 非常小,因而其平方可忽略,得

$$\frac{1}{\rho} = w'' \tag{6.4}$$

把式(6.4)代入式(6.2),得

$$w'' = \frac{M(x)}{EI} \tag{6.5}$$

式(6.5)称为弯曲梁的挠曲线微分方程。

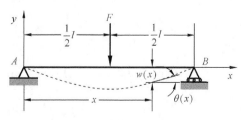

图 6.3

## 6.1 积 分 法

积分式(6.5)，得

$$w' = \int \frac{M(x)}{EI} \mathrm{d}x + C_1 \tag{6.6}$$

式中，$C_1$ 为积分常数。设 $\theta(x)$ 为变形后中性层切线与水平方向的夹角，如图 6.3 所示，考虑到夹角很小，则有

$$w' = \tan\theta \approx \theta \tag{6.7}$$

因此，式(6.6)可重写为

$$\theta = \int \frac{M(x)}{EI} \mathrm{d}x + C_1 \tag{6.8}$$

积分，得

$$w = \iint \left( \int \frac{M(x)}{EI} \mathrm{d}x \right) \mathrm{d}x + C_1 x + C_2 \tag{6.9}$$

式中，$C_2$ 也为积分常数。

积分常数 $C_1$ 和 $C_2$ 可通过边界条件确定。三类常用支撑及其边界条件如图 6.4 所示。

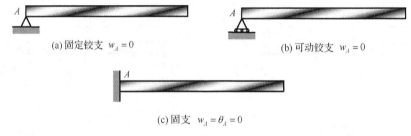

图 6.4

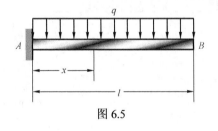

图 6.5

**例 6.1** 悬臂梁受如图 6.5 所示载荷作用，求：(1)梁的挠曲线方程；(2)自由端的挠度和转角。

**解** 考虑梁的 $x$ 截面，如图 6.5 所示，则弯矩为 $M(x) = -\frac{1}{2}q(l-x)^2$ 和挠曲线微分方程为 $w'' = -\frac{q(l-x)^2}{2EI}$。

利用积分法，得

$$\theta = \int \left[ -\frac{q(l-x)^2}{2EI} \right] \mathrm{d}x + C_1 = -\frac{q}{6EI}(3l^2 x - 3lx^2 + x^3) + C_1$$

$$w = \int \left\{ \int \left[ -\frac{q(l-x)^2}{2EI} \right] \mathrm{d}x \right\} \mathrm{d}x + C_1 x + C_2 = -\frac{q}{24EI}(6l^2 x^2 - 4lx^3 + x^4) + C_1 x + C_2$$

当 $x = 0$ 时，$\theta = w = 0$，则得 $C_1 = 0$ 和 $C_2 = 0$。因此挠曲线方程为

$$w = -\frac{q}{24EI}(6l^2 x^2 - 4lx^3 + x^4)$$

利用上述方程，得自由端挠度和转角分别为

$$w_B = w_{x=l} = -\frac{ql^4}{8EI}, \quad \theta_B = \theta_{x=l} = -\frac{ql^3}{6EI}$$

## 6.2 叠 加 法

如图 6.6(a)所示，当梁受多个载荷作用时，首先分别计算每个载荷单独作用时的挠度和转角，然后采用叠加原理，把对应各个载荷的挠度和转角相加，即可得到多个载荷同时作用时的挠度和转角。大部分工程手册都包含梁在各种载荷作用与支撑条件下的挠度和转角列表。

考虑如图 6.6(a)所示简支梁，受集中载荷和分布载荷同时作用，则梁在任意截面的挠度和转角等于如图 6.6(b)所示集中载荷与如图 6.6(c)所示分布载荷分别引起的挠度和转角之和。因此，受两个载荷同时作用时，梁的挠度和转角可表示为

$$w = w_F + w_q, \quad \theta = \theta_F + \theta_q \quad (6.10)$$

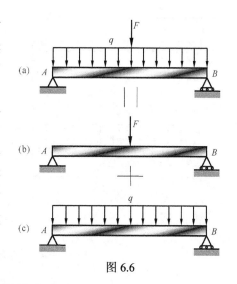

图 6.6

**例 6.2** 对于如图 6.7 所示梁和载荷，求截面 $C$ 的挠度和截面 $B$ 的转角。

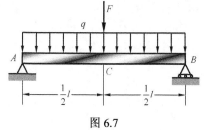

图 6.7

**解** 对集中载荷 $F$，查附录Ⅲ，得

$$(w_C)_F = -\frac{Fl^3}{48EI}, \quad (\theta_B)_F = \frac{Fl^2}{16EI}$$

同理，对分布载荷 $q$，查附录Ⅲ，得

$$(w_C)_q = -\frac{5ql^4}{384EI}, \quad (\theta_B)_q = \frac{ql^3}{24EI}$$

利用叠加原理，得截面 $C$ 的挠度和截面 $B$ 的转角为

$$w_C = (w_C)_F + (w_C)_q = -\left(\frac{Fl^3}{48EI} + \frac{5ql^4}{384EI}\right), \quad \theta_B = (\theta_B)_F + (\theta_B)_q = \frac{Fl^2}{16EI} + \frac{ql^3}{24EI}$$

## 6.3 静 不 定 梁

考虑如图 6.8 所示等截面梁，$A$ 端固支，$B$ 端可动铰支。梁有 4 个未知反力，但只有 3 个平衡条件，因此梁为 1 次静不定。

叠加法可以用于求解静不定梁在支撑处的反力。对如图 6.8 所示 1 次静不定梁，把 1 个反力(如 $B$ 处反

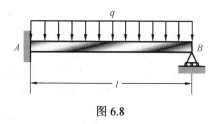

图 6.8

力)看成多余反力,并消除多余反力处的支撑。那么多余反力可以作为未知载荷,连同其余载荷,产生与原支撑相协调的变形。通过叠加给定载荷与多余反力分别产生的挠度和转角而得到消除多余支撑后的挠度和转角。

**例 6.3** 如图 6.9(a)所示梁受载荷作用,求可动铰支处的反力。

**解** 考虑 $B$ 处为多余反力,并释放 $B$ 处支撑,则多余反力 $R_B$ 可根据 $B$ 处挠度等于零而求得。

查附录Ⅲ,得 $B$ 处由多余反力 $R_B$ 引起的挠度为

$$(w_B)_{R_B} = \frac{R_B l^3}{3EI}$$

同理,$B$ 处由均布载荷 $q$ 引起的挠度为

$$(w_B)_q = -\frac{ql^4}{8EI}$$

利用叠加原理,得 $B$ 处总挠度为

$$w_B = (w_B)_{R_B} + (w_B)_q = \frac{R_B l^3}{3EI} - \frac{ql^4}{8EI}$$

因 $B$ 处总挠度为零,则得

$$\frac{R_B l^3}{3EI} - \frac{ql^4}{8EI} = 0$$

解上式,得 $B$ 处反力为

$$R_B = \frac{3}{8}ql$$

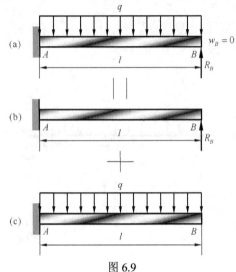

图 6.9

## 习 题

6.1 对于如图 6.10 所示载荷,求:(1)悬臂梁的挠曲线方程;(2)自由端的挠度和转角。

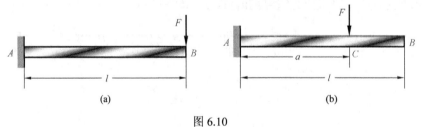

图 6.10

6.2 对于如图 6.11 所示载荷,求:(1)简支梁的挠曲线方程;(2)中间截面 $C$ 的挠度和截面 $B$ 的转角。

6.3 对于如图 6.12 所示梁和载荷,求:(1)中间截面 $C$ 的挠度;(2)截面 $B$ 的转角。

6.4 对于如图 6.13 所示悬臂梁和载荷,求截面 $B$ 的挠度和转角。

6.5 对于如图 6.14 所示梁和载荷,求可动铰支 $B$ 处反力。

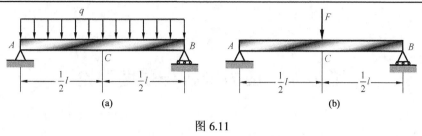

图 6.11

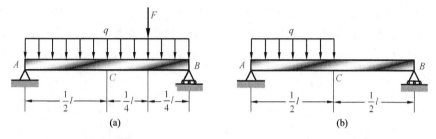

图 6.12

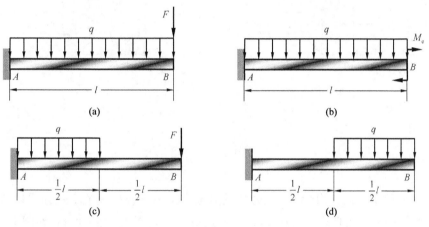

图 6.13

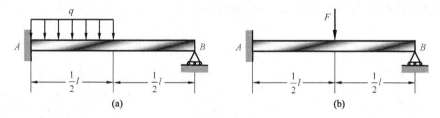

图 6.14

# 第7章 应力分析与强度理论

给定点 $P$ 的最一般应力状态可通过六个应力分量表示，其中三个分量 $\sigma_x$、$\sigma_y$ 和 $\sigma_z$ 为单元体面上的正应力，而其余三个分量 $\tau_{xy}(\tau_{yx}=\tau_{xy})$、$\tau_{yz}(\tau_{zy}=\tau_{yz})$ 和 $\tau_{zx}(\tau_{xz}=\tau_{zx})$ 为剪应力，如图 7.1 所示。如果坐标轴旋转，则相同的应力状态可通过不同的分量进行表示。应力状态的讨论主要涉及平面应力，即单元体的两个平行面上不受应力作用，如图 7.2 所示。

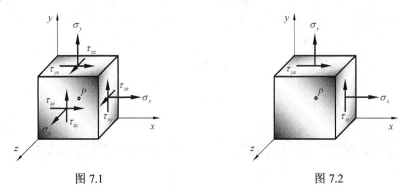

图 7.1　　　　　　　　　图 7.2

## 7.1　应　力　变　换

考虑点 $P$ 处于平面应力状态，应力分量由 $\sigma_x$、$\sigma_y$ 和 $\tau_x(\tau_y=\tau_x)$ 表示，如图 7.3(a) 所示。下面求单元体绕 $z$ 轴逆时针旋转 $\theta$ 后的应力分量 $\sigma_{x'}$、$\sigma_{y'}$ 和 $\tau_{x'}(\tau_{y'}=\tau_{x'})$，如图 7.3(b) 所示。

考虑表面分别垂直于 $x$、$y$ 和 $x'$ 轴的单元体，如图 7.4 所示。假设斜面面积为 $\Delta A$，则垂直面和水平面的面积分别等于 $\Delta A\cos\theta$ 和 $\Delta A\sin\theta$。

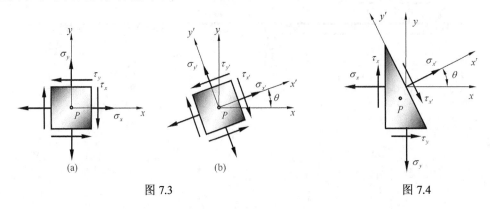

图 7.3　　　　　　　　　图 7.4

利用 $x'$ 和 $y'$ 方向的静平衡条件，得

$$\sum F_{x'}=0: \sigma_{x'}\Delta A - \sigma_x(\Delta A\cos\theta)\cos\theta + \tau_x(\Delta A\cos\theta)\sin\theta$$
$$- \sigma_y(\Delta A\sin\theta)\sin\theta + \tau_y(\Delta A\sin\theta)\cos\theta = 0$$
$$\sum F_{y'}=0: -\tau_{x'}\Delta A + \sigma_x(\Delta A\cos\theta)\sin\theta + \tau_x(\Delta A\cos\theta)\cos\theta$$
$$- \sigma_y(\Delta A\sin\theta)\cos\theta - \tau_y(\Delta A\sin\theta)\sin\theta = 0$$
(7.1)

解式(7.1)，并利用 $\tau_y = \tau_x$、$\cos^2\theta = \frac{1}{2}[1+\cos(2\theta)]$、$\sin^2\theta = \frac{1}{2}[1-\cos(2\theta)]$ 和 $2\sin\theta\cos\theta = \sin(2\theta)$，得

$$\sigma_{x'} = \frac{1}{2}(\sigma_x+\sigma_y) + \frac{1}{2}(\sigma_x-\sigma_y)\cos(2\theta) - \tau_x\sin(2\theta)$$
$$\tau_{x'} = \frac{1}{2}(\sigma_x-\sigma_y)\sin(2\theta) + \tau_x\cos(2\theta)$$
(7.2)

如果需要求 $\sigma_{y'}$，则用 $\theta + \frac{1}{2}\pi$ 代替 $\theta$ 代入式(7.2)，得

$$\sigma_{y'} = \frac{1}{2}(\sigma_x+\sigma_y) - \frac{1}{2}(\sigma_x-\sigma_y)\cos(2\theta) + \tau_x\sin(2\theta)$$
(7.3)

**例 7.1** 对于如图 7.5(a) 所示应力状态，求单元体顺时针旋转 20° 后的正应力和剪应力。

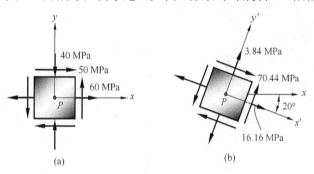

图 7.5

**解** 根据如图 7.5(a) 所示参考系，则有

$$\sigma_x = 60 \text{ MPa}, \quad \sigma_y = -40 \text{ MPa}, \quad \tau_x = -50 \text{ MPa}, \quad \theta = -20°$$

利用式(7.2)和式(7.3)，得

$$\sigma_{x'} = \frac{1}{2}(\sigma_x+\sigma_y) + \frac{1}{2}(\sigma_x-\sigma_y)\cos(2\theta) - \tau_x\sin(2\theta) = 16.16 \text{ MPa}$$
$$\sigma_{y'} = \frac{1}{2}(\sigma_x+\sigma_y) - \frac{1}{2}(\sigma_x-\sigma_y)\cos(2\theta) + \tau_x\sin(2\theta) = 3.84 \text{ MPa}$$
$$\tau_{x'} = \frac{1}{2}(\sigma_x-\sigma_y)\sin(2\theta) + \tau_x\cos(2\theta) = -70.44 \text{ MPa}$$

图 7.5(b) 为单元体顺时针旋转 20° 后的正应力和剪应力。

## 7.2 主应力

设 $\sigma_{avg} = \dfrac{1}{2}(\sigma_x + \sigma_y)$ 和 $R = \sqrt{\left[\dfrac{1}{2}(\sigma_x - \sigma_y)\right]^2 + \tau_x^2}$,则利用式(7.2),得

$$(\sigma_{x'} - \sigma_{avg})^2 + \tau_{x'}^2 = R^2 \tag{7.4}$$

式(7.4)表示圆心在 $C(\sigma_{avg}, 0)$、半径为 $R$ 的圆,如图 7.6 所示。该圆称为莫尔圆或应力圆。

应力圆与水平轴的交点 $A$ 和 $B$ 是特别感兴趣的点:$A$ 点对应最大正应力 $\sigma_{max}$,而 $B$ 点对应最小正应力 $\sigma_{min}$。同时,这两点对应的剪应力为零。通过在式(7.2)中设 $\tau_{x'} = 0$,则对应点 $A$ 和 $B$ 的 $\theta_p$ 值为

$$\tan(2\theta_p) = \dfrac{-2\tau_x}{\sigma_x - \sigma_y} \tag{7.5}$$

式(7.5)定义了相差 90° 的两个 $\theta_p$ 值。通过任何一个 $\theta_p$ 值即可确定对应于最大和最小正应力的单元体的方位,如图 7.7 所示。

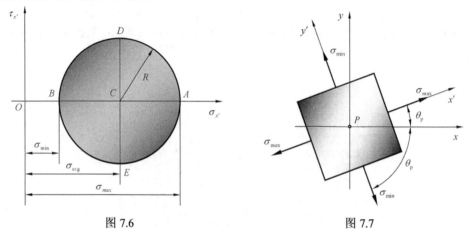

图 7.6　　　　　图 7.7

通过上述方法确定的包含单元体表面的平面称为主平面,主平面上的正应力 $\sigma_{max}$ 和 $\sigma_{min}$ 称为面内主应力。显然,主平面上没有剪应力。根据式(7.4)得

$$\sigma_{max,min} = \sigma_{avg} \pm R = \dfrac{1}{2}(\sigma_x + \sigma_y) \pm \sqrt{\left[\dfrac{1}{2}(\sigma_x - \sigma_y)\right]^2 + \tau_x^2} \tag{7.6}$$

均匀、各向同性材料的任何一点在平面应力状态下,都存在三个相互垂直的主应力。如果 $\sigma_{min} \geq 0$,则三个主应力可表示为 $\sigma_1 = \sigma_{max}$、$\sigma_2 = \sigma_{min}$ 和 $\sigma_3 = 0$;如果 $\sigma_{max} \geq 0 \geq \sigma_{min}$,则 $\sigma_1 = \sigma_{max}$、$\sigma_2 = 0$ 和 $\sigma_3 = \sigma_{min}$;如果 $\sigma_{max} \leq 0$,则 $\sigma_1 = 0$、$\sigma_2 = \sigma_{max}$ 和 $\sigma_3 = \sigma_{min}$。

**例 7.2**　对于如图 7.8(a)所示应力状态,求主应力和相应主平面。

**解**　参考如图 7.8(a)所示坐标系,得

$$\sigma_x = 60 \text{ MPa}, \quad \sigma_y = -40 \text{ MPa}, \quad \tau_x = -50 \text{ MPa}$$

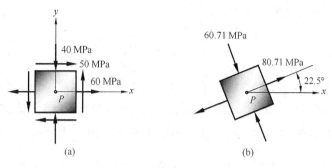

图 7.8

利用式(7.6)，得面内主应力为

$$\sigma_{\max} = \frac{1}{2}(\sigma_x + \sigma_y) + \sqrt{\left[\frac{1}{2}(\sigma_x - \sigma_y)\right]^2 + \tau_x^2} = 80.71 \text{ MPa}$$

$$\sigma_{\min} = \frac{1}{2}(\sigma_x + \sigma_y) - \sqrt{\left[\frac{1}{2}(\sigma_x - \sigma_y)\right]^2 + \tau_x^2} = -60.71 \text{ MPa}$$

利用式(7.5)，得面内主应力的方位为

$$\theta_p = \frac{1}{2}\arctan\frac{-2\tau_x}{\sigma_x - \sigma_y} = \begin{matrix} 22.5° & (\text{对应 } \sigma_{\max}) \\ -67.5° & (\text{对应 } \sigma_{\min}) \end{matrix}$$

主应力和相应主平面如图 7.8(b) 所示。因此，平面应力状态的三个主应力分别为 $\sigma_1 = 80.71$ MPa、$\sigma_2 = 0$ 和 $\sigma_3 = -60.71$ MPa。

## 7.3 最大剪应力

参考如图 7.6 所示应力圆，位于应力圆垂直方向直径上的 $D$ 点和 $E$ 点对应最大剪应力和最小剪应力。因 $D$ 和 $E$ 点的横坐标为 $\sigma_{\text{avg}} = \frac{1}{2}(\sigma_x + \sigma_y)$，故通过在式(7.2)中设 $\sigma_{x'} = \frac{1}{2}(\sigma_x + \sigma_y)$，则对应点 $D$ 和 $E$ 的 $\theta_s$ 值为

$$\tan(2\theta_s) = \frac{\sigma_x - \sigma_y}{2\tau_x} \tag{7.7}$$

式(7.7)定义了相差 90° 的两个 $\theta_s$ 值。通过任何一个 $\theta_s$ 值即可确定对应于最大面内剪应力的单元体的方位，如图 7.9 所示。

由式(7.4)，得

$$(\tau_{\max})_{\text{in-plane}} = R = \sqrt{\left[\frac{1}{2}(\sigma_x - \sigma_y)\right]^2 + \tau_x^2} \tag{7.8}$$

以及最大面内剪应力面上的正应力为

$$\sigma_{\text{avg}} = \frac{1}{2}(\sigma_x + \sigma_y) \tag{7.9}$$

比较式(7.5)和式(7.7)，注意到 $\tan(2\theta_s)$ 是 $\tan(2\theta_p)$ 的负倒数，即 $\tan(2\theta_s) \cdot \tan(2\theta_p) = -1$。这意味两个角度相差45°。因此，最大面内剪应力面与主平面成45°夹角。

应该注意，剪应力 $\tau_{\max} = \frac{1}{2}(\sigma_1 - \sigma_3)$ 可能会比由式(7.8)定义的 $(\tau_{\max})_{\text{in-plane}}$ 大。这种情况会发生在由式(7.6)定义的主应力具有相同正负号，即都是拉应力或都是压应力。

**例7.3** 对于如图7.10(a)所示应力状态，求：(1)最大面内剪应力；(2)最大面内剪应力面的方位；(3)最大面内剪应力面上的正应力。

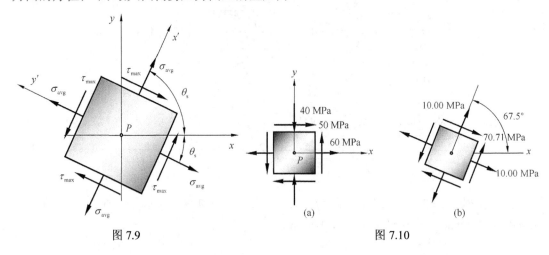

图 7.9　　　　　　　　　　图 7.10

**解** 对如图7.10(a)所示坐标系，有

$$\sigma_x = 60 \text{ MPa}, \quad \sigma_y = -40 \text{ MPa}, \quad \tau_x = -50 \text{ MPa}$$

(1) 利用式(7.8)，得最大面内剪应力为

$$(\tau_{\max})_{\text{in-plane}} = \sqrt{\left[\frac{1}{2}(\sigma_x - \sigma_y)\right]^2 + \tau_x^2} = 70.71 \text{ MPa}$$

(2) 利用式(7.7)，得最大面内剪应力面的方位为

$$\theta_s = \frac{1}{2}\arctan\frac{\sigma_x - \sigma_y}{2\tau_x} = \begin{matrix}67.5° \\ -22.5°\end{matrix}$$

(3) 利用式(7.9)，得最大面内剪应力面上的正应力为

$$\sigma_{\text{avg}} = \frac{1}{2}(\sigma_x + \sigma_y) = 10.00 \text{ MPa}$$

最大面内剪应力和最大面内剪应力面上的正应力及这些应力的方位如图7.10(b)所示。

## 7.4 压力容器

薄壁压力容器为平面应力提供了重要应用。薄壁压力容器的应力分析仅考虑两种常见的容器类型，即柱形压力容器和球形压力容器。

1. 柱形压力容器

如图 7.11 所示，柱形压力容器的内径为 $d$，壁厚为 $t$，内压为 $p$。下面求壁单元体上的应力，单元体的边分别平行和垂直于圆柱纵轴。

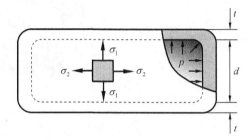

图 7.11

显然，没有剪应力作用于单元体，单元体上的正应力因此是主应力 $\sigma_1$ 和 $\sigma_2$。$\sigma_1$ 称为环向应力或周向应力，$\sigma_2$ 称为纵向应力或轴向应力。

为了求纵向应力 $\sigma_2$，考虑如图 7.12 所示自由体（或隔离体）的平衡，则有

$$\sum F_x = 0: \quad \sigma_2(\pi dt) - p\left(\frac{1}{4}\pi d^2\right) = 0 \tag{7.10}$$

解方程，得

$$\sigma_2 = \frac{pd}{4t} \tag{7.11}$$

为了求环向应力 $\sigma_1$，分离一部分容器，如图 7.13 所示。利用平衡条件，得

$$\sum F_z = 0: \quad \sigma_1(2t\Delta x) - p(d\Delta x) = 0 \tag{7.12}$$

解方程，得

$$\sigma_1 = \frac{pd}{2t} \tag{7.13}$$

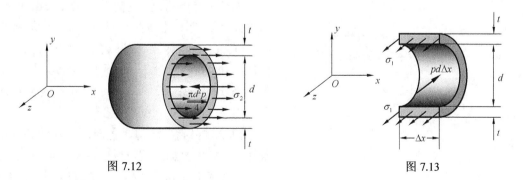

图 7.12　　　　　　　　　　　　图 7.13

利用式(7.11)和式(7.13)，得最大剪应力为

$$(\tau_{max})_{\text{in-plane}} = \frac{1}{2}(\sigma_1 - \sigma_2) = \frac{pd}{8t}, \quad \tau_{max} = \frac{1}{2}(\sigma_1 - \sigma_3) = \frac{pd}{4t} \tag{7.14}$$

### 2. 球形压力容器

如图 7.14(a)所示，球形压力容器的内径为 $d$，壁厚为 $t$，内压为 $p$。利用容器对称性，两个面内主应力一定相等，即 $\sigma_1 = \sigma_2$。

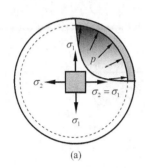 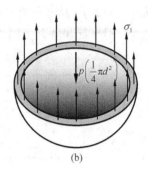

图 7.14

为了求应力 $\sigma_1$（或 $\sigma_2$），考虑如图 7.14(b)所示自由体。根据自由体在垂直方向的平衡，得

$$\sigma_1(\pi dt) - p\left(\frac{1}{4}\pi d^2\right) = 0 \tag{7.15}$$

解方程，得

$$\sigma_1 = \sigma_2 = \frac{pd}{4t} \tag{7.16}$$

利用式(7.16)，得最大剪应力为

$$(\tau_{max})_{\text{in-plane}} = \frac{1}{2}(\sigma_1 - \sigma_2) = 0, \quad \tau_{max} = \frac{1}{2}(\sigma_1 - \sigma_3) = \frac{pd}{8t} \tag{7.17}$$

## 7.5 广义胡克定律

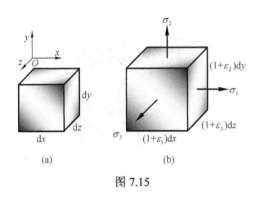

图 7.15

如图 7.15(a)所示，考虑边长为 $dx$、$dy$ 和 $dz$ 的单元体。若单元体受主应力 $\sigma_1$、$\sigma_2$ 和 $\sigma_3$ 作用，则单元体将变形为边长为 $(1+\varepsilon_1)dx$、$(1+\varepsilon_2)dy$ 和 $(1+\varepsilon_3)dz$ 的新的单元体，如图 7.15(b)所示。其中，$\varepsilon_1$、$\varepsilon_2$ 和 $\varepsilon_3$ 为对应于主应力 $\sigma_1$、$\sigma_2$ 和 $\sigma_3$ 的线应变。

假设材料满足均匀、各向同性和线弹性的条件，则利用叠加原理可以确定应力应变关系。首先考虑由每个正应力引起的沿 $x$ 方向的

线应变。当 $\sigma_1$ 作用时，单元体沿 $x$ 方向伸长，由 $\sigma_1$ 引起的线应变 $\varepsilon_1'$ 为

$$\varepsilon_1' = \frac{\sigma_1}{E} \tag{7.18}$$

作用 $\sigma_2$ 时，单元体在 $x$ 方向收缩，线应变 $\varepsilon_1''$ 表示为

$$\varepsilon_1'' = -\mu \frac{\sigma_2}{E} \tag{7.19}$$

同理，$\sigma_3$ 也引起 $x$ 方向收缩，线应变 $\varepsilon_1'''$ 写为

$$\varepsilon_1''' = -\mu \frac{\sigma_3}{E} \tag{7.20}$$

利用叠加原理，由 $\sigma_1$、$\sigma_2$ 和 $\sigma_3$ 共同引起的沿 $x$ 方向的线应变 $\varepsilon_1$ 为

$$\varepsilon_1 = \varepsilon_1' + \varepsilon_1'' + \varepsilon_1''' = \frac{1}{E}[\sigma_1 - \mu(\sigma_2 + \sigma_3)] \tag{7.21}$$

同理，通过叠加原理，可以得到由 $\sigma_1$、$\sigma_2$ 和 $\sigma_3$ 共同引起的沿 $y$ 和 $z$ 方向的线应变。因此三个主应力作用下的应力应变关系可表示为

$$\varepsilon_1 = \frac{1}{E}[\sigma_1 - \mu(\sigma_2 + \sigma_3)], \quad \varepsilon_2 = \frac{1}{E}[\sigma_2 - \mu(\sigma_3 + \sigma_1)], \quad \varepsilon_3 = \frac{1}{E}[\sigma_3 - \mu(\sigma_1 + \sigma_2)] \tag{7.22}$$

上述关系称为均匀、各向同性材料处于主应力状态下的广义胡克定律。只要应力不超过比例极限和变形满足小变形条件，上述结果均有效。

**例 7.4** 在大型钢压力容器的侧面划出边长为 40 mm 的正方形。加压后正方形处于二向应力，如图 7.16 所示。已知 $E$=200 GPa 和 $\mu$=0.30，求边 $AB$、边 $BC$ 和对角线 $AC$ 的长度改变。

**解** 已知 $\sigma_1 = 60$ MPa, $\sigma_2 = 30$ MPa, $\sigma_3 = 0$，则利用广义胡克定律，得

$$\varepsilon_1 = \frac{1}{E}[\sigma_1 - \mu(\sigma_2 + \sigma_3)] = 2.55 \times 10^{-4}, \quad \varepsilon_2 = \frac{1}{E}[\sigma_2 - \mu(\sigma_3 + \sigma_1)] = 6.0 \times 10^{-5}$$

利用 $\varepsilon_{AB} = \delta_{AB}/l_{AB} = \varepsilon_1$，得

$$\delta_{AB} = \varepsilon_1 l_{AB} = 1.02 \times 10^{-2} \text{ mm}$$

同理，得

$$\delta_{BC} = \varepsilon_2 l_{BC} = 2.4 \times 10^{-3} \text{ mm}$$

利用 $\delta_{AC} = \sqrt{(l_{AB} + \delta_{AB})^2 + (l_{BC} + \delta_{BC})^2} - \sqrt{l_{AB}^2 + l_{BC}^2}$，得

$$\delta_{AC} = 8.9 \times 10^{-3} \text{ mm}$$

对如图 7.17 所示一般应力状态，剪应力 $\tau_{xy}$、$\tau_{yz}$ 和 $\tau_{zx}$ 将把立方单元体变形为倾斜

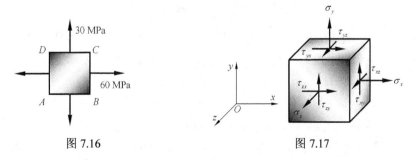

图 7.16　　　　　　　　　　　　　图 7.17

平行六面体。如果应力不超过比例极限，应用叠加原理，得一般应力状态下的广义胡克定律为

$$\varepsilon_x = \frac{1}{E}[\sigma_x - \mu(\sigma_y + \sigma_z)], \quad \varepsilon_y = \frac{1}{E}[\sigma_y - \mu(\sigma_z + \sigma_x)], \quad \varepsilon_z = \frac{1}{E}[\sigma_z - \mu(\sigma_x + \sigma_y)]$$
$$\gamma_{xy} = \frac{\tau_{xy}}{G}, \quad \gamma_{yz} = \frac{\tau_{yz}}{G}, \quad \gamma_{zx} = \frac{\tau_{zx}}{G}$$
(7.23)

## 7.6 强度理论

构件通常被设计为在预期加载条件下材料不能发生失效。材料失效通常是指屈服或断裂。对于塑性材料，失效通常是指屈服；而对于脆性材料，失效是指断裂。

塑性材料只要满足 $\sigma < \sigma_s$（$\sigma_s$ 为屈服强度）或脆性材料只要满足 $\sigma < \sigma_u$（$\sigma_u$ 为极限强度），则单向应力状态下的构件满足安全要求，如图 7.18 所示。

当构件处于平面应力状态或二向应力状态时，不可能通过试验预测构件是否失效，如图 7.19 所示。因此对于复杂应力状态，需要建立与材料失效有关的理论。这种理论称为强度理论或失效理论。

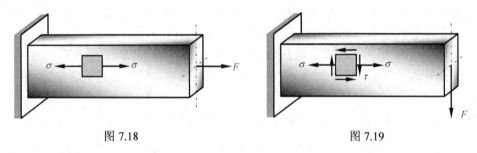

图 7.18　　　　　　　　　　图 7.19

存在两类强度理论：一类适用于脆性断裂，另一类适用于塑性屈服。

**1. 最大正应力理论**

该理论认为，当材料处于二向或三向应力状态时，只要最大正应力达到相同材料单向拉伸失效正应力值，材料即发生失效。该理论是基于事实：脆性断裂由最大正应力引起。

假设三个主应力由 $\sigma_1$、$\sigma_2$ 和 $\sigma_3$ 表示，并按照 $\sigma_1 \geq \sigma_2 \geq \sigma_3$ 进行排序，则失效准则可表示为

$$\sigma_{\text{eq}1} = \sigma_1 = \sigma_u \tag{7.24}$$

式中，$\sigma_{\text{eq}1}$ 为对应最大正应力理论的相当应力；$\sigma_u$ 为材料单向拉伸的极限强度。

假设安全因数为 $n$，则二向或三向应力状态下的强度条件可表示为

$$\sigma_{\text{eq}1} = \sigma_1 \leq \sigma_{\text{allow}} = \frac{\sigma_u}{n} \tag{7.25}$$

式中，$\sigma_{\text{allow}}$ 为许用应力。最大正应力理论与脆性断裂实验结果吻合。

**例 7.5**　如图 7.20 所示平面应力状态出现在由铸铁制成的构件中，已知铸铁的拉伸极

限强度为$\sigma_{ut}$=160 MPa，压缩极限强度为$\sigma_{uc}$=320 MPa。采用最大正应力理论，判断断裂是否发生。如果断裂不发生，求安全因数。

**解** 参考例 7.2，有

$$\sigma_1 = 80.71 \text{ MPa}, \quad \sigma_2 = 0, \quad \sigma_3 = -60.71 \text{ MPa}$$

因 $\sigma_{eq1} = \sigma_1 = 80.71 \text{ MPa} < \sigma_{ut} = 160 \text{ MPa}$，故断裂不发生，相应的安全因数为

$$n = \frac{\sigma_{ut}}{\sigma_{eq1}} = 1.98$$

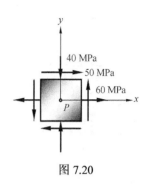

图 7.20

### 2. 最大线应变理论

该理论认为，当材料处于二向或三向应力状态时，只要最大线应变达到相同材料单向拉伸失效拉应变值，材料即发生失效。该理论是基于事实：脆性断裂由最大线应变引起。

失效准则可表示为

$$\varepsilon_1 = \frac{1}{E}[\sigma_1 - \mu(\sigma_2 - \sigma_3)] = \frac{1}{E}\sigma_b \tag{7.26}$$

式中，$\varepsilon_1$ 为最大线应变；$\mu$ 为泊松比。该准则也可表示为

$$\sigma_{eq2} = \sigma_1 - \mu(\sigma_2 - \sigma_3) = \sigma_b \tag{7.27}$$

式中，$\sigma_{eq2}$ 为对应最大线应变理论的相当应力。因此，二向或三向应力状态下的强度条件可表示为

$$\sigma_{eq2} = \sigma_1 - \mu(\sigma_2 - \sigma_3) \leqslant \sigma_{allow} = \frac{\sigma_b}{n} \tag{7.28}$$

式中，$\sigma_{allow}$ 为许用应力；$n$ 为安全因数。目前，最大线应变理论已经很少使用。

### 3. 最大剪应力理论

该理论认为，当材料处于二向或三向应力状态时，只要最大剪应力达到相同材料单向拉伸失效剪应力值，材料即发生失效。该理论是基于事实：塑性屈服由最大剪应力引起。

失效准则可表示为

$$\tau_{max} = \frac{1}{2}(\sigma_1 - \sigma_3) = \frac{1}{2}\sigma_s \tag{7.29}$$

式中，$\tau_{max}$ 为最大剪应力；$\sigma_s$ 为单向拉伸材料的屈服强度。该准则也可表示为

$$\sigma_{eq3} = \sigma_1 - \sigma_3 = \sigma_s \tag{7.30}$$

式中，$\sigma_{eq3}$ 为对应最大剪应力理论的相当应力。因此，二向或三向应力状态下的强度条件可表示为

$$\sigma_{eq3} = \sigma_1 - \sigma_3 \leqslant \sigma_{allow} = \frac{\sigma_s}{n} \tag{7.31}$$

式中，$\sigma_{allow}$ 为许用应力；$n$ 为安全因数。最大剪应力理论广泛应用于塑性屈服。

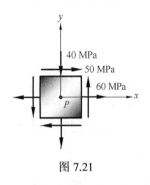

图 7.21

**例 7.6** 如图 7.21 所示平面应力状态出现在由钢制成的构件中,已知屈服强度 $\sigma_s$=310 MPa。采用最大剪应力理论,判断屈服是否发生。如果屈服不发生,求安全因数。

**解** 参考例 7.2,有

$$\sigma_1 = 80.71 \text{ MPa}, \quad \sigma_2 = 0, \quad \sigma_3 = -60.71 \text{ MPa}$$

因 $\sigma_{eq3} = \sigma_1 - \sigma_3 = 141.42 \text{ MPa} < \sigma_s = 310 \text{ MPa}$,故屈服不发生,相应的安全因数为

$$n = \frac{\sigma_s}{\sigma_{eq3}} = 2.19$$

**4. 最大畸变能理论**

该理论认为,当材料处于二向或三向应力状态时,只要畸变能密度(即单位体积畸变能)等于或超过相同材料单向拉伸屈服畸变能密度,材料即发生失效。该理论是基于事实:塑性屈服由与材料形状改变有关的最大畸变能引起。失效准则可表示为

$$u_d = \frac{(1+\mu)}{6E}[(\sigma_1-\sigma_2)^2+(\sigma_2-\sigma_3)^2+(\sigma_3-\sigma_1)^2]=\frac{(1+\mu)}{6E}2\sigma_s^2 \qquad (7.32)$$

式中,$u_d$ 为对应二向或三向应力状态的畸变能密度;$E$ 为弹性模量。上述准则也可表示为

$$\sigma_{eq4} = \sqrt{\frac{1}{2}[(\sigma_1-\sigma_2)^2+(\sigma_2-\sigma_3)^2+(\sigma_3-\sigma_1)^2]}=\sigma_s \qquad (7.33)$$

式中,$\sigma_{eq4}$ 为对应最大畸变能理论的相当应力。因此,二向或三向应力状态下的强度条件可表示为

$$\sigma_{eq4} = \sqrt{\frac{1}{2}[(\sigma_1-\sigma_2)^2+(\sigma_2-\sigma_3)^2+(\sigma_3-\sigma_1)^2]} \leq \sigma_{allow} = \frac{\sigma_s}{n} \qquad (7.34)$$

式中,$\sigma_{allow}$ 为许用应力;$n$ 为安全因数。最大畸变能理论与塑性屈服非常吻合。

**例 7.7** 如图 7.22 所示平面应力状态出现在由钢制成的构件中,已知屈服强度 $\sigma_s$=310 MPa。采用最大畸变能理论,判断屈服是否发生。如果屈服不发生,求安全因数。

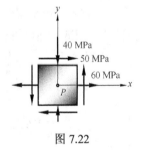

图 7.22

**解** 参考例 7.2,有

$$\sigma_1 = 80.71 \text{ MPa}, \quad \sigma_2 = 0, \quad \sigma_3 = -60.71 \text{ MPa}$$

因 $\sigma_{eq4} = \sqrt{\frac{1}{2}[(\sigma_1-\sigma_2)^2+(\sigma_2-\sigma_3)^2+(\sigma_3-\sigma_1)^2]} = 122.88 \text{ MPa} < \sigma_s$,故屈服不发生,相应的安全因数为

$$n = \frac{\sigma_s}{\sigma_{eq4}} = 2.52$$

# 习 题

7.1 对于如图 7.23 所示应力状态，求单元体逆时针旋转 30°后的正应力和剪应力。

7.2 对于如图 7.24 所示应力状态，求单元体顺时针旋转 15°后的正应力和剪应力。

7.3 对于如图 7.25 所示应力状态，求单元体顺时针旋转 55°后的正应力和剪应力。

7.4 对于如图 7.26 所示应力状态，求单元体逆时针旋转 45°后的正应力和剪应力。

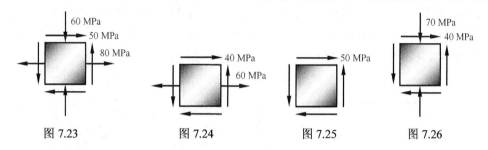

图 7.23　　　　图 7.24　　　　图 7.25　　　　图 7.26

7.5 对于如图 7.27 所示应力状态，求主应力和相应的主平面。

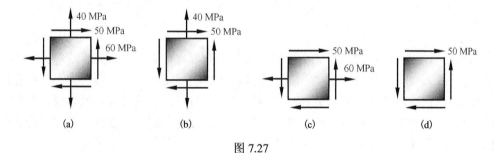

(a)　　　　(b)　　　　(c)　　　　(d)

图 7.27

7.6 对于如图 7.28 所示应力状态，求：(1)最大面内剪应力；(2)最大面内剪应力面的方位；(3)最大面内剪应力面上的正应力。

7.7 边长为 40 mm 的方板受如图 7.29 所示平面应力作用。已知 $E$=200 GPa 和 $\mu$=0.30，求边 $AB$、边 $BC$ 和对角线 $AC$ 的长度改变。

7.8 对于如图 7.30 所示平面应力状态出现在由钢制成的构件中，已知屈服强度 $\sigma_s$=310 MPa。采用最大剪应力理论，判断屈服是否发生。如果屈服不发生，求安全因数。

7.9 对于如图 7.31 所示平面应力状态出现在由钢制成的构件中，已知屈服强度 $\sigma_s$=310 MPa。采用最大畸变能理论，判断屈服是否发生。如果屈服不发生，求安全因数。

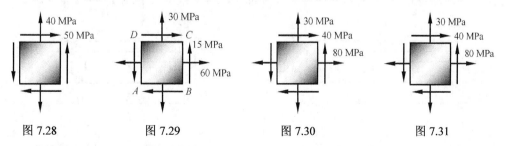

图 7.28　　　　图 7.29　　　　图 7.30　　　　图 7.31

# 第 8 章 组 合 载 荷

前面各章讨论了受单一内力作用的构件，例如，分析了拉压杆、扭转轴和弯曲梁，同时阐明了每一内力作用下的应力和应变求解方法。

然而，在许多结构中，构件常常受两种或两种以上的内力，这些同时作用于构件的内力称为组合载荷或组合加载。例如，图 8.1 中梁 ADB 的 AD 段同时承受弯矩和轴力的作用，图 8.2 中曲柄 ABC 的 AB 段同时承受扭矩和弯矩的作用。

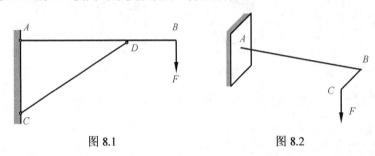

图 8.1　　　　　　　　图 8.2

组合载荷作用下构件的应力分布可以通过叠加原理进行确定。首先求解每个内力单独作用引起的应力分布，然后将这些分布应力进行叠加即可得到合成分布应力。只要在应力和载荷之间存在线性关系，那么叠加原理都能使用。另外，在使用叠加原理时，构件受载时在几何方面仅发生小变形以保证每个载荷所产生的应力之间相互独立。常用结构都满足上述条件，因此叠加原理广泛应用于工程领域。

## 8.1　偏心拉压杆

如果外载荷作用线通过截面形心，则轴向加载杆横截面上的应力可以假设均匀分布。这种加载方式称为中心加载。

现在考虑外载荷作用线不通过截面形心时的应力分布。当外载荷作用线不通过截面形心时，称为偏心加载，如图 8.3 所示。

假设外载荷作用于杆件对称面，设偏心距为 $e$，那么作用于任何横截面上的内力包括通过截面形心的轴力 $N=F$ 和杆件对称面内的弯矩 $M=Fe$，如图 8.4 所示。

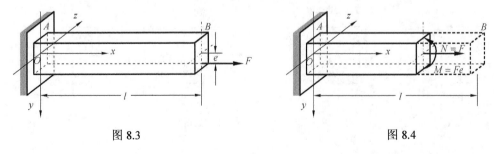

图 8.3　　　　　　　　图 8.4

如果材料满足线弹性条件，并发生小变形，那么由于偏心加载引起的应力分布可以通过叠加对应于轴力 $N$ 的均匀应力分布 $\sigma_N$ 和对应于弯矩 $M$ 的线性应力分布 $\sigma_M$ 获得

$$\sigma = \sigma_N + \sigma_M = \frac{N}{A} + \frac{My}{I} \tag{8.1}$$

式中，$A$ 为横截面面积；$I$ 为横截面的形心惯性矩；$y$ 为横截面上点到形心轴的距离。

根据横截面几何和外载荷偏心，合成应力可以为正和负，或有相同符号。根据式(8.1)，截面上存在一根线，满足 $\sigma = 0$。这根线表示截面的中性轴。需要注意的是中性轴与形心轴不再重合。

合成正应力在下表面有最大值 $\sigma_{max}$，在上表面有最小值 $\sigma_{min}$，这些应力可表示为

$$\sigma_{max,min} = \frac{N}{A} \pm \frac{My_{max}}{I} = \frac{N}{A} \pm \frac{M}{S} \tag{8.2}$$

式中，$y_{max}$ 为上、下表面点到形心轴的距离；$S = I/y_{max}$ 为相对形心轴的截面模量。

上述分析表明，截面下、上表面点具有最大、最小应力。这些点称为临界点或危险点。假设许用应力为 $\sigma_{allow}$，则偏心拉压杆的强度条件为

$$|\sigma_{max,min}| = \left|\frac{N}{A} \pm \frac{M}{S}\right| \leqslant \sigma_{allow} \tag{8.3}$$

**例 8.1**  $F = 50$ kN 的两个外载荷作用于 20a 工字钢梁的自由端（图 8.5）。已知 $e = 80$ mm，求梁下表面的应力。

**解**  梁截面上的轴力和弯矩分别表示为

$N = 2F = 100$ kN，$M = Fe = 4$ kN·m

查阅附录 II 型钢表，20a 工字钢的横截面面积和截面模量分别为

$A = 35.578$ cm$^2$，$S = 237$ cm$^3$

利用式(8.2)，得梁下表面的应力为

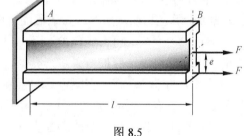

图 8.5

$$\sigma = \sigma_N + \sigma_M = \frac{N}{A} + \frac{M}{S} = 44.98 \text{ MPa}$$

## 8.2 横力弯曲工字梁

根据梁截面形状和弯矩取最大值处临界截面（危险截面）上的剪力值，最大应力可能不在截面的上、下表面，而是在截面上其他点。工字梁腹板和翼缘交界处有较大正应力与较大剪应力，当这两个应力同时作用时，在交界处产生的应力往往大于工字梁表面的应力。

考虑如图 8.6 所示受横向载荷作用的工字梁 $AB$，整个梁上剪力保持不变，而弯矩在固定端达到最大，因此危险截面位于固定端，并且危险截面上的内力可表示为

$$\begin{aligned} V &= F \\ M_{max} &= Fl \end{aligned} \tag{8.4}$$

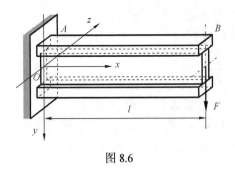

图 8.6

在危险截面上，剪力引起的剪应力在工字梁中性轴上达到最大值，并且最大剪应力可表示为

$$\tau_{max} = \frac{VQ'_{max}}{Ib} = \frac{V}{A_w} \tag{8.5}$$

式中，$I$ 为工字型截面对中性轴的惯性矩；$Q'_{max}$ 为中性轴上方（或下方）截面面积对中性轴的静矩；$b$ 为腹板宽度；$A_w$ 为腹板横截面面积。

在危险截面上，弯矩引起的正应力在上表面或下表面达到最大，最大应力可表示为

$$\sigma_{max} = \frac{M_{max} y_{max}}{I} = \frac{M_{max}}{S} \tag{8.6}$$

式中，$y_{max}$ 为上、下表面点到中性轴的距离；$S = I/y_{max}$ 为对中性轴的截面模量。

在危险截面的任何其他点，材料同时受正应力和剪应力作用，其值分别为

$$\sigma = \frac{M_{max} y}{I} \tag{8.7}$$

$$\tau = \frac{VQ'}{Ib} \tag{8.8}$$

式中，$y$ 为到中性轴的距离；$Q'$ 为需要计算应力的点的上侧或下侧的截面对中性轴的静矩；$b$ 为需要计算应力的点所在位置的截面宽度。

通过应力变换，可得到危险截面上任一点的主应力。对于工字梁，较大的剪应力和较大的主应力都出现在腹板与翼缘交界处，因此在交界处主应力会大于上、下表面处的应力和中性轴上的应力。在工字梁设计中特别要注意这种可能性，因此应该计算腹板和翼缘交界处 $a$ 点与 $b$ 点的主应力，如图 8.7 所示。

利用应力变换公式，危险截面上 $a$ 点和 $b$ 点的主应力可表示为

$$\begin{aligned} \sigma_{1,3} &= \frac{\sigma}{2} \pm \sqrt{\left(\frac{\sigma}{2}\right)^2 + \tau^2} \quad (a\text{点}) \\ \sigma_{1,3} &= -\frac{\sigma}{2} \pm \sqrt{\left(\frac{\sigma}{2}\right)^2 + \tau^2} \quad (b\text{点}) \end{aligned} \tag{8.9}$$

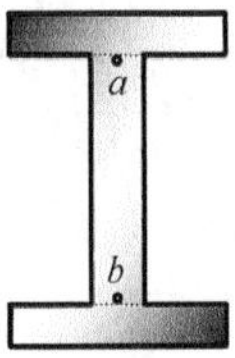

图 8.7

对于塑性材料，应该采用最大剪应力理论或最大畸变能理论进行工字梁设计。

如果采用最大剪应力理论，则 $a$ 点或 $b$ 点的相当应力可写为

$$\sigma_{eq3} = \sigma_1 - \sigma_3 = \sqrt{\sigma^2 + 4\tau^2} \tag{8.10}$$

式中，$\sigma_{eq3}$ 为对应最大剪应力理论的相当应力。假设材料许用应力为 $\sigma_{allow}$，则横力弯曲工字梁上交接点处的最大剪应力强度条件可表示为

$$\sigma_{eq3} = \sqrt{\sigma^2 + 4\tau^2} \leq \sigma_{allow} \tag{8.11}$$

如果采用最大畸变能理论，则横力弯曲工字梁上交接点处的强度条件可表示为

$$\sigma_{\mathrm{eq4}} = \sqrt{\frac{1}{2}[(\sigma_1-\sigma_2)^2+(\sigma_2-\sigma_3)^2+(\sigma_3-\sigma_1)^2]} = \sqrt{\sigma^2+3\tau^2} \leqslant \sigma_{\mathrm{allow}} \qquad (8.12)$$

式中，$\sigma_{\mathrm{eq4}}$ 为对应最大畸变能理论的相当应力。

**例 8.2** $F=150$ kN 的外载荷施加到工字钢梁的自由端，钢的屈服强度为 $\sigma_s=235$ MPa（图 8.8）。利用最大剪应力理论，确定屈服是否发生。如果屈服不发生，求安全因数。忽略倒角效应。

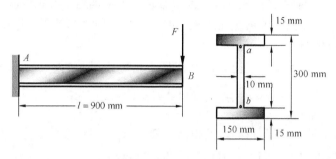

图 8.8

**解** 危险截面位于固定端，弯矩和剪力分别为

$$M_{\max} = Fl = 135 \text{ kN} \cdot \text{m}, \quad V = F = 150 \text{ kN}$$

截面惯性矩等于

$$I = \frac{1}{12}\times(150)\times(300)^3 - \frac{1}{12}\times(140)\times(270)^3 = 1.07865\times10^8 \text{ (mm}^4\text{)}$$

固定端截面上最大正应力和剪应力为

$$\sigma_{\max} = \frac{M_{\max} y_{\max}}{I} = 187.73 \text{ MPa}, \quad \tau_{\max} = \frac{V}{A_w} = 55.56 \text{ MPa}$$

固定端截面上 $a$ 点（或 $b$ 点）的正应力和剪应力为

$$\sigma_a = \frac{y_a}{y_{\max}}\sigma_{\max} = 168.96 \text{ MPa}, \quad \tau_a \approx \tau_{\max} = \frac{V}{A_w} = 55.56 \text{ MPa}$$

对应于最大剪应力理论的相当应力为

$$\sigma_{\mathrm{eq3}} = \sqrt{\sigma_a^2 + 4\tau_a^2} = 202.23 \text{ MPa}$$

由 $\sigma_{\mathrm{eq3}} > \sigma_{\max}$，得危险点位于 $a$ 点（或 $b$ 点）。因 $\sigma_{\mathrm{eq3}} < \sigma_s$，故不会发生屈服，相应的安全因数等于

$$n = \frac{\sigma_s}{\sigma_{\mathrm{eq3}}} = 1.16$$

## 8.3 拉压弯曲梁

考虑如图 8.9 所示矩形截面梁，只要材料保持线弹性和发生小变形，则可通过叠加原理求组合载荷作用下梁的合成应力分布。

危险截面位于固定端，危险截面上的内力为

$$V = F_2, \quad M_{max} = F_2 l, \quad N = F_1 \tag{8.13}$$

危险截面上由剪力引起的剪应力在矩形截面形心轴上将达到最大值，最大剪应力为

$$\tau_{max} = \frac{V Q'_{max}}{I b} = \frac{3V}{2A} \tag{8.14}$$

式中，$I$ 为截面对形心轴的惯性矩；$Q'_{max}$ 为形心轴上方截面对形心轴的静矩；$b$ 为矩形截面梁的宽度；$A$ 为矩形截面梁的横截面面积。剪力引起的应力与轴力或弯矩引起的应力相比非常小，因而可以忽略。

危险截面上由弯矩引起的正应力在上表面或下表面达到最大，最大值可表示为

$$\sigma_M = \frac{M_{max} y_{max}}{I} = \frac{M_{max}}{S} \tag{8.15}$$

式中，$y_{max}$ 为上、下表面点到形心轴的距离；$S = I/y_{max}$ 为对形心轴的截面模量。

危险截面上由轴力引起的正应力均匀分布，可表示为

$$\sigma_N = \frac{N}{A} \tag{8.16}$$

式中，$A$ 为矩形截面梁的横截面面积。

综上所述，危险截面的上、下表面点将有最大或最小应力，如图 8.10 所示。最大或最小值为

$$\sigma_{max,min} = \sigma_N \pm \sigma_M = \frac{N}{A} \pm \frac{M_{max}}{S} \tag{8.17}$$

如果许用应力为 $\sigma_{allow}$，则强度条件可表示为

$$\left| \sigma_{max,min} \right| = \left| \frac{N}{A} \pm \frac{M_{max}}{S} \right| \leq \sigma_{allow} \tag{8.18}$$

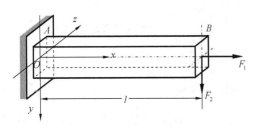

图 8.9

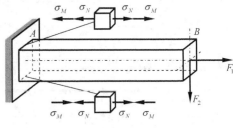

图 8.10

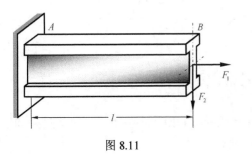

图 8.11

**例 8.3**　$F_1$=100 kN 和 $F_2$=4 kN 的两个外载荷作用于 20a 工字钢梁的自由端(图 8.11)。已知 $l$=1 m，求梁中最大拉应力。

**解**　最大拉应力出现在梁的固定端上表面。固定端的轴力和弯矩为

$$N = F_1 = 100 \text{ kN}, \quad M_{max} = F_2 l = 4 \text{ kN} \cdot \text{m}$$

查 20a 工字钢梁，得
$$A = 35.578 \text{ cm}^2, \quad S = 237 \text{ cm}^3$$
利用式(8.17)，得
$$\sigma_{\max}^{t} = \sigma_N + \sigma_M = \frac{N}{A} + \frac{M_{\max}}{S} = 44.98 \text{ MPa}$$

## 8.4 弯曲扭转轴

考虑如图 8.12 所示圆轴，设材料保持线弹性和发生小变形，那么可以通过叠加原理求组合载荷作用构件的合成应力。

危险截面 $A$ 处内力为
$$V = F, \quad M_{\max} = Fl, \quad T = M_e \tag{8.19}$$

危险截面上由剪力引起的剪应力在圆截面中性轴上有最大值为
$$\tau_V = \frac{V Q'_{\max}}{I b} = \frac{4V}{3A} \tag{8.20}$$

式中，$I$ 为截面对中性轴的惯性矩；$Q'_{\max}$ 为中性轴上方截面对中性轴的静矩；$b$ 为圆轴直径；$A$ 为圆轴横截面面积。剪力引起的应力可以忽略不计。因此，仅需要考虑由扭矩和弯矩引起的应力。

危险截面上由弯矩引起的正应力在上、下表面有最大绝对值为
$$\sigma_M = \frac{M_{\max} y_{\max}}{I} = \frac{M_{\max}}{S} \tag{8.21}$$

式中，$y_{\max}$ 为上、下表面点到中性轴的距离；$S = I/y_{\max}$ 为截面模量。

危险截面上由扭矩引起的剪应力在圆轴外表面达到最大值为
$$\tau_T = \frac{T_{\max} \rho_{\max}}{I_p} = \frac{T_{\max}}{S_p} \tag{8.22}$$

式中，$I_p$ 为圆截面对形心的极惯性矩；$\rho_{\max}$ 为圆截面半径；$S_p = I_p/\rho_{\max}$ 为极截面模量。

显然，危险截面上危险点 $P_1$ 和 $P_2$ 处于平面应力状态，如图 8.13 所示。

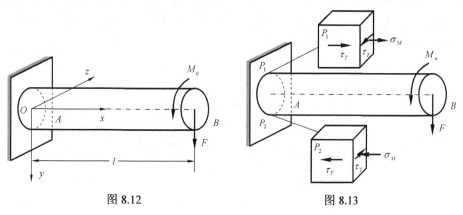

图 8.12　　　　　　图 8.13

危险点 $P_1$ 和 $P_2$ 的主应力可分别表示为

$$\sigma_{1,3} = \frac{\sigma_M}{2} \pm \sqrt{\left(\frac{\sigma_M}{2}\right)^2 + \tau_T^2} \quad (P_1 \text{点})$$

$$\sigma_{1,3} = -\frac{\sigma_M}{2} \pm \sqrt{\left(\frac{\sigma_M}{2}\right)^2 + \tau_T^2} \quad (P_2 \text{点})$$
(8.23)

弯扭组合作用的轴通常由塑性材料制成，因此应该采用最大剪应力理论或最大畸变能理论进行轴的设计。

对塑性材料，如果采用最大剪应力理论，则危险点 $P_1$ 和 $P_2$ 的相当应力为

$$\sigma_{eq3} = \sigma_1 - \sigma_3 = \sqrt{\sigma_M^2 + 4\tau_T^2} \tag{8.24}$$

式中，$\sigma_{eq3}$ 为对应最大剪应力理论的相当应力。

利用式(8.21)和式(8.22)，以及 $S_p = 2S$，得

$$\sigma_{eq3} = \frac{\sqrt{M_{max}^2 + T_{max}^2}}{S} \tag{8.25}$$

式中，对直径为 $d$ 的实心圆轴，$S = \frac{1}{32}\pi d^3$；对外径为 $D$、内径为 $d$ 的空心圆轴，$S = \frac{1}{32}\pi D^3(1-\alpha^4)$，其中 $\alpha = d/D$。

假设许用应力为 $\sigma_{allow}$，则最大剪应力强度条件可表示为

$$\sigma_{eq3} = \frac{\sqrt{M_{max}^2 + T_{max}^2}}{S} \leqslant \sigma_{allow} \tag{8.26}$$

对于塑性材料，如果采用最大畸变能理论，则强度条件为

$$\sigma_{eq4} = \sqrt{\frac{1}{2}[(\sigma_1-\sigma_2)^2 + (\sigma_2-\sigma_3)^2 + (\sigma_3-\sigma_1)^2]} = \sqrt{\sigma_M^2 + 3\tau_T^2} = \frac{\sqrt{M_{max}^2 + \frac{3}{4}T_{max}^2}}{S} \leqslant \sigma_{allow} \tag{8.27}$$

式中，$\sigma_{eq4}$ 为对应最大畸变能理论的相当应力。

**例 8.4** 如图 8.14 所示钢管，屈服强度 $\sigma_s$=325 MPa，长度 $l$=200 mm，外径 $D$=80 mm，壁厚 $t$=5 mm，受载荷 $F$=6 kN 和 $M_e$=1.5 kN·m 作用。采用最大剪应力理论，判断屈服是否发生。如果不发生屈服，求安全因数。

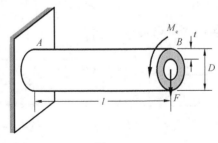

图 8.14

**解** 危险点位于固定端上、下表面点。固定端的弯矩和扭矩为

$$M_{\max} = Fl = 1.2 \text{ kN·m}, \quad T = M_e = 1.5 \text{ kN·m}$$

利用式(8.26)，得

$$\sigma_{eq3} = \frac{\sqrt{M_{\max}^2 + T^2}}{S} = \frac{\sqrt{M_{\max}^2 + T^2}}{\frac{1}{32}\pi D^3 \{1-[(D-2t)/D]^4\}} = 92.35 \text{ MPa}$$

因 $\sigma_{eq3} < \sigma_s$，故不会发生屈服，相应的安全因数为

$$n = \frac{\sigma_s}{\sigma_{eq3}} = 3.52$$

## 习　题

8.1　$F=80$ kN 的外载荷作用于22a工字钢梁的自由端(图8.15)。已知偏心距 $e=80$ mm，求梁上、下表面的应力。

8.2　$F=80$ kN 的三个外载荷作用于 22a 工字钢梁的自由端(图 8.16)。已知偏心距 $e=80$ mm，求梁中最大拉应力、压应力。

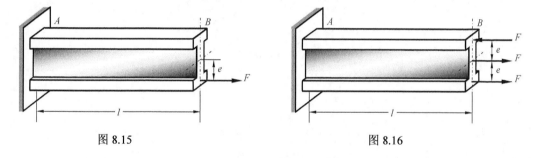

图 8.15　　　　　　　　　　　图 8.16

8.3　$F=180$ kN 的外载荷作用于宽缘工字钢梁的自由端(图8.17)。已知屈服强度 $\sigma_s=235$ MPa，采用最大畸变能理论判断屈服是否发生。如果不发生屈服，求安全因数。

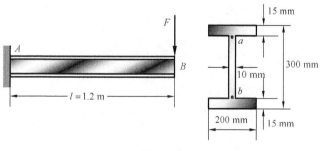

图 8.17

8.4　外载荷 $F_1=100$ kN 和 $F_2=20$ kN 作用于36a工字钢梁的自由端(图8.18)。已知屈服强度 $\sigma_s=235$ MPa，利用最大剪应力理论，确定屈服是否发生。如果屈服不发生，求安全因数。

8.5 如图 8.19 所示钢管，屈服强度 $\sigma_s=325$ MPa，长度 $l=200$ mm，外径 $D=80$ mm，壁厚 $t=5$ mm，承受载荷 $F=6$ kN 和 $M_e=1.5$ kN·m 作用。采用最大畸变能理论，判断屈服是否发生。如果屈服不发生，求安全因数。

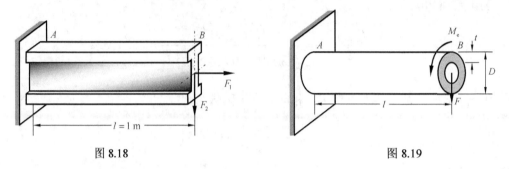

图 8.18　　　　　　　　　　图 8.19

8.6 悬臂工字钢梁承受如图 8.20 所示外载荷。已知 $F=200$ kN 和 $\sigma_{\text{allow}}=180$ MPa，求：(1) 梁中最大正应力；(2) 腹板和翼缘交界处最大主应力；(3) 工字钢梁是否合适。

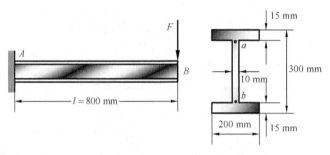

图 8.20

# 第9章 压杆稳定

进行构件设计时，必须要保证所设计的构件满足强度、刚度和稳定要求。前面各章讨论了构件的强度和刚度。本章将讨论构件的稳定，即构件承受轴向压缩载荷而不发生突然横向挠曲。

承受轴向压缩的细长构件称为压杆。压杆失效是通过失稳(即突然横向挠曲)而发生的。压杆失稳甚至在最大应力远小于屈服强度和极限强度时也能发生。

压杆失稳常常导致结构突然失效，因此需要特别关注压杆设计问题，以便所设计的压杆能够安全支撑给定载荷而不发生失稳现象。

本章仅考虑理想压杆。理想压杆是指受载前压杆的轴线为直线，压杆由均匀材料制成，载荷通过截面形心沿轴线进行加载。

## 9.1 两端铰支细长压杆临界载荷

压杆失稳前所能承受的最大轴向压力称为临界载荷，用 $F_{cr}$ 表示。图 9.1(a)中压杆两端铰支受轴向载荷作用。压杆在受载后会发生失稳，即发生突然横向挠曲，如图 9.1(b)所示。

考虑如图 9.1(c)所示 $AC$ 段平衡，则 $C$ 截面上的弯矩可表示为 $M = -Fw$。利用 $w'' = M/EI$，得

$$w'' + \frac{F}{EI}w = 0 \qquad (9.1)$$

式(9.1)是常系数线性二阶齐次微分方程。设

$$k^2 = \frac{F}{EI} \qquad (9.2)$$

则

$$w'' + k^2 w = 0 \qquad (9.3)$$

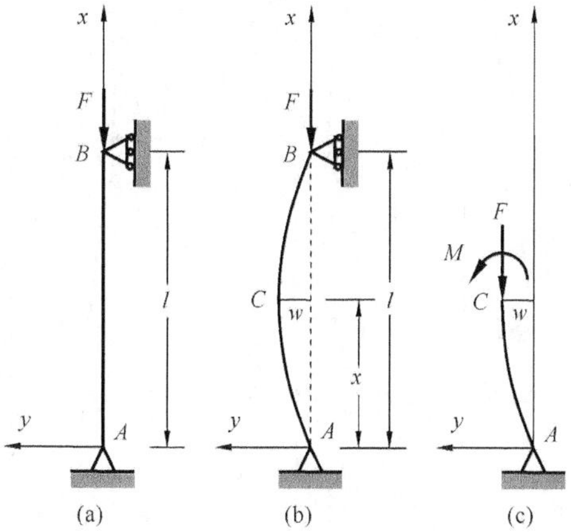

上述方程的通解可表示为

图 9.1

$$w = A\sin(kx) + B\cos(kx) \qquad (9.4)$$

利用 $x=0, w=0$，得 $B=0$；再利用 $x=l, w=0$，得

$$A\sin(kl) = 0 \qquad (9.5)$$

上述方程满足的条件是 $A=0$ 或 $\sin(kx)=0$。如果 $A=0$，得到 $w \equiv 0$，即压杆始终是直杆。如果 $\sin(kx)=0$，得 $kl=n\pi$，即

$$F = \frac{n^2\pi^2 EI}{l^2} \qquad (9.6)$$

式(9.6)的最小值对应于 $n=1$，因此得临界载荷为

$$F_{cr} = \frac{\pi^2 EI}{l^2} \tag{9.7}$$

式(9.7)称为欧拉公式。对于圆形或方形横截面，相对于任意形心轴的惯性矩 $I$ 都相同，因此压杆会围绕任意形心轴而发生失稳。然而，对于其他截面形状，式(9.7)中的惯性矩是指最小惯性矩，即取 $I = I_{min}$ 计算临界载荷。

## 9.2 其他支撑细长压杆临界载荷

欧拉公式通过两端铰支压杆推导而来，下面确定其他支撑条件下的临界载荷。考虑如图 9.2(a) 所示一端固支一端自由压杆。该压杆等效于图 9.2(b) 中的两端铰支压杆的上半部分，因此图 9.2(a) 中压杆的临界载荷与图 9.2(b) 中压杆相同，其临界载荷可取二倍杆长通过欧拉公式进行计算而得到。由此可得临界载荷为

$$F_{cr} = \frac{\pi^2 EI}{(2l)^2} = \frac{\pi^2 EI}{(\mu l)^2} = \frac{\pi^2 EI}{l_e^2} \tag{9.8}$$

式中，$\mu$ 为有效长度因数；$l_e = \mu l$ 为有效长度。式(9.8)是欧拉公式的延伸，因此也称为欧拉公式。

对应各种支撑的有效长度因数和有效长度如图 9.3 所示。

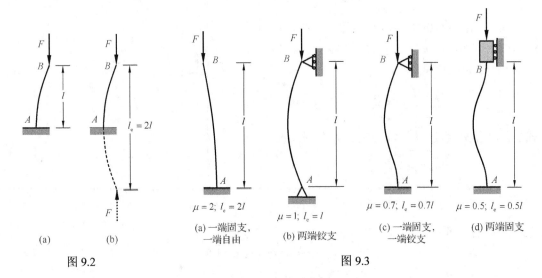

(a)    (b)

图 9.2

(a) 一端固支，一端自由    (b) 两端铰支    (c) 一端固支，一端铰支    (d) 两端固支

图 9.3

## 9.3 细长压杆临界应力

对应临界载荷的应力称为临界应力，由 $\sigma_{cr}$ 表示。利用欧拉公式并设 $i^2 = I/A$（其中 $A$ 为横截面面积，$i$ 为惯性半径），得

$$\sigma_{cr} = \frac{F_{cr}}{A} = \frac{\pi^2 E}{(\mu l/i)^2} = \frac{\pi^2 E}{\lambda^2} \tag{9.9}$$

式中，$\lambda = \mu l/i$ 为压杆的长细比。式(9.9)也称为欧拉公式。欧拉公式表明，临界应力正比于弹性模量，而反比于长细比的平方。

欧拉公式是在临界应力不超过比例极限的前提之下推导而来的，即

$$\sigma_{cr} = \frac{\pi^2 E}{\lambda^2} \leqslant \sigma_p \tag{9.10}$$

定义 $\lambda_p = \sqrt{\pi^2 E/\sigma_p}$，则上述关系可表示为

$$\lambda \geqslant \lambda_p \tag{9.11}$$

式中，$\lambda_p$ 为临界长细比。式(9.11)表明，欧拉公式仅当 $\lambda \geqslant \lambda_p$ 时才成立。对于低碳钢(Q235)，$E=206$ GPa 和 $\sigma_p=200$ MPa，则 $\lambda_p=100.8$；对于铝合金，$E=70$ GPa 和 $\sigma_p=175$ MPa，则 $\lambda_p=62.8$。

**例 9.1** 外径和内径分别为 30 mm 和 20 mm 的细长压杆由钢管制成，如图 9.4 所示。已知 $l=1.5$ m 和 $E=200$ GPa，求该支撑条件下临界载荷和临界应力。

**解** 从中间截面把压杆截开，则压杆分为 $BC$ 和 $AC$ 两段，每段都是一端固支、一端自由的压杆。因此，图9.4中压杆的临界载荷就等于任意一段压杆的临界载荷，利用 $\mu_{BC} = \mu_{AC} = 2$ 和 $l_{BC} = l_{AC} = \frac{1}{2} l$，得

$$F_{cr} = (F_{cr})_{BC} = (F_{cr})_{AC} = \frac{\pi^2 EI}{(\mu_{AC} l_{AC})^2}$$

$$= \frac{\pi^2 E \left[ \frac{1}{64} \pi (D^4 - d^4) \right]}{l^2} = 27.99 \text{ kN}$$

利用式(9.9)，临界应力为

$$\sigma_{cr} = \frac{F_{cr}}{A} = \frac{F_{cr}}{\frac{1}{4} \pi (D^2 - d^2)} = 71.28 \text{ MPa}$$

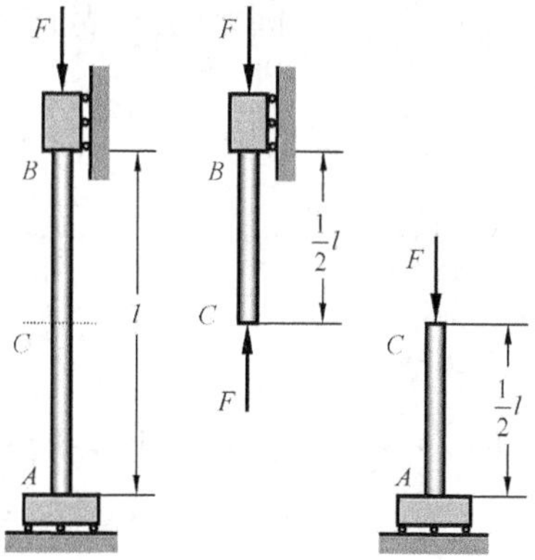

图 9.4

## 9.4 中长压杆临界应力

前面所讨论的压杆都假设在受载前轴线为直线，压杆由均匀材料制成，应力不超过比例极限。然而，实际压杆达不到这样的理想情况，因此实际上压杆的设计往往采用经验公式。

对于细长压杆，长细比较大，压杆失效由欧拉公式确定，临界应力与弹性模量有关，而与屈服强度或极限强度无关。

对于粗短压杆，压杆失效由屈服或断裂引起。对塑性材料，临界应力等于屈服强度；而对脆性材料，临界应力等于极限强度。

对于中长压杆，压杆失效与弹性模量和屈服强度(或极限强度)有关。中长压杆的失效是非常复杂的现象，因而通常采用经验公式进行中心受载中长压杆的设计。最简单的经验

公式是线性公式，可表示为

$$\sigma_{cr} = a - b\lambda \tag{9.12}$$

式中，$a$ 和 $b$ 为与所用材料有关的常数。对于低碳钢，$a=304$ MPa 和 $b=1.12$ MPa；对铸铁，$a=332.2$ MPa 和 $b=1.454$ MPa。

**例 9.2** 20a 工字钢压杆在自由端承受中心压缩载荷（图 9.5）。已知 $\sigma_p=200$ MPa、$E=206$ GPa、$a=304$ MPa 和 $b=1.12$，求压杆长度分别为 (1) $l=1.2$ m、(2) $l=1.0$ m 时的临界载荷。

**解** 对给定压杆，有效长度因数 $\mu=2.0$，临界长细比为

$$\lambda_p = \sqrt{\frac{\pi^2 E}{\sigma_p}} = 101$$

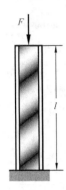

图 9.5

查阅附录 II 型钢表，20a 工字钢压杆的横截面面积和最小惯性矩分别为

$$A = 35.578 \text{ cm}^2, \quad I_{min} = 158 \text{ cm}^4$$

(1) 长细比为

$$\lambda = \frac{\mu l}{i} = \frac{\mu l}{\sqrt{I_{min}/A}} = 114$$

因 $\lambda > \lambda_p$，故应该选用欧拉公式求解临界载荷，即

$$F_{cr} = A\sigma_{cr} = A\frac{\pi^2 E}{\lambda^2} = 557 \text{ kN}$$

(2) 长细比为

$$\lambda = \frac{\mu l}{\sqrt{I_{min}/A}} = 95$$

因 $\lambda < \lambda_p$，故应该选用经验公式求解临界载荷，即

$$F_{cr} = A\sigma_{cr} = A(a - b\lambda) = 703 \text{ kN}$$

## 9.5 压杆设计

直线经验公式通常用于中长压杆的设计，而欧拉公式则用于细长压杆的设计。因此，用于细长、中长和粗短压杆设计的三个临界应力公式表示如下：

$$\begin{aligned}
\sigma_{cr} &= \frac{\pi^2 E}{\lambda^2} & (\lambda \geqslant \lambda_p) \\
\sigma_{cr} &= a - b\lambda & (\lambda_s \leqslant \lambda < \lambda_p) \\
\sigma_{cr} &= \sigma_s & (\lambda < \lambda_s)
\end{aligned} \tag{9.13}$$

式中，$\lambda_s = (a - \sigma_s)/b$；$\lambda_p = \sqrt{\pi^2 E/\sigma_p}$。

临界应力 $\sigma_{cr}$ 与安全因数 $n$ 的比值定义为许用应力 $\sigma_{allow}$。如果压杆稳定，则中心受载压杆的工作应力 $\sigma$ 应该满足如下条件：

$$\sigma \leq \sigma_{\text{allow}} = \frac{\sigma_{\text{cr}}}{n} \tag{9.14}$$

**例 9.3** 外径和壁厚分别为 $D=120$ mm 和 $t=15$ mm 的空心圆形钢杆在自由端承受中心压缩载荷 $F$ 作用(图 9.6)。求压杆长度分别为 1.8 m、1.2 m 和 0.6 m 时的临界载荷。已知 $\sigma_p=280$ MPa、$\sigma_s=350$ MPa、$E=210$ GPa、$a=461$ MPa 和 $b=2.568$ MPa。

**解** 对如图 9.6 所示压杆,有效长度因数 $\mu = 2.0$。对应比例极限和屈服强度的临界长细比分别等于

$$\lambda_p = \sqrt{\frac{\pi^2 E}{\sigma_p}} = 86, \quad \lambda_s = \frac{a - \sigma_s}{b} = 43$$

(1) 长细比为

$$\lambda = \frac{\mu l}{i} = \frac{\mu l}{\sqrt{I/A}} = \frac{\mu l}{\sqrt{\left[\frac{1}{64}\pi(D^4 - d^4)\right]/\left[\frac{1}{4}\pi(D^2 - d^2)\right]}} = \frac{4\mu l}{\sqrt{D^2 + d^2}} = 96$$

因 $\lambda > \lambda_p$,故应采用欧拉公式计算临界载荷。相应的临界载荷为

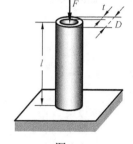

图 9.6

$$F_{\text{cr}} = A\sigma_{\text{cr}} = \frac{1}{4}\pi(D^2 - d^2)\frac{\pi^2 E}{\lambda^2} = 1113 \text{ kN}$$

(2) 长细比为

$$\lambda = \frac{4\mu l}{\sqrt{D^2 + d^2}} = 64$$

因 $\lambda_s < \lambda < \lambda_p$,故应采用经验公式计算临界载荷。相应的临界载荷为

$$F_{\text{cr}} = A\sigma_{\text{cr}} = \frac{1}{4}\pi(D^2 - d^2)(a - b\lambda) = 1468 \text{ kN}$$

(3) 长细比为

$$\lambda = \frac{4\mu l}{\sqrt{D^2 + d^2}} = 32$$

因 $\lambda < \lambda_s$,故应采用强度公式计算临界载荷。相应的临界载荷为

$$F_{\text{cr}} = A\sigma_{\text{cr}} = \frac{1}{4}\pi(D^2 - d^2)\sigma_s = 1732 \text{ kN}$$

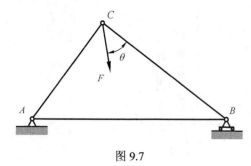

图 9.7

**例 9.4** 平面结构 $ABC$ 由杆件 $AB$、$AC$ 和 $BC$ 分别在节点 $A$、$B$ 和 $C$ 处铰接而成,杆件 $AB$、$AC$ 和 $BC$ 的直径均为 32 mm,长度分别为 1 m、0.6 m 和 0.8 m,节点 $C$ 受集中力 $F$ 作用(图 9.7)。已知 $\sigma_p = 280$ MPa、$\sigma_s = 350$ MPa、$E = 210$ GPa、$a = 461$ MPa、$b = 2.568$ MPa 和 $0 < \theta < 90°$,试求临界载荷 $F_{\text{cr}}$ 及其对应临界角 $\theta_{\text{cr}}$ (注:仅考虑面内失稳)。

**解** $\lambda_p = \sqrt{\dfrac{\pi^2 E}{\sigma_p}} = 86$，$\lambda_s = \dfrac{a-\sigma_s}{b} = 43$，$\lambda_{AC} = \dfrac{4\mu l_{AC}}{d} = 75$，$\lambda_{BC} = \dfrac{4\mu l_{BC}}{d} = 100$

由于 $\lambda_s < \lambda_{AC} < \lambda_p < \lambda_{BC}$，即 AC 和 BC 分别为中长杆和细长杆，故应该采用经验公式和欧拉公式分别计算 AC 杆和 BC 杆的临界载荷，即

$$(F_{AC})_{cr} = \left(\dfrac{1}{4}\pi d^2\right)(a - b\lambda_{AC}) = 215.86 \text{ kN}，\quad (F_{BC})_{cr} = \left(\dfrac{1}{4}\pi d^2\right)\dfrac{\pi^2 E}{\lambda_{BC}^2} = 166.69 \text{ kN}$$

所以临界载荷和相应临界角可表示为

$$F_{cr} = \sqrt{(F_{AC})_{cr}^2 + (F_{BC})_{cr}^2} = 272.73 \text{ kN}，\quad \theta_{cr} = \arctan\dfrac{(F_{AC})_{cr}}{(F_{BC})_{cr}} = 52.32°$$

## 习 题

**9.1** 四根细长压杆由直径 30 mm 的钢杆制成(图 9.8)。已知 $l$=1.5 m 和 $E$=200 GPa，求每一种支撑条件下的临界载荷 $F_{cr}$。

**9.2** 四根细长压杆由外径 30 mm 和壁厚 5 mm 的铝管制成(图 9.9)。利用 $l$=1.2 m、$E$=70 GPa 和安全因数 $n$=2.3，求每一种支撑条件下的许可载荷 $F_{\text{allow}}$。

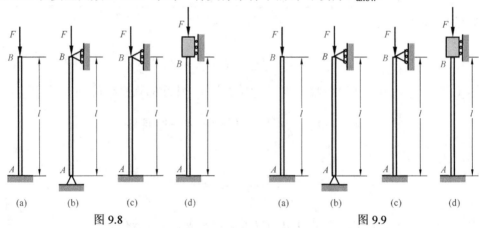

图 9.8　　　　　　图 9.9

**9.3** 直径 $d$=150 mm 的实心圆钢杆在自由端承受中心压缩载荷(图 9.10)，求杆长分别为 2.0 m、1.2 m 和 0.5 m 时的临界载荷。已知 $\sigma_p$=280 MPa、$\sigma_s$=350 MPa、$E$=210 GPa、$a$=461 MPa 和 $b$=2.568 MPa。

**9.4** 25a 工字钢压杆在自由端承受中心压缩载荷(图 9.11)。已知 $\sigma_p$=200 MPa、$E$=206 GPa、$a$=304 MPa 和 $b$=1.12 MPa，求杆长分别为 $l$= 1.5 m 和 $l$=1.0 m 时的临界载荷。

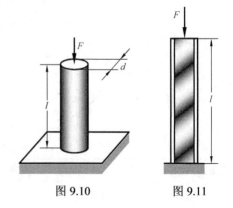

图 9.10　　　图 9.11

# 第 10 章　非对称弯曲

前面讨论的弯曲仅限于梁至少有一个纵向对称面，并且载荷作用于该纵向对称面内，如图 10.1 所示。

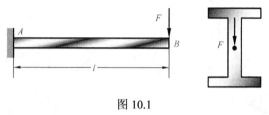

图 10.1

本章将讨论其余两种一般情形：如图 10.2 所示梁没有纵向对称面，或如图 10.3 所示载荷并不作用于纵向对称面。

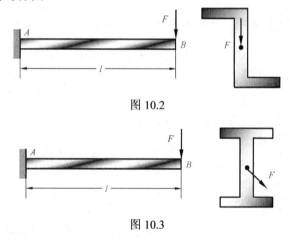

图 10.2

图 10.3

## 10.1　非对称纯弯曲

假设外力偶作用于 xy 面，如图 10.4 所示。

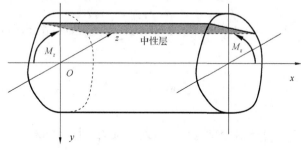

图 10.4

从如图 10.4 所示梁截取线元 $\mathrm{d}x$，如图 10.5 所示，则纵向线应变可表示为

$$\varepsilon = \frac{\eta}{\rho} \tag{10.1}$$

式中，$\eta$ 为面元 $\mathrm{d}A$ 到中性层的距离；$\rho$ 为中性层的曲率半径。

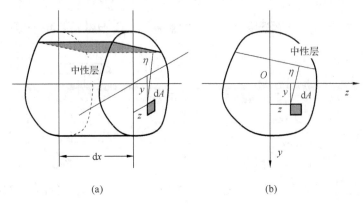

图 10.5

假设正应力低于比例极限，并且单向胡克定律适用，即 $\sigma = E\varepsilon$，则根据式(10.1)，有

$$\sigma = \frac{E}{\rho}\eta \tag{10.2}$$

式中，$E$ 为弹性模量。

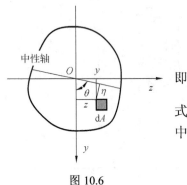

图 10.6

根据 $x$ 方向的平衡，有

$$\sum F_x = 0: N = \int \sigma \mathrm{d}A = \frac{E}{\rho}\int \eta \mathrm{d}A = 0 \tag{10.3}$$

即

$$\int \eta \mathrm{d}A = 0 \tag{10.4}$$

式(10.4)表明，对非对称纯弯曲梁，只要应力不超过弹性范围，中性轴则通过截面形心，如图 10.6 所示。

设 $\theta$ 为中性轴与 $y$ 轴的夹角，如图 10.6 所示，则有

$$\eta = y\sin\theta - z\cos\theta \tag{10.5}$$

根据绕 $y$ 轴的平衡条件，有

$$\sum M_y = 0: M_y = \int z\sigma \mathrm{d}A = \frac{E}{\rho}\int z(y\sin\theta - z\cos\theta)\mathrm{d}A = 0 \tag{10.6}$$

或

$$\tan\theta = \frac{I_y}{I_{yz}} \tag{10.7}$$

式中，$I_y = \int z^2 \mathrm{d}A$ 和 $I_{yz} = \int yz \mathrm{d}A$ 分别为横截面对 $y$ 轴的惯性矩和对 $y$，$z$ 轴的惯性积。

根据绕 $z$ 轴的平衡条件，有

$$\sum M_z = 0: \quad M_z = \int y\sigma \mathrm{d}A = \frac{E}{\rho}\int y(y\sin\theta - z\cos\theta)\mathrm{d}A \tag{10.8}$$

或

$$\frac{1}{\rho} = \frac{M_z}{E\int(y^2\sin\theta - yz\cos\theta)\mathrm{d}A} = \frac{M_z}{E(I_z\sin\theta - I_{yz}\cos\theta)} \tag{10.9}$$

式中，$I_z = \int y^2 \mathrm{d}A$ 为横截面对 $z$ 轴的惯性矩。把式(10.5)中的 $\eta$ 和式(10.9)中的 $1/\rho$ 代入式(10.2)，得

$$\sigma = \frac{M_z(yI_y - zI_{yz})}{I_yI_z - I_{yz}^2} \tag{10.10}$$

如果外力偶分别作用于 $xy$ 平面和 $xz$ 平面，如图 10.7 所示，则有

$$\sigma = \frac{M_z(yI_y - zI_{yz})}{I_yI_z - I_{yz}^2} + \frac{M_y(zI_z - yI_{yz})}{I_yI_z - I_{yz}^2} \tag{10.11}$$

根据式(10.11)，中性轴的取向可表示为

$$\tan\theta = -\frac{M_zI_y - M_yI_{yz}}{M_yI_z - M_zI_{yz}} \tag{10.12}$$

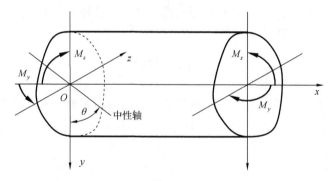

图 10.7

如果外力偶分别作用于形心主惯性面，即 $I_{yz}=0$，则式(10.11)和式(10.12)可简化为

$$\sigma = \frac{M_zy}{I_z} + \frac{M_yz}{I_y}, \quad \tan\theta = -\frac{M_zI_y}{M_yI_z} \tag{10.13}$$

当外力偶并不作用于任何形心主惯性面时，首先把外力偶分解到形心主惯性面，然后用式(10.13)计算每个力偶分量产生的正应力，最后利用叠加原理，确定各点的合成应力。

如果外力偶仅作用于一个形心主惯性面，如 $xy$ 平面，那么式(10.13)可写为

$$\sigma = \frac{M_zy}{I_z}, \quad \theta = 90° \tag{10.14}$$

式(10.14)表明，当外力偶作用于形心主惯性面时，挠度分布是平面曲线，并与形心主惯性面重合。这类弯曲称为平面弯曲。

## 10.2 非对称横力弯曲

从非对称纯弯曲导出的正应力公式仍然适用于非对称横力弯曲，只要非对称横力弯曲梁满足细长条件。然而，对非对称横力弯曲，如图 10.8 所示，该梁在通过截面形心的外力作用下，既要发生弯曲，又要发生扭转。

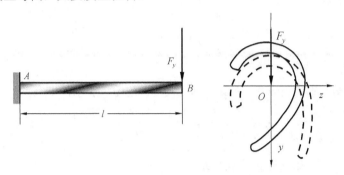

图 10.8

在外力作用下梁只发生弯曲而不发生扭转，这是否可能？如果可能，外力应该作用于何处？如图 10.9 所示，设 $y$ 和 $z$ 轴是形心主惯性轴，并设外力作用于点 $S$ 时薄壁梁只弯不扭，那么截面上任意点的剪应力可表示为

$$\tau = \frac{V_y Q_z'}{I_z t} \tag{10.15}$$

式中，$Q_z'$ 为阴影面积对形心主惯性轴 $z$ 的静矩；$I_z$ 为截面对形心主惯性轴 $z$ 的惯性矩；$V_y$ 为截面上沿 $y$ 轴的剪力；$t$ 为薄壁梁的厚度。

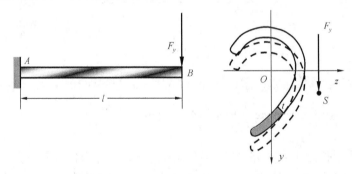

图 10.9

当外力作用于点 $S$ 时，梁将会只弯不扭。点 $S$ 通常称为剪力中心(剪心)或弯曲中心(弯心)。对薄壁梁而言，剪心通常不与截面形心重合。设 $a_z$ 为截面上外力作用线到任意参考点 $a$ 的距离，由图 10.10，得

$$F_y a_z = \int r \tau dA \tag{10.16}$$

式中，$r$ 为面元 $dA$ 上的合力作用线到参考点 $a$ 的距离。

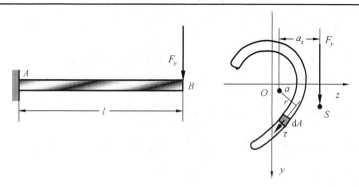

图 10.10

如图 10.11 所示,当外力沿任意方向时,只要外力作用于剪心,则薄壁梁也不发生扭转,因为外力可分解为分量 $F_y$ 和 $F_z$,而这两分量都不会引起扭转。

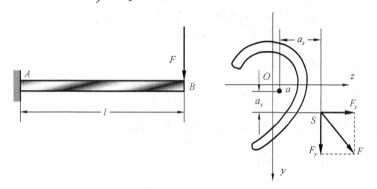

图 10.11

**例 10.1** 如图 10.12 所示,型号为 160×160×16 的角钢悬臂梁自由端受外力作用。已知 $F = 15 \text{ kN}$,$l = 1 \text{ m}$,$E = 200 \text{ GPa}$,试求最大正应力和最大挠度。

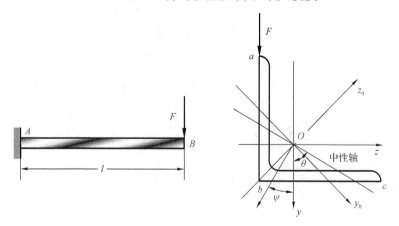

图 10.12

**解** (1)查附录 II 型钢表,有

$$I_y = I_z = 1175.08 \text{ cm}^4, \quad I_{y_0} = 484.59 \text{ cm}^4, \quad I_{z_0} = 1865.57 \text{ cm}^4$$

相对 $Oxyz$ 坐标系的惯性积可表示为

$$I_{yz} = \frac{I_{y_0} - I_{z_0}}{2}\sin(2\alpha) + I_{y_0 z_0}\cos(2\alpha) = 690.49 \text{ cm}^4$$

中性轴的取向可表示为
$$\theta = \arctan\frac{I_y}{I_{yz}} = 59.53°$$

显然，最大拉应力位于截面 $A$ 上点 $a$ 处，最大压应力位于截面 $A$ 上点 $b$ 处。利用 $\sigma = \dfrac{M_z(yI_y - zI_{yz})}{I_y I_z - I_{yz}^2}$，得为

$$\sigma_a = \frac{(-15)\times[(-11.45)\times(1175.08) - (-4.55)\times(690.49)]}{1175.08^2 - 690.49^2} = 171.11(\text{MPa})$$

$$\sigma_b = \frac{(-15)\times[(4.55)\times(1175.08) - (-4.55)\times(690.49)]}{1175.08^2 - 690.49^2} = -140.84(\text{MPa})$$

上述计算表明，最大正应力为
$$\sigma_{\max} = \sigma_a = 171.11 \text{ MPa}$$

(2) 最大正应力也可通过外力向形心主惯性面分解进行计算。根据 $\sigma = \dfrac{M_{z_0} y_0}{I_{z_0}} + \dfrac{M_{y_0} z_0}{I_{y_0}}$，可得截面 $A$ 上点 $a$ 处最大拉应力为

$$\sigma_a = \frac{Fl}{\sqrt{2}}\left[-\frac{(y_0)_a}{I_{z_0}} + \frac{(z_0)_a}{I_{y_0}}\right] = \frac{15}{\sqrt{2}}\times\left(-\frac{-11.31}{1865.57} + \frac{4.88}{484.59}\right) = 171.11(\text{MPa})$$

截面 $A$ 上点 $b$ 处最大压应力为

$$\sigma_b = \frac{Fl}{\sqrt{2}}\left[-\frac{(y_0)_b}{I_{z_0}} + \frac{(z_0)_b}{I_{y_0}}\right] = \frac{15}{\sqrt{2}}\times\left(-\frac{0}{1865.57} + \frac{-6.43}{484.59}\right) = -140.74(\text{MPa})$$

因此，最大正应力为
$$\sigma_{\max} = \sigma_a = 171.11 \text{ MPa}$$

(3) 通过把外力分解到形心主惯性面，进行最大挠度计算。位于形心主惯性面的最大挠度分量分别为

$$w_{y_0} = \frac{(F/\sqrt{2})l^3}{3EI_{z_0}} = 0.95 \text{ mm}, \quad w_{z_0} = -\frac{(F/\sqrt{2})l^3}{3EI_{y_0}} = -3.65 \text{ mm}$$

显然，最大挠度为
$$w_{\max} = \sqrt{w_{y_0}^2 + w_{z_0}^2} = 3.77 \text{ mm}$$

最大挠度方向为
$$\psi = \arctan\frac{(w_{y_0}/\sqrt{2}) + (w_{z_0}/\sqrt{2})}{(w_{y_0}/\sqrt{2}) - (w_{z_0}/\sqrt{2})} = -30.41°$$

**例 10.2** 如图 10.13 所示，厚度为 $t$（$t \ll h$，$t \ll b$）的薄壁悬臂梁在剪心处受外力作用。试求剪应力分布，并确定剪心位置。

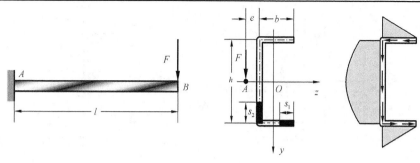

图 10.13

**解** (1)横截面对 $z$ 轴的惯性矩为

$$I_z = \frac{1}{12}th^3 + 2\left[\frac{1}{12}bt^3 + bt\left(\frac{h}{2}\right)^2\right]$$

忽略 $t^3$ 项，得

$$I_z = \frac{1}{12}th^2(h+6b)$$

假设外力作用于剪心，则梁在横力作用下不会发生扭转。上、下翼缘剪应力为

$$\tau = \frac{V_y Q_z'}{I_z t} = \frac{6Fs_1}{th(h+6b)}$$

腹板剪应力为

$$\tau = \frac{V_y Q_z'}{I_z t} = \frac{6F(hb + hs_2 - s_2^2)}{th^2(h+6b)}$$

最大剪应力位于中性轴，数值为

$$\tau_{\max} = \frac{3F(h+4b)}{2th(h+6b)}$$

(2) 设 $e$ 为剪心到腹板中线的距离，则有

$$Fe = \int_0^b h\tau t \mathrm{d}s_1 = \int_0^b \frac{6Fs_1}{h+6b}\mathrm{d}s_1 = \frac{3Fb^2}{h+6b}$$

即

$$e = \frac{3b^2}{h+6b}$$

## 习　题

10.1　厚度为 $t\,(t \ll a)$ 的薄壁简支梁在中间截面受力 $F$ 作用(图 10.14)。试求载荷作用处横截面上正应力分布，并确定最大正应力的位置和数值。

10.2　厚度为 $t\,(t \ll a)$ 的薄壁悬臂梁受均布载荷 $q$ 作用(图 10.15)。试求固定端横截面上正应力分布。

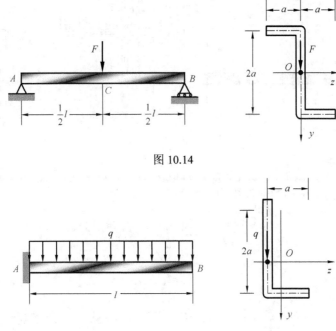

图 10.14

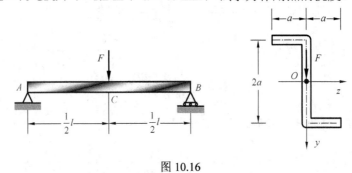

图 10.15

**10.3** 厚度 $t = 2$ mm 的薄壁简支梁在中间截面受力 $F$ 作用（图 10.16）。已知 $F = 1.5$ kN、$E = 70$ GPa、$l = 1.2$ m、$a = 40$ mm，试求力作用点的挠度。

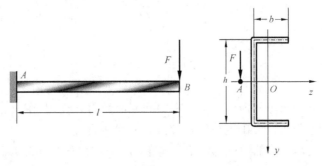

图 10.16

**10.4** 厚度为 $t$（$t \ll h$，$t \ll b$）的薄壁悬臂梁在剪心处受外力作用（图 10.17）。已知弹性模量 $E$，试求固定端截面正应力、剪应力分布和梁自由端挠度。

# 第 11 章 能 量 方 法

## 11.1 外 功

### 1. 外力功

如图 11.1 所示,若作用于物体的外力从零缓慢增加到终值 $F$,则力作用点沿力方向的相应位移也缓慢从零增加到终值 $\delta$。假设材料处于线弹性范围,那么外力对物体所做功可表示为

$$W_\mathrm{e} = \frac{1}{2}F\delta \tag{11.1}$$

### 2. 外力偶功

如图 11.2 所示,当作用于物体的外力偶从零缓慢增加到终值 $M_\mathrm{e}$ 时,力偶的相应角位移也缓慢从零增加到终值 $\theta$。如果材料处于线弹性范围,则外力偶所做功可表示为

$$W_\mathrm{e} = \frac{1}{2}M_\mathrm{e}\theta \tag{11.2}$$

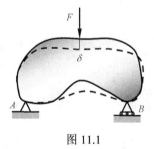

图 11.1

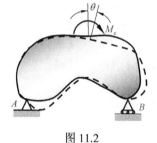

图 11.2

### 3. 多外载荷功

如图 11.3 所示,物体受外力 $F_1, F_2, \cdots, F_i, \cdots, F_n$ 作用,若外力由零开始缓慢作用于物体,直至达到终值,则力作用点沿力方向的相应位移也从零缓慢增加到终值。假设材料处于线弹性范围,则外力所做功可表示为

$$W_\mathrm{e} = \sum \frac{1}{2}F_i \Delta_i \tag{11.3}$$

式中,$\Delta_i$ 为力 $F_1, F_2, \cdots, F_i, \cdots, F_n$ 的函数,即 $\Delta_i = \Delta_i(F_1, F_2, \cdots, F_i, \cdots, F_n)$ $(i=1,2,\cdots,n)$。应该注意,$F_i$ 为广义力,$\Delta_i$ 为相应的广义位移。

## 4. 组合载荷功

如图 11.4 所示，细长圆轴受组合载荷作用。如果是小变形，则组合载荷所做功可表示为

$$W_e = \sum \frac{1}{2} F_i \delta_i \tag{11.4}$$

式中，$\delta_i$ 为 $F_i$ 的函数，即 $\delta_i = \delta_i(F_i)$ $(i=1,2,3)$。

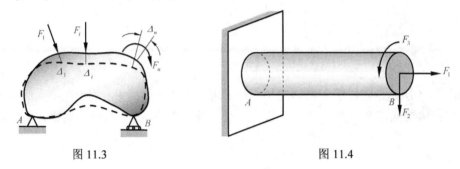

图 11.3　　　　　　　　　图 11.4

## 11.2　应变能密度

### 1. 单向应力应变能密度

如图 11.5 所示，当构件受单向应力时，应变能密度等于应力应变曲线下方的面积，并可表示为

$$u = \int \sigma d\varepsilon \tag{11.5}$$

式中，$\varepsilon$ 为与 $\sigma$ 对应的线应变。若 $\sigma$ 位于线弹性范围，则有 $\sigma = E\varepsilon$，其中 $E$ 为材料的弹性模量。把 $\sigma$ 代入式(11.5)，得

$$u = \frac{1}{2} E\varepsilon^2 = \frac{1}{2}\sigma\varepsilon = \frac{\sigma^2}{2E} \tag{11.6}$$

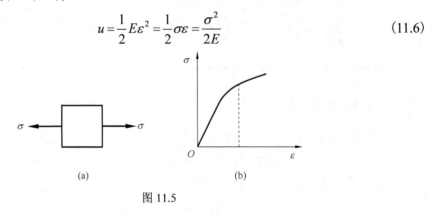

(a)　　　　　　(b)

图 11.5

### 2. 纯剪应力应变能密度

若构件受纯剪应力，如图 11.6 所示，则应变能密度可表示为

$$u = \int \tau \mathrm{d}\gamma \tag{11.7}$$

式中，$\gamma$ 为与 $\tau$ 对应的剪应变。根据式(11.7)，应变能密度 $u$ 等于应力应变曲线下方的面积。若 $\tau$ 位于线弹性范围，即 $\tau = G\gamma$，其中 $G$ 为材料的剪切弹性模量，则有

$$u = \frac{1}{2}G\gamma^2 = \frac{1}{2}\tau\gamma = \frac{\tau^2}{2G} \tag{11.8}$$

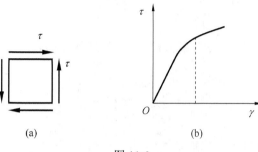

图 11.6

## 11.3 应 变 能

**1. 轴向拉压应变能**

轴向拉杆受缓慢增加的与杆轴线重合的外力 $F$ 作用，如图 11.7 所示。

根据式(11.6)，杆的应变能等于

$$V_\varepsilon = \int \frac{\sigma^2}{2E} A \mathrm{d}x \tag{11.9}$$

利用 $\sigma = \dfrac{N}{A}$，其中，$N$ 为轴力，$A$ 为横截面面积，得

$$V_\varepsilon = \int \frac{N^2}{2EA} \mathrm{d}x \tag{11.10}$$

若面积 $A$ 均匀，轴力 $N$ 不变，则式(11.10)可重写为

$$V_\varepsilon = \frac{N^2 l}{2EA} \tag{11.11}$$

**2. 扭转应变能**

圆轴受缓慢增加的外力偶 $M_\mathrm{e}$ 作用，外力偶作用面垂直纵轴，如图 11.8 所示。

根据式(11.8)，轴的应变能可表示为

$$V_\varepsilon = \int \left( \int \frac{\tau^2}{2G} \mathrm{d}A \right) \mathrm{d}x \tag{11.12}$$

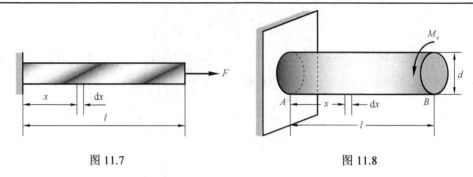

图 11.7　　　　　　　　　　图 11.8

因 $\tau = \dfrac{T\rho}{I_p}$，其中，$T$ 为扭矩，$I_p$ 为极惯性矩，$\rho$ 为到截面形心的距离，则有

$$V_\varepsilon = \int \left( \dfrac{T^2}{2GI_p^2} \int \rho^2 dA \right) dx \tag{11.13}$$

利用 $I_p = \int \rho^2 dA$，则式(11.13)可重写为

$$V_\varepsilon = \int \dfrac{T^2}{2GI_p} dx \tag{11.14}$$

如果极惯性矩 $I_p$ 和扭矩 $T$ 均为常数，则式(11.14)可表示为

$$V_\varepsilon = \dfrac{T^2 l}{2GI_p} \tag{11.15}$$

### 3. 弯曲应变能

弯曲梁受缓慢增加的外力 $F$ 作用，外力作用于纵向对称面，如图 11.9 所示。对细长梁，与正应力相比，剪应力引起的应变能通常很小，可忽略不计，因此梁的应变能可表示为

$$V_\varepsilon = \int \left( \int \dfrac{\sigma^2}{2E} dA \right) dx \tag{11.16}$$

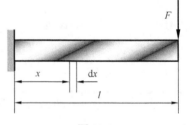

图 11.9

利用 $\sigma = \dfrac{My}{I}$，其中 $M$ 为弯矩，$I$ 为惯性矩，$y$ 为到中性轴的距离，则有

$$V_\varepsilon = \int \left( \dfrac{M^2}{2EI^2} \int y^2 dA \right) dx \tag{11.17}$$

把 $I = \int y^2 dA$ 代入式(11.17)，则式(11.17)可重写为

$$V_\varepsilon = \int_0^l \dfrac{M^2}{2EI} dx \tag{11.18}$$

对纯弯曲梁,如图 11.10 所示,弯矩 $M$ 为常数,如果惯性矩 $I$ 也保持不变,则式(11.18)可表示为

$$V_\varepsilon = \frac{M^2 l}{2EI} \tag{11.19}$$

**4. 组合载荷应变能**

受组合载荷作用的圆截面杆件如图 11.11 所示。对细长杆件,剪应力引起的应变能可忽略不计,则组合载荷作用下杆件的应变能可表示为

$$V_\varepsilon = \int \frac{N^2}{2EA} dx + \int \frac{T^2}{2GI_p} dx + \int \frac{M^2}{2EI} dx \tag{11.20}$$

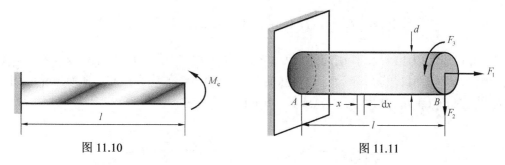

图 11.10　　　　　　　　图 11.11

## 11.4　功 能 原 理

如果不计能量损失,则根据能量守恒,外力对物体所做功将会全部转化为物体应变能,数学上可表示为

$$V_\varepsilon = W_e \tag{11.21}$$

式中,$V_\varepsilon$ 为应变能;$W_e$ 为外功。式(11.21)称为功能原理,可用于计算结构变形。

**例 11.1**　已知弯曲刚度 $EI$,利用功能原理计算悬臂梁 $AB$ 自由端 $B$ 处挠度(图 11.12)。

**解**　外功和应变能分别为

$$W_e = \frac{1}{2} F \delta_B, \quad V_\varepsilon = \int_0^l \frac{M^2}{2EI} dx = \int_0^l \frac{(Fx)^2}{2EI} dx = \frac{F^2 l^3}{6EI}$$

利用功能原理,即 $W_e = V_\varepsilon$,得

$$\frac{1}{2} F \delta_B = \frac{F^2 l^3}{6EI}, \quad \delta_B = \frac{Fl^3}{3EI}$$

**例 11.2**　利用功能原理,计算悬臂梁 $AB$ 自由端 $B$ 处转角(图 11.13)。已知弯曲刚度 $EI$。

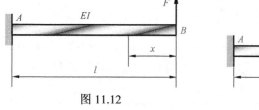　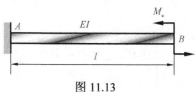

图 11.12　　　　　　　　图 11.13

**解** $W_e = \dfrac{1}{2}M_e\theta_B$, $V_\varepsilon = \displaystyle\int_0^l \dfrac{M^2}{2EI}\,\mathrm{d}x = \int_0^l \dfrac{M_e^2}{2EI}\,\mathrm{d}x = \dfrac{M_e^2 l}{2EI}$

根据 $W_e = V_\varepsilon$，得

$$\dfrac{1}{2}M_e\theta_B = \dfrac{M_e^2 l}{2EI}, \quad \theta_B = \dfrac{M_e l}{EI}$$

## 11.5 互 等 定 理

考虑如图 11.14 所示线弹性体受广义力作用，设先加 $F_1$ 和 $F_3$，再加 $F_2$ 和 $F_4$，则物体的应变能可表示为

$$V_{\varepsilon 1} = \dfrac{1}{2}F_1\Delta_1 + \dfrac{1}{2}F_3\Delta_3 + \dfrac{1}{2}F_2\Delta_2 + \dfrac{1}{2}F_4\Delta_4 + F_1\Delta_1' + F_3\Delta_3' \tag{11.22}$$

式中，$\Delta_1$（或 $\Delta_3$）为由 $F_1$ 和 $F_3$ 引起的 $F_1$（或 $F_3$）作用点沿 $F_1$（或 $F_3$）方向的广义位移；$\Delta_2$（或 $\Delta_4$）为由 $F_2$ 和 $F_4$ 引起的 $F_2$（或 $F_4$）作用点沿 $F_2$（或 $F_4$）方向的广义位移；$\Delta_1'$（或 $\Delta_3'$）为由 $F_2$ 和 $F_4$ 引起的 $F_1$（或 $F_3$）作用点沿 $F_1$（或 $F_3$）方向的广义位移。

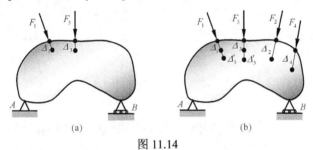

图 11.14

若广义力 $F_2$ 和 $F_4$ 首先作用于物体，广义力 $F_1$ 和 $F_3$ 再作用于物体，则物体的应变能为

$$V_{\varepsilon 2} = \dfrac{1}{2}F_2\Delta_2 + \dfrac{1}{2}F_4\Delta_4 + \dfrac{1}{2}F_1\Delta_1 + \dfrac{1}{2}F_3\Delta_3 + F_2\Delta_2' + F_4\Delta_4' \tag{11.23}$$

式中，$\Delta_2'$（或 $\Delta_4'$）为由 $F_1$ 和 $F_3$ 引起的 $F_2$（或 $F_4$）作用点沿 $F_2$（或 $F_4$）方向的广义位移。

比较式(11.22)和式(11.23)，并利用 $V_{\varepsilon 1} = V_{\varepsilon 2}$，得

$$F_1\Delta_1' + F_3\Delta_3' = F_2\Delta_2' + F_4\Delta_4' \tag{11.24}$$

由此得到结论，第 1 组广义力在第 2 组广义力引起的广义位移上所做功等于第 2 组广义力在第 1 组广义力引起的广义位移上所做功。上述关系称为功的互等定理。

假设 $F_3 = F_4 = 0$ 和 $F_1 = F_2$，则式(11.24)可简化为

$$\Delta_1' = \Delta_2' \tag{11.25}$$

由此得到结论，若两个广义力数值相等，则由第 2 个广义力引起的第 1 个广义力作用点沿第 1 个广义力方向的广义位移等于由第 1 个广义力引起的第 2 个广义力作用点沿第 2 个广义力方向的广义位移。该关系称为位移互等定理。

**例 11.3** 如图 11.15(a)所示，线弹性球受等值反向径向拉伸载荷作用。已知球的弹性模量 $E$、泊松比 $\mu$、直径 $d$，求球的体积变化。

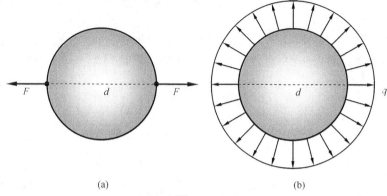

图 11.15

**解** 假设球受三向均匀拉伸应力作用，如图 11.15(b)所示，则有

$$\varepsilon_1 = \frac{1}{E}[\sigma_1 - \mu(\sigma_2 + \sigma_3)] = \frac{1-2\mu}{E}q$$

利用功的互等定理，得

$$F(\varepsilon_1 d) = q\Delta V$$

即

$$\Delta V = \frac{F(\varepsilon_1 d)}{q} = \frac{1-2\mu}{E}Fd$$

## 11.6 卡氏定理

如图 11.16 所示，变形体受外力 $F_1, F_2, \cdots, F_i, \cdots, F_n$ 作用。若作用于物体的外力由零开始缓慢增加到终值，则力作用点沿力方向的相应位移也从零缓慢增加到终值。如果材料处于线弹性范围，则作用于物体的外力所做功可表示为

$$W_e = \sum \frac{1}{2}F_i \Delta_i \tag{11.26}$$

式中，$\Delta_i$ 为外力 $F_1, F_2, \cdots, F_i, \cdots, F_n$ 的线性齐次函数。式(11.26) 表明，外功 $W_e$ 为外力 $F_1, F_2, \cdots, F_i, \cdots, F_n$ 的二次齐次函数。利用功能原理，即 $W_e = V_\varepsilon$，应变能也是外力 $F_1, F_2, \cdots, F_i, \cdots, F_n$ 的二次齐次函数，并可表示为

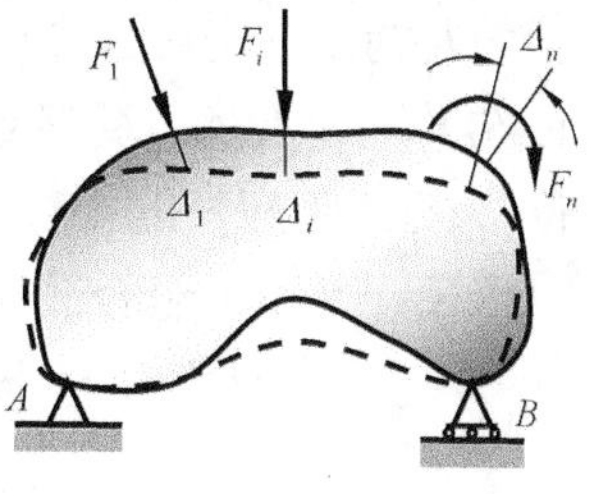

图 11.16

$$V_\varepsilon = V_\varepsilon(F_1, F_2, \cdots, F_i, \cdots, F_n) \tag{11.27}$$

如果任何一个外力，如外力 $F_i$，增加 $\mathrm{d}F_i$，而其余外力保持不变，那么应变能也将增加 $\mathrm{d}V_\varepsilon$，并可表示为

$$\mathrm{d}V_\varepsilon = \frac{\partial V_\varepsilon}{\partial F_i}\mathrm{d}F_i \tag{11.28}$$

因此，总应变能可写为

$$V_{\varepsilon 1} = V_{\varepsilon} + \mathrm{d}V_{\varepsilon} = \sum \frac{1}{2} F_i \Delta_i + \frac{\partial V_{\varepsilon}}{\partial F_i} \mathrm{d}F_i \tag{11.29}$$

总应变能 $V_{\varepsilon 1}$ 应该与外力加载次序无关，例如，可以先加 $\mathrm{d}F_i$，再加 $F_1, F_2, \cdots, F_i, \cdots, F_n$，那么总应变能应该不变。当先加 $\mathrm{d}F_i$ 时，外力 $\mathrm{d}F_i$ 所做功 $\mathrm{d}W'_e$ 可写为

$$\mathrm{d}W'_e = \frac{1}{2} \mathrm{d}F_i \mathrm{d}\Delta_i \tag{11.30}$$

式中，$\mathrm{d}\Delta_i$ 为对应 $\mathrm{d}F_i$ 的位移。若再加 $F_1, F_2, \cdots, F_i, \cdots, F_n$，则外力 $F_1, F_2, \cdots, F_i, \cdots, F_n$ 和 $\mathrm{d}F_i$ 所做功可表示为

$$W'_e = \sum \frac{1}{2} F_i \Delta_i + \mathrm{d}F_i \Delta_i \tag{11.31}$$

因此，总功为

$$W_{e2} = W'_e + \mathrm{d}W'_e = \sum \frac{1}{2} F_i \Delta_i + \mathrm{d}F_i \Delta_i + \frac{1}{2} \mathrm{d}F_i \mathrm{d}\Delta_i \tag{11.32}$$

根据功能原理，总应变能为

$$V_{\varepsilon 2} = W_{e2} = \sum \frac{1}{2} F_i \Delta_i + \mathrm{d}F_i \Delta_i + \frac{1}{2} \mathrm{d}F_i \mathrm{d}\Delta_i \tag{11.33}$$

利用 $V_{\varepsilon 1} = V_{\varepsilon 2}$，则有

$$\sum \frac{1}{2} F_i \Delta_i + \frac{\partial V_{\varepsilon}}{\partial F_i} \mathrm{d}F_i = \sum \frac{1}{2} F_i \Delta_i + \mathrm{d}F_i \Delta_i + \frac{1}{2} \mathrm{d}F_i \mathrm{d}\Delta_i \tag{11.34}$$

忽略高阶小量，得

$$\Delta_i = \frac{\partial V_{\varepsilon}}{\partial F_i} \tag{11.35}$$

式中，$F_i$ 为广义力；$\Delta_i$ 为相应广义位移。如果 $F_i$ 是力偶，则 $\Delta_i$ 为与力偶对应的转角。式 (11.35) 称为卡氏第二定理或卡氏定理。该定理可表述为：广义力 $F_i$ 作用点处沿 $F_i$ 方向的广义位移 $\Delta_i$ 等于应变能相对广义力 $F_i$ 的一阶偏导数。

卡氏定理用于组合载荷，则可表示为

$$\Delta_i = \frac{\partial V_{\varepsilon}}{\partial F_i} = \int \frac{N}{EA} \frac{\partial N}{\partial F_i} \mathrm{d}x + \int \frac{T}{GI_\mathrm{p}} \frac{\partial T}{\partial F_i} \mathrm{d}x + \int \frac{M}{EI} \frac{\partial M}{\partial F_i} \mathrm{d}x \tag{11.36}$$

**例 11.4** 已知弯曲刚度 $EI$，利用卡氏定理求悬臂梁自由端挠度（图 11.17）。

**解** 应变能为

$$V_{\varepsilon} = \int_0^l \frac{M^2}{2EI} \mathrm{d}x = \int_0^l \frac{(Fx)^2}{2EI} \mathrm{d}x = \frac{F^2 l^3}{6EI}$$

利用卡氏定理，得

$$\delta_B = \frac{\partial V_{\varepsilon}}{\partial F} = \frac{\partial}{\partial F} \left( \frac{F^2 l^3}{6EI} \right) = \frac{Fl^3}{3EI}$$

**例 11.5** 已知弯曲刚度 $EI$，利用卡氏定理，求悬臂梁自由端转角（图 11.18）。

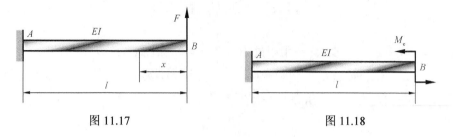

图 11.17　　　　　　　　图 11.18

**解** 应变能为

$$V_\varepsilon = \int_0^l \frac{M^2}{2EI} dx = \int_0^l \frac{M_e^2}{2EI} dx = \frac{M_e^2 l}{2EI}$$

利用 $\Delta_i = \dfrac{\partial V_\varepsilon}{\partial F_i}$，得

$$\theta_B = \frac{\partial V_\varepsilon}{\partial M_e} = \frac{\partial}{\partial M_e}\left(\frac{M_e^2 l}{2EI}\right) = \frac{M_e l}{EI}$$

卡氏定理能用于求外力作用处的位移。如果需求无外力作用处的位移，则需要在待求位移处施加虚拟力。应变能相对虚拟力求导，进而得到虚拟力处位移，该位移由实际力和虚拟力共同产生。通过设置虚拟力等于零，即可求得实际力产生的位移。

**例 11.6** 已知弯曲刚度 $EI$，利用卡氏定理求悬臂梁自由端转角，如图 11.19(a)所示。

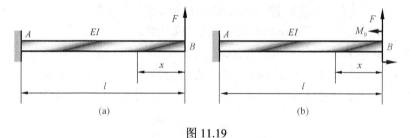

图 11.19

**解** 在梁自由端施加虚拟力偶 $M_0$，如图 11.19(b)所示，则截面 $x$ 处弯矩为

$$M = Fx + M_0 \quad (0 \leqslant x < l)$$

因此，应变能可表示为

$$V_\varepsilon = \int_0^l \frac{M^2}{2EI} dx = \int_0^l \frac{(Fx + M_0)^2}{2EI} dx = \frac{\frac{1}{3}F^2 l^3 + FM_0 l^2 + M_0^2 l}{2EI}$$

利用卡氏定理，即 $\Delta_i = \dfrac{\partial V_\varepsilon}{\partial F_i}$，得

$$\theta_B = \left(\frac{\partial V_\varepsilon}{\partial M_0}\right)_{M_0=0} = \left(\frac{Fl^2 + 2M_0 l}{2EI}\right)_{M_0=0} = \frac{Fl^2}{2EI}$$

## 11.7 虚功原理

变形体在载荷 $F$、$q$ 和 $M_e$ 作用下处于平衡状态，如图 11.20 所示。

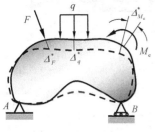

图 11.20

当因其他因素，如载荷变化或温度波动，在物体内引起虚变形时，则由载荷 $F$、$q$ 和 $M_e$ 在虚变形上所做虚功可表示为

$$W_e = F\Delta_F^* + \int q\Delta_q^* \mathrm{d}x + M_e \Delta_{M_e}^* \tag{11.37}$$

式中，$\Delta_F^*$、$\Delta_q^*$ 和 $\Delta_{M_e}^*$ 分别为 $F$、$q$ 和 $M_e$ 作用点沿 $F$、$q$ 和 $M_e$ 方向的虚位移。

若物体发生虚变形，则作用于物体的载荷 $F$、$q$ 和 $M_e$ 产生的内力所做虚功可表示为

$$W_i = \int N\mathrm{d}\delta^* + \int M\mathrm{d}\theta^* + \int V\mathrm{d}\lambda^* + \int T\mathrm{d}\varphi^* \tag{11.38}$$

式中，$N$、$M$、$V$ 和 $T$ 分别为由载荷 $F$、$q$ 和 $M_e$ 产生的轴力、弯矩、剪力和扭矩；$\delta^*$、$\theta^*$、$\lambda^*$ 和 $\varphi^*$ 分别为与 $N$、$M$、$V$ 和 $T$ 对应的虚变形。

利用 $W_i = W_e$，得

$$\int N\mathrm{d}\delta^* + \int M\mathrm{d}\theta^* + \int V\mathrm{d}\lambda^* + \int T\mathrm{d}\varphi^* = F\Delta_F^* + \int q\Delta_q^* \mathrm{d}x + M_e \Delta_{M_e}^* \tag{11.39}$$

这就是虚功原理，可表述如下：内力在虚变形上所做虚功等于外力在虚位移上所做虚功。虚功原理既可用于线性弹性体，也可用于非线性弹性体。

**例 11.7** 如图 11.21(a)所示，桁架 $ABC$ 由杆 $AC$ 和 $BC$ 构成，每杆截面积均为常数 $A$，在节点 $C$ 受集中力 $F$ 作用。杆 $AC$ 和 $BC$ 的应力应变关系分别是 $\sigma = E\varepsilon$ 和 $\sigma^2 = k\varepsilon$，其中 $E$ 和 $k$ 均为常数。利用虚功原理，求 $C$ 处水平和垂直位移分量(注：结构稳定性无须考虑)。

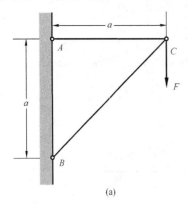

图 11.21

**解** 如图 11.21(b)所示，当方向向右的水平力 $F_0$ 作用于桁架节点 $C$ 时，杆 $AC$ 和 $BC$ 中产生的内力可表示为

$$N_{AC} = F_0, \quad N_{BC} = 0$$

当原载荷 $F$ 作用于结构时，在 $AC$ 和 $BC$ 中产生的变形可作为虚变形，并可表示为

$$\delta_{AC}^* = \frac{Fa}{EA}, \quad \delta_{BC}^* = \left[ -\frac{(-\sqrt{2}F)^2}{kA^2} \right](\sqrt{2}a) = -\frac{2\sqrt{2}F^2 a}{kA^2}$$

在 $C$ 处产生的位移可作为虚位移，并为

$$\delta_C^* = \Delta_H$$

式中，$\Delta_H$ 为 $C$ 处待求水平位移分量。

利用虚功原理，得

$$F_0 \Delta_H = N_{AC} \frac{Fa}{EA} + N_{BC} \left( -\frac{2\sqrt{2}F^2 a}{kA^2} \right)$$

即

$$\Delta_H = \frac{Fa}{EA}$$

如果方向向下的垂直力作用于桁架节点 $C$，同理可得 $C$ 处垂直位移分量为

$$\Delta_V = \frac{Fa}{EA} + (-\sqrt{2}) \left( -\frac{2\sqrt{2}F^2 a}{kA^2} \right) = \frac{Fa}{EA} + \frac{4F^2 a}{kA^2}$$

## 11.8 单位载荷法

变形体在载荷 $F$、$q$ 和 $M_e$ 作用下处于平衡状态，如图 11.22(a) 所示。

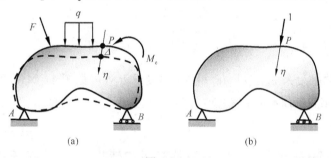

图 11.22

为求 $P$ 处沿 $\eta$ 方向的位移 $\Delta$，假设单位载荷沿 $\eta$ 方向作用于物体 $P$ 处，如图 11.22(b) 所示，然后原载荷也作用于物体。如果把原载荷引起的变形作为虚变形，那么根据虚功原理，得

$$\Delta = \int \bar{N} d\delta + \int \bar{M} d\theta + \int \bar{V} d\lambda + \int \bar{T} d\varphi \tag{11.40}$$

式中，$\bar{N}$、$\bar{M}$、$\bar{V}$ 和 $\bar{T}$ 分别为由单位载荷引起的轴力、弯矩、剪力和扭矩；$\delta$、$\theta$、$\lambda$ 和 $\varphi$ 分别为由载荷 $F$、$q$ 和 $M_e$ 引起的与轴力、弯矩、剪力和扭矩对应的变形。上述关系称为单位载荷法，可用于确定结构上任意点沿任意方向的位移。如果转角需要确定，那么单位力偶需要施加于待求转角的位置。单位载荷法可用于线性弹性体和非线性弹性体。

对受组合载荷作用的线弹性构件，式(11.40)可重写为

$$\Delta = \int \frac{N(x)\bar{N}(x)}{EA} dx + \int \frac{M(x)\bar{M}(x)}{EI} dx + \int \frac{T(x)\bar{T}(x)}{GI_p} dx \tag{11.41}$$

式中，$N$、$M$ 和 $T$ 分别为由组合载荷引起的轴力、弯矩和扭矩；$EA$、$EI$ 和 $GI_p$ 分别为组合载荷作用构件的拉压刚度、弯曲刚度和扭转刚度。

**例 11.8** 已知弯曲刚度 $EI$，利用单位载荷法求悬臂梁自由端转角，如图 11.23(a)所示。

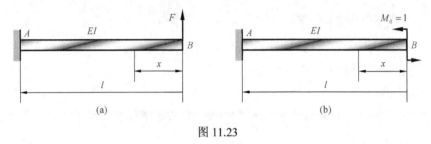

图 11.23

**解** 如图 11.23(a)所示，原载荷弯矩可表示为

$$M(x) = Fx \quad (0 \leqslant x < l)$$

如图 11.23(b)所示，单位载荷弯矩为

$$\bar{M}(x) = M_0 = 1 \quad (0 < x < l)$$

根据单位载荷法，得

$$\theta_B = \int_0^l \frac{M(x)\bar{M}(x)}{EI}\mathrm{d}x = \frac{Fl^2}{2EI}$$

## 11.9  能量方法应用

1. 用于静定结构

**例 11.9** 弯曲刚度为 $EI$ 的杆 $AB$ 和 $BC$ 焊接于 $B$ 处，如图 11.24(a)所示。采用单位载荷法，求点 $C$ 处垂直挠度和截面 $C$ 处转角。

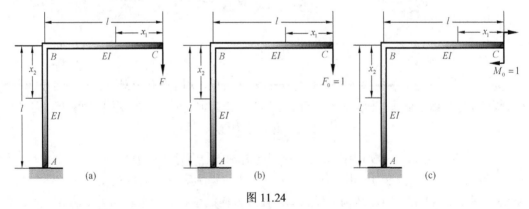

图 11.24

**解** 图 11.24(a)中原载荷引起的弯矩可表示为

$$M(x_1) = -Fx_1 \quad (0 \leqslant x_1 \leqslant l), \quad M(x_2) = -Fl \quad (0 \leqslant x_2 < l)$$

为求点 $C$ 的垂直挠度，在该点垂直方向施加单位载荷 $F_0 = 1$，如图 11.24(b)所示，则

单位载荷引起的弯矩为

$$\bar{M}(x_1) = -x_1 \ (0 \leqslant x_1 \leqslant l), \quad \bar{M}(x_2) = -l \ (0 \leqslant x_2 < l)$$

利用单位载荷法，得

$$(w_C)_y = \int_0^l \frac{M(x_1)\bar{M}(x_1)}{EI} dx_1 + \int_0^l \frac{M(x_2)\bar{M}(x_2)}{EI} dx_2 = \frac{4Fl^3}{3EI}$$

为求截面 $C$ 的转角，在该截面施加单位力偶 $M_0 = 1$，如图 11.24(c)所示，显然单位力偶引起的弯矩为

$$\bar{M}(x_1) = -1 \ (0 \leqslant x_1 \leqslant l), \quad \bar{M}(x_2) = -1 \ (0 \leqslant x_2 < l)$$

利用单位载荷法，得

$$\theta_C = \int_0^l \frac{M(x_1)\bar{M}(x_1)}{EI} dx_1 + \int_0^l \frac{M(x_2)\bar{M}(x_2)}{EI} dx_2 = \frac{3Fl^2}{2EI}$$

2. 用于静不定结构

**例 11.10** 弯曲刚度为 $EI$ 的杆件 $AB$ 和 $BC$ 焊接于 $B$ 处，如图 11.25(a)所示，点 $A$ 固定，点 $C$ 铰支。采用单位载荷法，求点 $C$ 处反力。

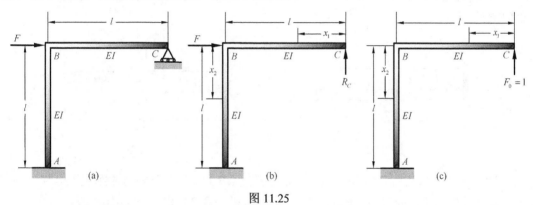

图 11.25

**解** 图 11.25(a)为 1 次静不定结构。设 $R_C$ 为点 $C$ 处多余反力，消除相应支撑，如图 11.25(b)所示。

参考图 11.25(b)，弯矩表示为

$$M(x_1) = R_C x_1 \ (0 \leqslant x_1 \leqslant l), \quad M(x_2) = R_C l - F x_2 \ (0 \leqslant x_2 < l)$$

为求点 $C$ 处垂直挠度，在该点垂直方向作用单位载荷 $F_0 = 1$，如图 11.25(c)所示，单位载荷引起的弯矩表示为

$$\bar{M}(x_1) = x_1 \ (0 \leqslant x_1 \leqslant l), \quad \bar{M}(x_2) = l \ (0 \leqslant x_2 < l)$$

利用单位载荷法，得

$$(w_C)_y = \int_0^l \frac{M(x_1)\bar{M}(x_1)}{EI} dx_1 + \int_0^l \frac{M(x_2)\bar{M}(x_2)}{EI} dx_2 = \frac{(8R_C - 3F)l^3}{6EI}$$

因 $(w_C)_y = 0$，则得

$$R_C = \frac{3}{8}F$$

# 习 题

11.1 利用功能原理,求悬臂梁自由端挠度(图 11.26)。已知 $AB$ 段弯曲刚度 $2EI$,$BC$ 段弯曲刚度 $EI$。

11.2 利用功能原理,求简支梁中间截面挠度(图 11.27)。已知 $CD$ 段弯曲刚度 $2EI$,$AC$ 和 $BD$ 段弯曲刚度 $EI$。

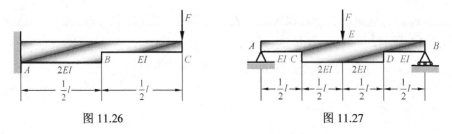

图 11.26　　　　　图 11.27

11.3 利用卡氏定理,求悬臂梁自由端挠度和转角(图 11.28)。已知弯曲刚度 $EI$。

11.4 利用卡氏定理,求悬臂梁自由端挠度(图 11.29)。已知 $AB$ 段弯曲刚度 $2EI$,$BC$ 段弯曲刚度 $EI$。

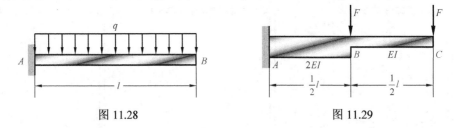

图 11.28　　　　　图 11.29

11.5 利用卡氏定理,求简支梁中间截面挠度(图 11.30)。已知弯曲刚度 $EI$。

11.6 利用卡氏定理,求点 $B$ 处挠度和转角(图 11.31)。已知弯曲刚度 $EI$。

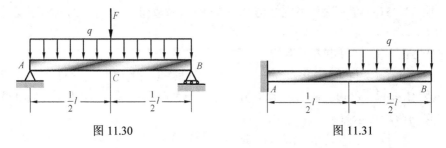

图 11.30　　　　　图 11.31

11.7 利用虚功原理,求悬臂梁自由端挠度和转角(图 11.32)。已知应力应变关系 $\sigma^2 = k\varepsilon$,其中 $k$ 是常数。

11.8 利用虚功原理,求简支梁中间截面挠度(图 11.33)。已知应力应变关系 $\sigma^2 = k\varepsilon$,其中 $k$ 是常数。

11.9 利用单位载荷法,求悬臂梁自由端挠度和转角(图 11.34)。已知弯曲刚度 $EI$。

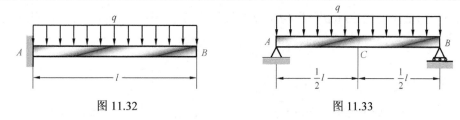

图 11.32　　　　　　　　　图 11.33

11.10　利用单位载荷法,求悬臂梁自由端挠度和转角(图 11.35)。已知 AB 段弯曲刚度 2EI,BC 段弯曲刚度 EI。

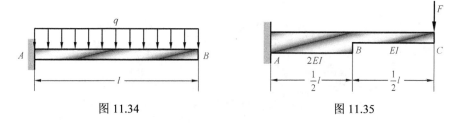

图 11.34　　　　　　　　　图 11.35

11.11　利用单位载荷法,求简支梁中间截面挠度(图 11.36)。已知弯曲刚度 EI。

11.12　利用单位载荷法,求简支梁中间截面 C 处挠度和截面 B 处转角(图 11.37)。已知弯曲刚度 EI。

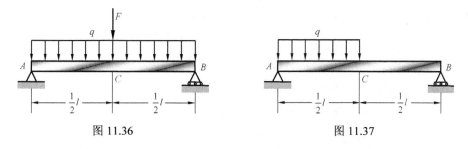

图 11.36　　　　　　　　　图 11.37

11.13　利用单位载荷法,求点 A 处水平和垂直挠度(图 11.38)。已知弯曲刚度 EI。

11.14　弯曲刚度为 EI 的杆件 AB 和 BC 焊接于 B 处,点 A 固定,点 C 铰支(图 11.39)。采用单位载荷法,求点 C 处反力。

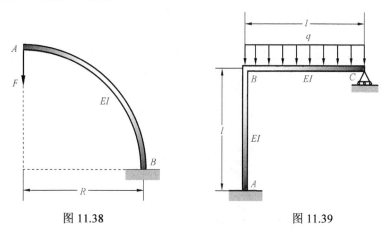

图 11.38　　　　　　　　　图 11.39

11.15 弯曲刚度为 $EI$ 的杆件 $AB$ 和 $BC$ 焊接于 $B$ 处(图 11.40)。采用单位载荷法,求点 $C$ 处水平和垂直挠度以及转角。

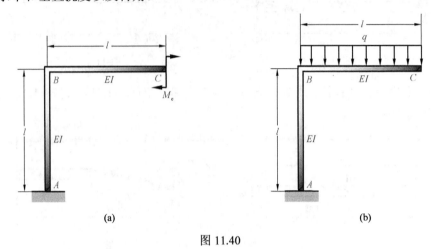

图 11.40

# 第 12 章 冲 击 载 荷

前面讨论的载荷都是静载荷,即载荷缓慢作用于构件直至终值,达到终值后保持不变。然而,也有载荷是动载荷,即载荷突然作用于构件。这些动载荷称为冲击载荷。当物体撞击构件时,撞击期间在构件内将会产生很大应力。

## 12.1 垂 直 冲 击

如图 12.1 所示物块弹簧系统,重量为 $P$、动能为 $T$ 的物块撞击不计质量的弹簧,假设物块停止运动时弹簧压缩 $\Delta_d$,如果不计冲击期间能量损失,那么能量守恒表明,物块动能和势能将会完全转化为弹簧内的应变能,或者说,动能和势能之和等于弹簧自由端移动 $\Delta_d$ 所做功。

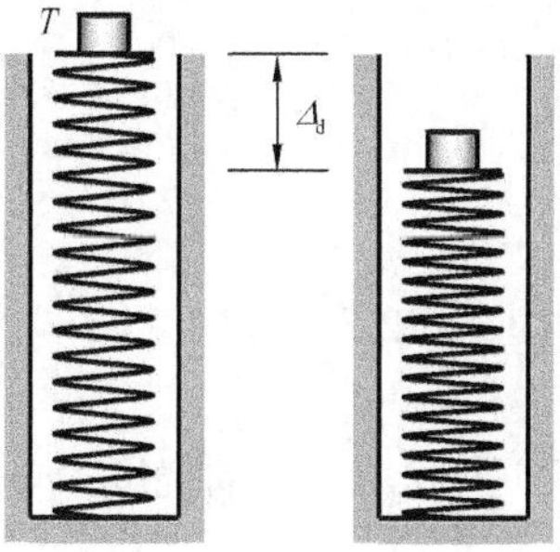

根据能量守恒,有

$$T + P\Delta_d = \frac{1}{2}F_d\Delta_d \tag{12.1}$$

式中,$F_d$ 为当弹簧压缩 $\Delta_d$ 时在物块和弹簧之间产生的作用力。假设弹簧处于线弹性范围,即 $\dfrac{F_d}{\Delta_d} = \dfrac{P}{\Delta_{st}}$,其中 $\Delta_{st}$ 为当物块静止作用于弹簧自由端时在弹簧自由端产生的静位移,则有

图 12.1

$$\frac{F_d}{P} = \frac{\Delta_d}{\Delta_{st}} = K_d \tag{12.2}$$

式中,$K_d$ 为动荷因数。把式(12.2)代入式(12.1),得

$$\frac{1}{2}K_d^2 - K_d - \frac{T}{P\Delta_{st}} = 0 \tag{12.3}$$

求解 $K_d$,得

$$K_d = 1 + \sqrt{1 + \frac{2T}{P\Delta_{st}}} \tag{12.4}$$

假设物块由静止释放,下落距离 $h$ 后撞击弹簧,那么动荷因数可表示为

$$K_d = 1 + \sqrt{1 + \frac{2h}{\Delta_{st}}} \tag{12.5}$$

$K_d$ 确定后,弹簧的动应力和动变形可分别表示为

$$\sigma_d = K_d\sigma_{st}, \quad \delta_d = K_d\delta_{st} \tag{12.6}$$

式中,$\sigma_{st}$ 和 $\delta_{st}$ 分别为当载荷静止作用于弹簧时而产生的静应力和静变形。

如果物块从刚接触弹簧时由静止状态突然释放，即 $h=0$，那么根据式(12.4)或式(12.5)，动荷因数 $K_d=2$ 和动应力 $\sigma_d=2\sigma_{st}$，即当物块从弹簧顶端由静止突然下落(突加载)，动应力是物块静止放在弹簧上(静加载)引起的静应力的两倍。

**例 12.1** 重量为 $P$ 的物块，初始静止，从高度 $h$ 处落下，撞击简支梁中点(图 12.2)。已知弹性模量 $E=105$ GPa，求：(1)梁的最大挠度；(2)梁中最大正应力。

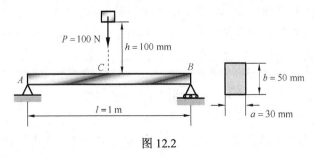

图 12.2

**解** (1)查询附录 III 挠度曲线表，得

$$\Delta_{st}=\frac{Pl^3}{48EI}=\frac{Pl^3}{48E\left(\frac{1}{12}ab^3\right)}=6.35\times10^{-2} \text{ mm}$$

利用式(12.5)，得

$$K_d=1+\sqrt{1+\frac{2h}{\Delta_{st}}}=57.13$$

因此，最大动挠度为

$$\Delta_d=K_d\Delta_{st}=3.63 \text{ mm}$$

(2)根据第 5 章正应力公式，得最大静应力为

$$\sigma_{st}=\frac{M}{W}=\frac{\frac{1}{4}Pl}{\frac{1}{6}ab^2}=2.00 \text{ MPa}$$

利用式(12.6)，得最大动应力为

$$\sigma_d=K_d\sigma_{st}=114.26 \text{ MPa}$$

**例 12.2** 质量为 $m$ 的环圈从如图 12.3 所示位置由静止释放，撞击固定在垂直杆 $AB$(直径为 $d$)自由端 $B$ 处的平板。已知 $E=70$ GPa，求：(1)杆的最大伸长；(2)杆内最大正应力 ($g=9.81$ m/s$^2$)。

**解** (1)杆的最大伸长为

$$\Delta_{st}=\frac{mgl}{EA}=\frac{mgl}{E\left(\frac{1}{4}\pi d^2\right)}=8.56\times10^{-3} \text{ mm}$$

$$K_d=1+\sqrt{1+\frac{2h}{\Delta_{st}}}=217.17, \quad \Delta_d=K_d\Delta_{st}=1.86 \text{ mm}$$

(2) 杆内最大正应力为

$$\sigma_{st} = \frac{mg}{A} = \frac{mg}{\frac{1}{4}\pi d^2} = 0.50 \text{ MPa}, \quad \sigma_d = K_d \sigma_{st} = 108.58 \text{ MPa}$$

**例 12.3** 平面机构 ABCD 由物块 A 和 D、光滑定滑轮 B 和 C、弹性绳索 ABCD 构成，绳索长度为 0.6 m，横截面面积为 50 mm²，物块 A 和 D 的重量均为 120 N，分别向下和向上匀速运动，速率均为 56 mm/s（图 12.4）。已知 E = 10 GPa，假设物块 D 突然卡住而停止运动，试求绳索最大应力。

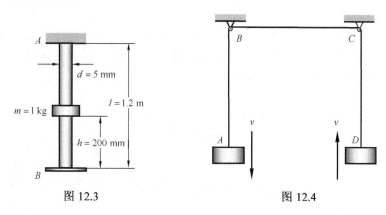

图 12.3    图 12.4

**解**

$$\Delta_{st} = \frac{P_A l}{EA} = 144 \text{ μm}, \quad K_d = 1 + \sqrt{\frac{2T_A}{P_A \Delta_{st}}} = 2.49$$

$$\sigma_{st} = \frac{P_A}{A} = 2.4 \text{ MPa}, \quad \sigma_d = K_d \sigma_{st} = 5.98 \text{ MPa}$$

## 12.2 水 平 冲 击

如图 12.5 所示，物块弹簧系统置于光滑水平面。重量为 P、动能为 T 的物块沿水平面滑动而撞击弹簧，假设物块达到最大位移时弹簧压缩量等于 $\Delta_d$。如果不计冲击能量损失和弹簧质量，并设弹簧处于线弹性范围，则物块运动静止时其动能将完全转化为弹簧的应变能。

根据能量守恒，得

$$T = \frac{1}{2} F_d \Delta_d = \frac{1}{2} P \Delta_{st} K_d^2 \quad (12.7)$$

式(12.7)的正根可表示为

$$K_d = \sqrt{\frac{2T}{P \Delta_{st}}} \quad (12.8)$$

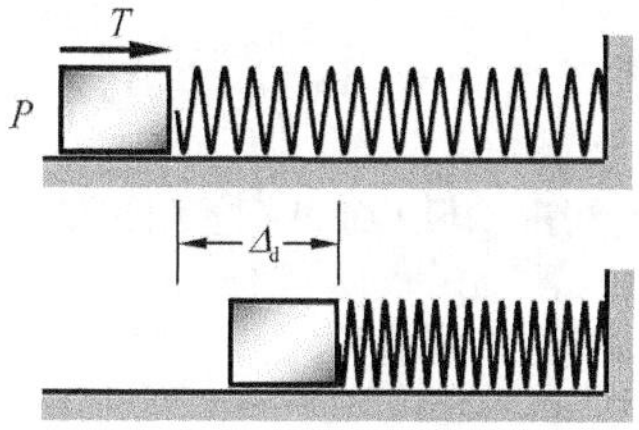

图 12.5

式中，$\Delta_{st}$ 为当与物块重量 P 相等的等效力作用于弹簧自由端时在弹簧自由端产生的静位

移。利用 $T = \dfrac{1}{2}\dfrac{P}{g}v^2$，得

$$K_{\mathrm{d}} = \sqrt{\dfrac{v^2}{g\Delta_{\mathrm{st}}}} \tag{12.9}$$

根据式(12.8)或式(12.9)，弹簧中动应力和弹簧的动变形分别为

$$\sigma_{\mathrm{d}} = K_{\mathrm{d}}\sigma_{\mathrm{st}}, \quad \delta_{\mathrm{d}} = K_{\mathrm{d}}\delta_{\mathrm{st}} \tag{12.10}$$

式中，$\sigma_{\mathrm{st}}$ 和 $\delta_{\mathrm{st}}$ 分别为当与物块重量 $P$ 相等的等效力静止作用于弹簧时产生的弹簧静应力和静变形。

**例 12.4** 质量为 $m$ 的小球向右运动以速度 $v$ 正碰立柱 $AB$（直径为 $d$）自由端 $A$ 处（图 12.6）。已知 $E=200$ GPa，求：(1)立柱最大挠度；(2)立柱最大正应力。

**解** (1)立柱最大挠度为

$$\Delta_{\mathrm{st}} = \dfrac{mgl^3}{3EI} = \dfrac{mgl^3}{3E\left(\dfrac{1}{64}\pi d^4\right)} = 6.58 \text{ mm}$$

$$K_{\mathrm{d}} = \sqrt{\dfrac{v^2}{g\Delta_{\mathrm{st}}}} = 3.94, \quad \Delta_{\mathrm{d}} = K_{\mathrm{d}}\Delta_{\mathrm{st}} = 25.9 \text{ mm}$$

(2)立柱最大正应力为

$$\sigma_{\mathrm{st}} = \dfrac{M}{W} = \dfrac{mgl}{\dfrac{1}{32}\pi d^3} = 29.6 \text{ MPa}, \quad \sigma_{\mathrm{d}} = K_{\mathrm{d}}\sigma_{\mathrm{st}} = 117 \text{ MPa}$$

**例 12.5** 质量为 $m$ 的小球以速度 $v$ 正碰杆 $AB$ 自由端 $A$（图 12.7）。已知 $E=100$ GPa，求：(1)点 $A$ 处位移；(2)杆内最大正应力。

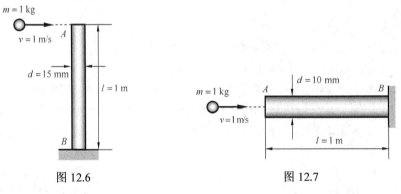

图 12.6　　　　　　图 12.7

**解** (1)点 $A$ 处位移为

$$\Delta_{\mathrm{st}} = \dfrac{mgl}{EA} = \dfrac{mgl}{E\left(\dfrac{1}{4}\pi d^2\right)} = 1.25 \times 10^{-6} \text{ m}$$

$$K_{\mathrm{d}} = \sqrt{\dfrac{v^2}{g\Delta_{\mathrm{st}}}} = 285.57, \quad \Delta_{\mathrm{d}} = K_{\mathrm{d}}\Delta_{\mathrm{st}} = 0.357 \text{ mm}$$

(2) 杆内最大正应力为

$$\sigma_{st} = \frac{mg}{A} = \frac{mg}{\frac{1}{4}\pi d^2} = 0.125 \text{ MPa}, \quad \sigma_d = K_d \sigma_{st} = 35.7 \text{ MPa}$$

# 习　题

12.1　质量为 $m$ 的物块，初始静止，从高度 $h$ 处落下，撞击悬臂梁自由端(图 12.8)。已知弹性模量 $E$=206 GPa，求：(1)梁的最大挠度；(2)梁中最大正应力。

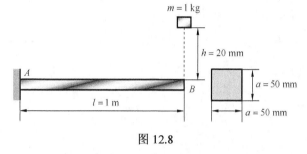

图 12.8

12.2　质量为 $m$ 的物块，初始静止，从高度 $h$ 处落下，撞击悬臂梁中点(图 12.9)。已知弹性模量 $E$=206 GPa，求：(1)梁的最大挠度；(2)梁中最大正应力。

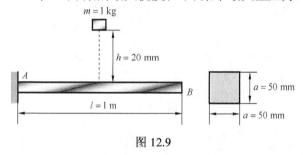

图 12.9

12.3　质量为 $m$ 的物块，初始静止，从高度 $h$ 处落下，撞击直径为 $d$ 的外伸梁自由端(图 12.10)。已知弹性模量 $E$=73 GPa，求：(1)点 $C$ 处挠度；(2)梁中最大正应力。

12.4　质量为 $m$ 的圆柱从如图 12.11 所示位置由静止释放，撞击垂直杆 $AB$(直径为 $d$)的自由端 $A$。已知 $E$=200 GPa，求杆中最大正应力。

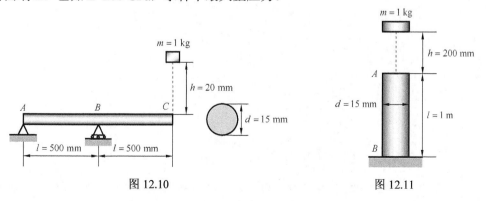

图 12.10　　　　　　　　　图 12.11

12.5 质量为 $m$ 的小球以速度 $v$ 正碰立柱 $AB$(直径为 $d$)的中点 $C$(图 12.12)。已知 $E=200$ GPa，求：(1)点 $A$ 的挠度；(2)立柱中最大正应力。

12.6 质量为 $m$ 的物块以速度 $v$ 正碰非均匀杆 $ABC$ 的自由端 $A$(图 12.13)。已知 $E=100$ GPa，$d_1=10$ mm，$d_2=20$ mm，求：(1)点 $A$ 处位移；(2)杆中最大正应力。

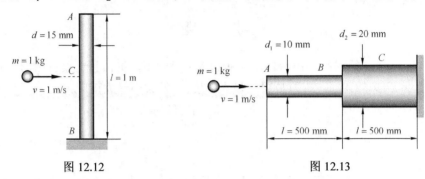

图 12.12　　　　　　图 12.13

# 第 13 章 静不定结构

结构分为静定和静不定。仅仅通过静力平衡方程即能完全确定所有反力和内力的结构称为静定结构，而仅仅通过静力平衡方程不能完全确定所有反力或内力的结构称为静不定结构。静不定结构的主要优点是强度比相应静定结构高，刚度比相应静定结构大。求解静不定结构，既要考虑静力平衡条件，又要考虑变形协调条件。

## 13.1 静 不 定

对静不定结构，反力需要作为多余力，则结构即外静不定；内力需要作为多余力，则结构即内静不定。静不定结构也可以同时包含外静不定和内静不定。

1. 外静不定

图 13.1 是典型的外静不定结构，结构有 4 个反力分量，只有 3 个独立平衡方程，因此有 1 个反力分量不能求解。超出平衡方程数的未知反力数称为静不定度，显然图 13.1 为 1 次外静不定。

2. 内静不定

对内静不定结构，利用平衡方程不可能求出所有内力，图 13.2 为 3 次内静不定。

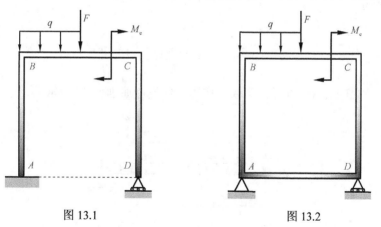

图 13.1　　　　　　　　图 13.2

## 13.2 力法分析静不定结构

力法可用于求解静不定结构。力法取多余力（多余反力或多余内力）作为未知量，利用位移协调条件确定多余力。基本思想可通过图 13.3(a)静不定梁进行说明，梁有 4 个未知反力,但只有3个独立平衡方程,因此是1次静不定。取未知反力,如B处反力$X_1$,作为多余反力并消除B处相应约束,那么原静不定梁可简化为如图13.3(b)所示稳定静定梁。

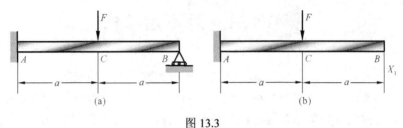

图13.3

B处沿$X_1$方向的位移$\Delta_1$可认为是两个独立位移的叠加,即

$$\Delta_1 = \Delta_{1X_1} + \Delta_{1F} \tag{13.1}$$

式中,$\Delta_{1X_1}$为由多余力$X_1$在B处产生的沿$X_1$方向的位移;$\Delta_{1F}$为由梁上外载荷F在B处产生的沿$X_1$方向的位移。应该注意,不可能在求出多余力$X_1$之前求解位移$\Delta_{1X_1}$,然而,利用叠加原理,$\Delta_{1X_1}$可表示为

$$\Delta_{1X_1} = \delta_{11} X_1 \tag{13.2}$$

式中,$\delta_{11}$为由B处沿$X_1$方向单位载荷在B处产生的沿$X_1$方向的位移。根据位移协调条件,B处沿$X_1$方向的位移$\Delta_1$必须等于零,即

$$\delta_{11} X_1 + \Delta_{1F} = 0 \tag{13.3}$$

式(13.3)称为力法正则方程,其中系数$\delta_{11}$和$\Delta_{1F}$可表示为

$$\delta_{11} = \int \frac{\overline{M}_1^2 \mathrm{d}x}{EI}, \quad \Delta_{1F} = \int \frac{M_F \overline{M}_1 \mathrm{d}x}{EI} \tag{13.4}$$

式中,$\overline{M}_1$为B处沿$X_1$方向单位载荷产生的弯矩;$M_F$为由梁上外载荷F产生的弯矩;EI是梁的弯曲刚度。

将式(13.4)代入式(13.3),得

$$X_1 = -\frac{\Delta_{1F}}{\delta_{11}} = -\int \frac{M_F \overline{M}_1 \mathrm{d}x}{EI} \bigg/ \int \frac{\overline{M}_1^2 \mathrm{d}x}{EI} \tag{13.5}$$

**例13.1** 如图13.4(a)所示梁受载荷作用,假设弯曲刚度EI为常数,求可动铰支座处反力。

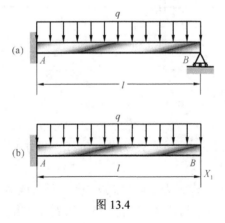

图13.4

**解** 原梁是1次静不定问题，取 $B$ 处可动铰支座作为多余约束，并进行释放，如图13.4(b)所示，根据 $B$ 处位移协调条件，即梁在该处挠度为零，则力法正则方程可表示为

$$\delta_{11}X_1 + \Delta_{1F} = 0$$

式中，$X_1$ 为多余反力。根据附录 III，$B$ 处沿 $X_1$ 方向单位载荷在 $B$ 处产生的挠度为

$$\delta_{11} = \frac{l^3}{3EI}$$

均布外载荷 $q$ 在 $B$ 处产生的挠度为 $\quad \Delta_{1F} = -\frac{ql^4}{8EI}$

把 $\delta_{11}$ 和 $\Delta_{1F}$ 代入 $\delta_{11}X_1 + \Delta_{1F} = 0$，得

$$\frac{l^3}{3EI}X_1 - \frac{ql^4}{8EI} = 0$$

解出 $X_1$，得 $B$ 处多余反力为 $\quad X_1 = \frac{3}{8}ql$

**例 13.2** 弯曲刚度为 $EI$ 的杆 $AB$ 和 $BC$ 在截面 $B$ 处通过焊接构成刚架，如图13.5(a)所示，用力法求 $C$ 处反力。

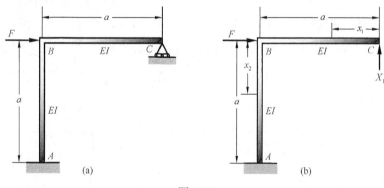

图 13.5

**解** 图13.5(a)为1次静不定结构。选 $C$ 处约束力 $X_1$ 作为多余反力，并消除相应约束，如图13.5(b)所示。由外载荷 $F$ 产生的弯矩可表示为

$$M_F(x_1) = 0 \quad (0 \leq x_1 \leq a), \quad M_F(x_2) = -Fx_2 \quad (0 \leq x_2 < a)$$

由 $C$ 处沿 $X_1$ 方向单位载荷产生的弯矩为

$$\overline{M}_1(x_1) = x_1 \quad (0 \leq x_1 \leq a), \quad \overline{M}_1(x_2) = a \quad (0 \leq x_2 < a)$$

根据式(13.7)，得

$$\delta_{11} = \int_0^a \frac{[\overline{M}_1(x_1)]^2}{EI}dx_1 + \int_0^a \frac{[\overline{M}_1(x_2)]^2}{EI}dx_2 = \frac{4a^3}{3EI}, \quad \Delta_{1F} = \int_0^a \frac{M_F(x_2)\overline{M}_1(x_2)}{EI}dx_2 = -\frac{Fa^3}{2EI}$$

利用力法正则方程 $\delta_{11}X_1 + \Delta_{1F} = 0$，有

$$X_1 = \frac{3}{8}F$$

**例 13.3** 弯曲刚度为 $EI$ 的梁 $AB$ 和拉伸刚度为 $EA$ 的杆 $BC$ 铰接于 $B$,如图 13.6(a)所示,梁 $A$ 端固定,杆 $C$ 端铰接。已知 $EA=3EI/(2a^2)$,采用力法求 $C$ 处反力。

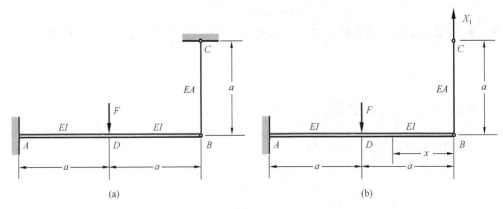

图 13.6

**解** 图 13.6(a)是 1 次静不定结构,取 $C$ 处 $X_1$ 作为多余反力,消除相应约束,如图 13.6(b)所示。

外载荷 $F$ 在梁 $AB$ 中产生的弯矩和在杆 $BC$ 中产生的轴力可分别表示为

$$M_F = 0 \quad (0 \le x \le a), \quad M_F = -F(x-a) \quad (a \le x < 2a), \quad N_F = 0$$

$C$ 处沿 $X_1$ 方向单位载荷在梁 $AB$ 中产生的弯矩和在杆 $BC$ 中产生的轴力分别为

$$\bar{M}_1(x) = x \quad (0 \le x < 2a), \quad \bar{N}_1 = 1$$

利用式(13.7),得

$$\delta_{11} = \int_0^{2a} \frac{\bar{M}_1^2}{EI} dx + \int_0^a \frac{\bar{N}_1^2}{EA} dx = \frac{8a^3}{3EI} + \frac{a}{EA} = \frac{10a^3}{3EI}, \quad \Delta_{1F} = \int_a^{2a} \frac{M_F \bar{M}_1}{EI} dx = -\frac{5Fa^3}{6EI}$$

根据力法正则方程 $\delta_{11} X_1 + \Delta_{1F} = 0$,有

$$X_1 = \frac{1}{4} F$$

对 $n$ 次静不定结构,假设 $X_1, X_2, \cdots, X_n$ 是多余力,那么静不定结构的力法正则方程为

$$\begin{cases} \delta_{11} X_1 + \delta_{12} X_2 + \cdots + \delta_{1n} X_n + \Delta_{1F} = 0 \\ \delta_{21} X_1 + \delta_{22} X_2 + \cdots + \delta_{2n} X_n + \Delta_{2F} = 0 \\ \vdots \\ \delta_{n1} X_1 + \delta_{n2} X_2 + \cdots + \delta_{nn} X_n + \Delta_{nF} = 0 \end{cases} \quad (13.6)$$

式中,$\delta_{ij}$ 为由 $X_j$ 处沿 $X_j$ 方向单位载荷在 $X_i$ 处产生的沿 $X_i$ 方向的位移;$\Delta_{iF}$ 为由结构上外载荷在 $X_i$ 处产生的沿 $X_i$ 方向的位移。$\delta_{ij}$ 和 $\Delta_{iF}$ 可分别表示为

$$\delta_{ij} = \int_l \frac{\bar{M}_i \bar{M}_j dx}{EI} \quad (i,j=1,2,\cdots,n), \quad \Delta_{iF} = \int \frac{M_F \bar{M}_i dx}{EI} \quad (i=1,2,\cdots,n) \quad (13.7)$$

式中,$\bar{M}_i$ 为 $X_i$ 处沿 $X_i$ 方向单位载荷产生的弯矩;$M_F$ 为静不定结构上外载荷产生的弯矩。

**例 13.4** 弯曲刚度为 $EI$ 的杆 $AB$ 和 $BC$ 在截面 $B$ 处通过焊接形成刚架,如图 13.7(a) 所示,刚架两端固定,受均布载荷作用,用力法求 $C$ 处反力。

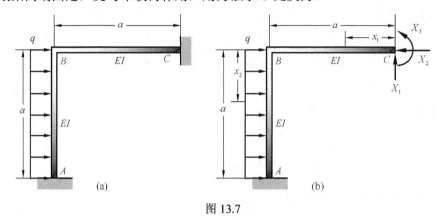

图 13.7

**解** 图 13.7(a) 为 3 次静不定刚架。选 $C$ 处约束力 $X_1$、$X_2$ 和 $X_3$ 作为多余反力,消除相应约束,如图 13.7(b) 所示。

均布载荷 $q$ 产生的弯矩可表示为

$$M_q(x_1) = 0 \ (0 < x_1 \leqslant a), \quad M_q(x_2) = -\frac{1}{2}qx_2^2 \ (0 \leqslant x_2 < a)$$

由 $C$ 处沿 $X_1$、$X_2$ 和 $X_3$ 方向单位载荷产生的弯矩分别可表示为

$$\bar{M}_1(x_1) = x_1 \ (0 < x_1 \leqslant a), \quad \bar{M}_1(x_2) = a \ (0 \leqslant x_2 < a)$$

$$\bar{M}_2(x_1) = 0 \ (0 < x_1 \leqslant a), \quad \bar{M}_2(x_2) = x_2 \ (0 \leqslant x_2 < a)$$

$$\bar{M}_3(x_1) = 1 \ (0 < x_1 \leqslant a), \quad \bar{M}_3(x_2) = 1 \ (0 \leqslant x_2 < a)$$

根据式(13.7),得

$$\delta_{11} = \frac{4a^3}{3EI}, \ \delta_{12} = \frac{a^3}{2EI}, \ \delta_{13} = \frac{3a^2}{2EI}, \ \delta_{22} = \frac{a^3}{3EI}, \ \delta_{23} = \frac{a^2}{2EI}, \ \delta_{33} = \frac{2a}{EI}$$

$$\Delta_{1F} = -\frac{qa^4}{6EI}, \ \Delta_{2F} = -\frac{qa^4}{8EI}, \ \Delta_{3F} = -\frac{qa^3}{6EI}$$

解力法正则方程

$$8aX_1 + 3aX_2 + 9X_3 = qa^2$$

$$12aX_1 + 8aX_2 + 12X_3 = 3qa^2$$

$$9aX_1 + 3aX_2 + 12X_3 = qa^2$$

得

$$\begin{bmatrix} X_1 \\ X_2 \\ X_3 \end{bmatrix} = \mathrm{inv}\left(\begin{bmatrix} 8a & 3a & 9 \\ 12a & 8a & 12 \\ 9a & 3a & 12 \end{bmatrix}\right)\begin{bmatrix} qa^2 \\ 3qa^2 \\ qa^2 \end{bmatrix} = \frac{1}{48}\begin{bmatrix} -3qa \\ 21qa \\ qa^2 \end{bmatrix}$$

式中,inv 表示矩阵求逆。

采用力法进行静不定结构分析的步骤简要总结如下：①确定静不定度，释放多余约束，获得稳定静定结构；②通过能量方法或叠加方法，计算 $\delta_{ij}$ 和 $\Delta_{iF}$；③利用位移协调，根据力法正则方程，计算 $X_i$。

## 13.3 力法分析对称静不定结构

### 1. 对称结构受对称载荷作用

对称载荷作用于对称静不定结构，则对称面上反对称内力(扭矩和剪力)恒等于零。如图 13.8 所示结构和载荷均对称于结构中间截面 $E$，因此截面 $E$ 上剪力为零。

### 2. 对称结构受反对称载荷作用

反对称载荷作用于对称静不定结构，则对称面上对称内力(轴力和弯矩)恒等于零。如图 13.9 所示结构对称而载荷反对称于结构中间截面 $E$，因此截面 $E$ 上轴力和弯矩均为零。

### 3. 对称结构受一般载荷作用

一般载荷作用于对称静不定结构，如图 13.10(a) 所示，可分为两个对称静不定结构的叠加，一个受对称载荷作用，如图 13.10(b) 所示；另一个受反对称载荷作用，如图 13.10(c) 所示。

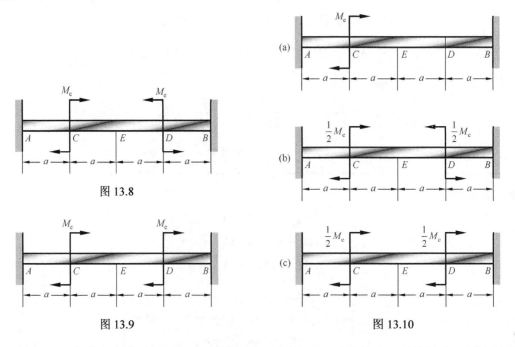

图 13.8

图 13.9

图 13.10

**例 13.5** 弯曲刚度为 $EI$ 的梁 $AB$ 两端固定，如图 13.11(a) 所示。不计轴力，用力法求中间截面 $E$ 处内力。

**解** 根据结构对称和载荷对称,中间截面 $E$ 处剪力为零。若轴力也不计,则弯矩将是中间截面 $E$ 处唯一内力。如果结构从中间截开,则原结构是 1 次静不定。取 $E$ 处弯矩 $X_1$ 作为多余,并利用 $E$ 处转角等于零,如图 13.11(b)所示,则有 $\delta_{11}X_1 + \Delta_{1F} = 0$。

外载荷 $F$ 在 $AE$ 段产生的弯矩可表示为

$$M_F = 0 \ (0 \leqslant x \leqslant a), \quad M_F = -F(x-a) \ (a \leqslant x < 2a)$$

$X_1$ 处单位载荷在 $AE$ 段产生的弯矩可表示为

$$\overline{M}_1(x) = 1 \ (0 \leqslant x < 2a)$$

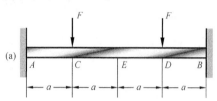

根据式(13.7),得

$$\delta_{11} = \int_0^{2a} \frac{\overline{M}_1^2}{EI} dx = \frac{2a}{EI}$$

$$\Delta_{1F} = \int_a^{2a} \frac{M_F \overline{M}_1}{EI} dx = -\frac{Fa^2}{2EI}$$

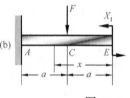

利用 $\delta_{11}X_1 + \Delta_{1F} = 0$,得

$$X_1 = \frac{1}{4}F$$

图 13.11

**例 13.6** 如图 13.12(a)所示,静不定刚架 $ABC$ 在 $A$ 和 $C$ 处固定。已知弯曲刚度 $EI$ 为常数,用力法求截面 $B$ 处内力。

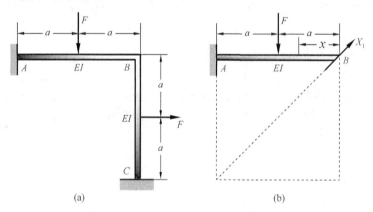

图 13.12

**解** 利用结构对称和载荷反对称,截面 $B$ 处轴力和弯矩均为零。剪力将是该处唯一内力。如果结构从截面 $B$ 处截开,则原结构是 1 次静不定。取 $B$ 处剪力 $X_1$ 作为多余,如图 13.12(b)所示,则有 $\delta_{11}X_1 + \Delta_{1F} = 0$。

外载荷 $F$ 在 $AB$ 段产生的弯矩可表示为

$$M_F = 0 \ (0 \leqslant x \leqslant a), \quad M_F = -F(x-a) \ (a \leqslant x < 2a)$$

$X_1$ 处单位载荷在 $AB$ 段产生的弯矩可表示为

$$\overline{M}_1(x) = \frac{1}{\sqrt{2}}x \quad (0 \leqslant x < 2a)$$

根据式(13.7)，有

$$\delta_{11} = \int_0^{2a} \frac{\overline{M}_1^2}{EI} \mathrm{d}x = \frac{4a^3}{3EI}, \quad \Delta_{1F} = \int_a^{2a} \frac{M_F \overline{M}_1}{EI} \mathrm{d}x = -\frac{5\sqrt{2}Fa^3}{12EI}$$

利用 $\delta_{11}X_1 + \Delta_{1F} = 0$，得

$$X_1 = \frac{5\sqrt{2}}{16}F$$

## 习　题

13.1　梁受如图 13.13 所示载荷作用，假设弯曲刚度 $EI$ 为常数，求可动铰支座处反力。

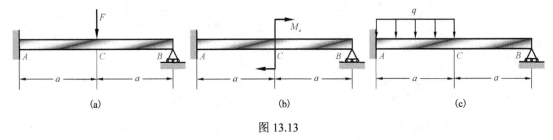

图 13.13

13.2　弯曲刚度为 $EI$ 的杆 $AB$ 和 $BC$ 焊接于 $B$ 处形成刚架(图 13.14)，用力法求 $C$ 处反力。

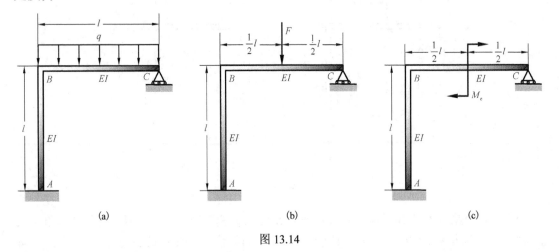

图 13.14

13.3　弯曲刚度为 $EI$ 的梁 $AB$ 和拉伸刚度为 $EA$ 的杆 $BC$ 铰接于 $B$(图 13.15)，已知 $EA = 3EI/(10a^2)$，采用力法求 $C$ 处反力。

13.4　弯曲刚度为 $EI$ 的梁 $AB$ 两端固定(图 13.16)，不计轴力，用力法求中间截面 $E$ 处内力。

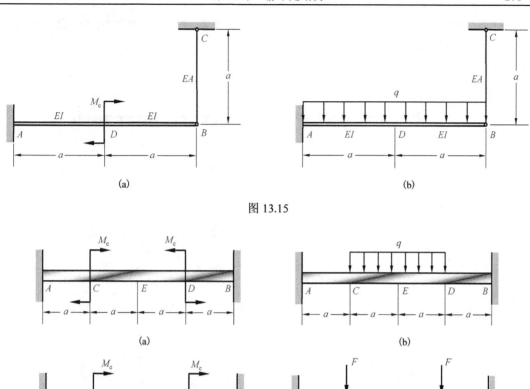

图 13.15

图 13.16

**13.5** 静不定刚架 $ABC$ 在 $A$ 和 $C$ 处固定(图 13.17)。已知弯曲刚度 $EI$ 为常数,用力法求截面 $B$ 处内力。

**13.6** 超静定刚架,两端固定,受均布载荷 $q$ 作用(图 13.18)。已知刚架弯曲刚度 $EI$ 为常量,试用力法求截面 $C$ 处内力(提示:已知截面 $C$ 处轴力和剪力均等于零)。

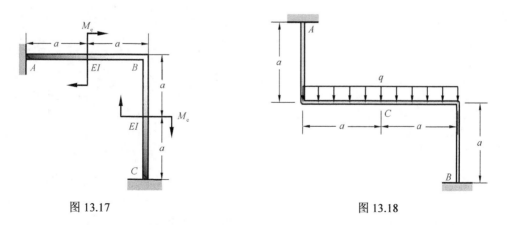

图 13.17        图 13.18

**13.7** 静不定刚架 $ABC$ 在 $A$ 处固定、$C$ 处铰支(图 13.19)。已知弯曲刚度 $EI$ 为常数，用力法求截面 $C$ 处反力。

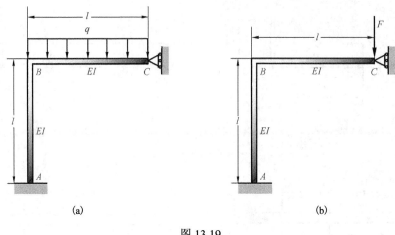

图 13.19

# 附录 I 截 面 性 质

## I.1 静 矩

考虑如图 I.1 所示阴影截面 $A$，截面对 $z$ 轴和 $y$ 轴的静矩分别定义为

$$Q_z = \int y \mathrm{d}A, \quad Q_y = \int z \mathrm{d}A \tag{I.1}$$

在国际单位制中，静矩 $Q_z$ 和 $Q_y$ 的单位为 $\mathrm{m}^3$。

截面的形心定义为
$$\bar{y} = \frac{\int y \mathrm{d}A}{A}, \quad \bar{z} = \frac{\int z \mathrm{d}A}{A} \tag{I.2}$$

比较方程(I.1)和方程(I.2)，得
$$Q_z = A\bar{y}, \quad Q_y = A\bar{z} \tag{I.3}$$

对于如图 I.2 所示组合截面，其由截面 $A_1$ 和 $A_2$ 组成，则组合截面对 $z$ 轴和 $y$ 轴的静矩分别表示为

$$Q_z = \sum (Q_z)_i = \sum A_i \bar{y}_i, \quad Q_y = \sum (Q_y)_i = \sum A_i \bar{z}_i \tag{I.4}$$

组合截面的形心可写为

$$\bar{y} = \frac{\sum A_i \bar{y}_i}{\sum A_i}, \quad \bar{z} = \frac{\sum A_i \bar{z}_i}{\sum A_i} \tag{I.5}$$

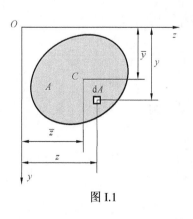

图 I.1

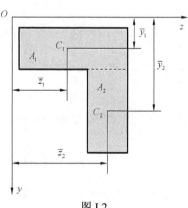

图 I.2

## I.2 惯性矩与极惯性矩

**1. 惯性矩**

考虑如图 I.3 所示阴影截面 $A$，截面对 $z$ 轴和 $y$ 轴的惯性矩分别定义为

$$I_z = \int y^2 \mathrm{d}A, \quad I_y = \int z^2 \mathrm{d}A \tag{I.6}$$

在国际单位制中，惯性矩 $I_z$ 和 $I_y$ 的单位为 $\mathrm{m}^4$。

**2. 极惯性矩**

考虑如图 I.4 所示阴影截面 $A$，截面对原点 $O$ 的极惯性矩定义为

$$I_\mathrm{p} = \int \rho^2 \mathrm{d}A \tag{I.7}$$

在国际单位制中，极惯性矩 $I_\mathrm{p}$ 的单位为 $\mathrm{m}^4$。利用 $\rho^2 = y^2 + z^2$，得

$$I_\mathrm{p} = \int \rho^2 \mathrm{d}A = \int y^2 \mathrm{d}A + \int z^2 \mathrm{d}A = I_z + I_y \tag{I.8}$$

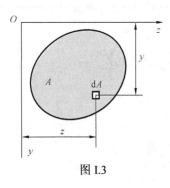

图 I.3

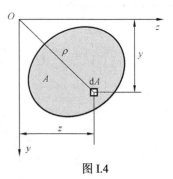

图 I.4

## I.3 惯性半径与极惯性半径

**1. 惯性半径**

考虑如图 I.3 所示阴影截面 $A$，截面对 $z$ 轴和 $y$ 轴的惯性半径分别定义为

$$i_z = \sqrt{\frac{I_z}{A}}, \quad i_y = \sqrt{\frac{I_y}{A}} \tag{I.9}$$

在国际单位制中，惯性半径 $i_z$ 和 $i_y$ 的单位为 $\mathrm{m}$。

**2. 极惯性半径**

考虑如图 I.4 所示阴影截面 $A$，截面对原点 $O$ 的极惯性半径定义为

$$i_p = \sqrt{\frac{I_p}{A}} \tag{I.10}$$

在国际单位制中,极惯性半径 $i_p$ 的单位为 m。

## I.4 惯性积

考虑如图 I.3 所示阴影截面 $A$,截面的惯性积定义为

$$I_{zy} = \int zy \, dA \tag{I.11}$$

在国际单位制中,惯性积 $I_{zy}$ 的单位为 $m^4$。

## I.5 平行移轴定理

考虑如图 I.5 所示阴影截面 $A$,截面对 $z$ 轴和 $z_C$ 轴的惯性矩分别表示为

$$I_z = \int y^2 dA, \quad I_{z_C} = \int y_C^2 dA \tag{I.12}$$

代入 $y = y_C + b$,有

$$I_z = \int y^2 dA = \int (y_C + b)^2 dA = \int y_C^2 dA + 2b \int y_C dA + b^2 \int dA \tag{I.13}$$

利用 $\int y_C dA = 0$,则方程(I.13)可简化为

$$I_z = I_{z_C} + Ab^2 \tag{I.14}$$

同理,得 $\quad I_y = I_{y_C} + Aa^2, \quad I_{zy} = I_{z_C y_C} + Aab, \quad I_p = (I_p)_C + A(a^2 + b^2) \tag{I.15}$

方程(I.14)和方程(I.15)所表示的关系称为平行移轴定理。当已知截面对形心轴的惯性矩和极惯性矩时,上述关系常用于计算截面对任意轴的惯性矩和极惯性矩。

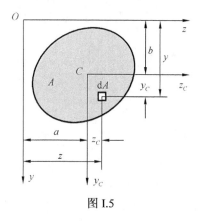

图 I.5

# I.6 常用截面几何性质

| 截面类型 | | 惯性矩和截面模量 |
|---|---|---|
| 矩形截面 | | $A = bh$<br>$I_z = \dfrac{1}{12}bh^3, \quad S_z = \dfrac{1}{6}bh^2$<br>$I_{z'} = I_z + Aa^2$ |
| 实心圆截面 | | $A = \dfrac{1}{4}\pi d^2$<br>$I_z = \dfrac{1}{64}\pi d^4, \quad S_z = \dfrac{1}{32}\pi d^3$<br>$(I_p)_C = \dfrac{1}{32}\pi d^4, \quad (S_p)_C = \dfrac{1}{16}\pi d^3$<br>$I_{z'} = I_z + Aa^2$ |
| 空心圆截面 | | $A = \dfrac{1}{4}\pi D^2(1-\alpha^2) \ (\alpha = d/D)$<br>$I_z = \dfrac{1}{64}\pi D^4(1-\alpha^4), \quad S_z = \dfrac{1}{32}\pi D^3(1-\alpha^4)$<br>$(I_p)_C = \dfrac{1}{32}\pi D^4(1-\alpha^4), \quad (S_p)_C = \dfrac{1}{16}\pi D^3(1-\alpha^4)$<br>$I_{z'} = I_z + Aa^2$ |

# 附录Ⅱ 型 钢

## Ⅱ.1 工 字 钢

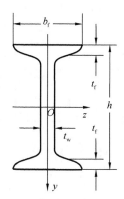

| 型号 | 单位长度质量/(kg/m) | 高度 $h$/mm | 翼缘 宽度 $b_f$/mm | 翼缘 厚度 $t_f$/mm | 腹板 厚度 $t_w$/mm | 面积 $A$/cm² | 惯性矩 $I_z$/cm⁴ | 惯性矩 $I_y$/cm⁴ | 截面模量 $S_z$/cm³ | 截面模量 $S_y$/cm³ |
|---|---|---|---|---|---|---|---|---|---|---|
| 10 | 11.261 | 100 | 68 | 7.6 | 4.5 | 14.345 | 245 | 33.0 | 49.0 | 9.72 |
| 12.6 | 14.223 | 126 | 74 | 8.4 | 5.0 | 18.118 | 488 | 46.9 | 77.5 | 12.7 |
| 14 | 16.890 | 140 | 80 | 9.1 | 5.5 | 21.516 | 712 | 64.4 | 102 | 16.1 |
| 16 | 20.513 | 160 | 88 | 9.9 | 6.0 | 26.131 | 1130 | 93.1 | 141 | 21.2 |
| 18 | 24.143 | 180 | 94 | 10.7 | 6.5 | 30.756 | 1660 | 122 | 185 | 26.0 |
| 20a | 27.929 | 200 | 100 | 11.4 | 7.0 | 35.578 | 2370 | 158 | 237 | 31.5 |
| 20b | 31.069 | 200 | 102 | 11.4 | 9.0 | 39.578 | 2500 | 169 | 250 | 33.1 |
| 22a | 33.070 | 220 | 110 | 12.3 | 7.5 | 42.128 | 3400 | 225 | 309 | 40.9 |
| 22b | 36.524 | 220 | 112 | 12.3 | 9.5 | 46.528 | 3570 | 239 | 325 | 42.7 |
| 25a | 38.105 | 250 | 116 | 13.0 | 8.0 | 48.541 | 5020 | 280 | 402 | 48.3 |
| 25b | 42.030 | 250 | 118 | 13.0 | 10.0 | 53.541 | 52800 | 309 | 423 | 52.4 |
| 28a | 43.492 | 280 | 122 | 13.7 | 8.5 | 55.404 | 7110 | 345 | 508 | 56.6 |
| 28b | 47.888 | 280 | 124 | 13.7 | 10.5 | 61.004 | 7480 | 379 | 534 | 61.2 |
| 32a | 52.717 | 320 | 130 | 15.0 | 9.5 | 67.156 | 11100 | 460 | 692 | 70.8 |
| 32b | 57.741 | 320 | 132 | 15.0 | 11.5 | 73.556 | 11600 | 502 | 723 | 76.0 |
| 32c | 62.765 | 320 | 134 | 15.0 | 13.5 | 79.956 | 12200 | 544 | 760 | 81.2 |
| 36a | 60.037 | 360 | 136 | 15.8 | 10.0 | 76.480 | 15800 | 552 | 875 | 81.2 |
| 36b | 65.689 | 360 | 138 | 15.8 | 12.0 | 83.680 | 16500 | 582 | 919 | 84.3 |
| 36c | 71.341 | 360 | 140 | 15.8 | 14.0 | 90.880 | 17300 | 612 | 962 | 87.4 |
| 40a | 67.598 | 400 | 142 | 16.5 | 10.5 | 86.112 | 21700 | 660 | 1090 | 93.2 |
| 40b | 73.878 | 400 | 144 | 16.5 | 12.5 | 94.112 | 22800 | 692 | 1140 | 96.2 |
| 40c | 80.158 | 400 | 146 | 16.5 | 14.5 | 102.112 | 23900 | 727 | 1190 | 99.6 |

续表

| 型号 | 单位长度质量 /(kg/m) | 高度 h/mm | 翼缘 宽度 $b_f$/mm | 翼缘 厚度 $t_f$/mm | 腹板 厚度 $t_w$/mm | 面积 A/cm² | 惯性矩 $I_z$/cm⁴ | 惯性矩 $I_y$/cm⁴ | 截面模量 $S_z$/cm³ | 截面模量 $S_y$/cm³ |
|---|---|---|---|---|---|---|---|---|---|---|
| 45a | 80.420 | 450 | 150 | 18.0 | 11.5 | 102.446 | 32200 | 855 | 1430 | 114 |
| 45b | 87.485 | 450 | 152 | 18.0 | 13.5 | 111.446 | 33800 | 894 | 1500 | 118 |
| 45c | 94.550 | 450 | 154 | 18.0 | 15.5 | 120.116 | 35300 | 938 | 1570 | 122 |
| 50a | 93.654 | 500 | 158 | 20.0 | 12.0 | 119.304 | 46500 | 1120 | 1860 | 142 |
| 50b | 101.504 | 500 | 160 | 20.0 | 14.0 | 129.304 | 48600 | 1170 | 1940 | 146 |
| 50c | 109.354 | 500 | 162 | 20.0 | 16.0 | 139.304 | 50600 | 1220 | 2080 | 151 |
| 56a | 106.316 | 560 | 166 | 21.0 | 12.5 | 135.435 | 65600 | 1370 | 2340 | 165 |
| 56b | 115.108 | 560 | 168 | 21.0 | 14.5 | 146.635 | 68500 | 1490 | 2450 | 174 |
| 56c | 123.900 | 560 | 170 | 21.0 | 16.5 | 157.835 | 71400 | 1560 | 2550 | 183 |
| 63a | 121.407 | 630 | 176 | 22.0 | 13.0 | 154.658 | 93900 | 1700 | 2980 | 193 |
| 63b | 131.298 | 630 | 178 | 22.0 | 15.0 | 167.258 | 98100 | 1810 | 3160 | 204 |
| 63c | 141.189 | 630 | 180 | 22.0 | 17.0 | 179.858 | 102000 | 1920 | 3300 | 214 |

## II.2 槽 钢

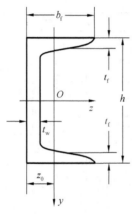

| 型号 | 单位长度质量 /(kg/m) | 高度 h/mm | 翼缘 宽度 $b_f$/mm | 翼缘 厚度 $t_f$/mm | 腹板 厚度 $t_w$/mm | 面积 A/cm² | 惯性矩 $I_z$/cm⁴ | 惯性矩 $I_y$/cm⁴ | 截面模量 $S_z$/cm³ | 截面模量 $S_y$/cm³ | $z_0$/cm |
|---|---|---|---|---|---|---|---|---|---|---|---|
| 5 | 5.438 | 50 | 37 | 7 | 4.5 | 6.928 | 26.0 | 8.3 | 10.4 | 3.55 | 1.35 |
| 6.3 | 6.634 | 63 | 40 | 7.5 | 4.8 | 8.451 | 50.8 | 11.9 | 16.1 | 4.50 | 1.36 |
| 8 | 8.045 | 80 | 43 | 8 | 5.0 | 10.248 | 101 | 16.6 | 25.3 | 5.79 | 1.43 |
| 10 | 10.007 | 100 | 48 | 8.5 | 5.3 | 42.748 | 198 | 25.6 | 39.7 | 7.8 | 1.52 |
| 12.6 | 12.318 | 126 | 53 | 9 | 5.5 | 15.692 | 391 | 38.0 | 62.1 | 10.2 | 1.59 |
| 14a | 14.535 | 140 | 58 | 9.5 | 6.0 | 18.516 | 564 | 53.2 | 80.5 | 13.0 | 1.71 |
| 14b | 16.733 | 140 | 60 | 9.5 | 8.0 | 21.316 | 609 | 61.1 | 87.1 | 14.1 | 1.67 |
| 16a | 17.240 | 160 | 63 | 10 | 6.5 | 21.962 | 866 | 73.3 | 108 | 16.3 | 1.80 |
| 16 | 19.752 | 160 | 65 | 10 | 8.5 | 25.162 | 935 | 83.4 | 117 | 17.6 | 1.75 |
| 18a | 20.174 | 180 | 68 | 10.5 | 7.0 | 25.699 | 1270 | 98.6 | 141 | 20.0 | 1.88 |
| 18 | 23.000 | 180 | 70 | 10.5 | 9.0 | 29.299 | 1370 | 111 | 152 | 21.5 | 1.84 |

续表

| 型号 | 单位长度质量 /(kg/m) | 高度 $h$/mm | 翼缘 宽度 $b_f$/mm | 翼缘 厚度 $t_f$/mm | 腹板 厚度 $t_w$/mm | 面积 $A$/cm$^2$ | 惯性矩 $I_z$/cm$^4$ | 惯性矩 $I_y$/cm$^4$ | 截面模量 $S_z$/cm$^3$ | 截面模量 $S_y$/cm$^3$ | $z_0$/cm |
|---|---|---|---|---|---|---|---|---|---|---|---|
| 20a | 22.637 | 200 | 73 | 11 | 7.0 | 28.837 | 1780 | 128 | 178 | 24.2 | 2.01 |
| 20 | 25.777 | 200 | 75 | 11 | 9.0 | 32.837 | 1910 | 144 | 191 | 25.9 | 1.95 |
| 22a | 24.999 | 220 | 77 | 11.5 | 7.0 | 31.846 | 2390 | 158 | 218 | 28.2 | 2.10 |
| 22 | 28.453 | 220 | 79 | 11.5 | 9.0 | 36.246 | 2570 | 176 | 234 | 30.1 | 2.03 |
| 25a | 27.410 | 250 | 78 | 12 | 7.0 | 34.917 | 3370 | 176 | 270 | 30.6 | 2.07 |
| 25b | 31.335 | 250 | 80 | 12 | 9.0 | 39.917 | 3530 | 196 | 282 | 32.7 | 1.98 |
| 25c | 35.260 | 250 | 82 | 12 | 11.0 | 44.917 | 3690 | 218 | 295 | 35.9 | 1.92 |
| 28a | 31.427 | 280 | 82 | 12.5 | 7.5 | 40.034 | 4760 | 218 | 340 | 35.7 | 2.10 |
| 28b | 35.823 | 280 | 84 | 12.5 | 9.5 | 45.634 | 5130 | 242 | 366 | 37.9 | 2.02 |
| 28c | 40.219 | 280 | 86 | 12.5 | 11.5 | 51.234 | 5500 | 268 | 393 | 40.3 | 1.95 |
| 32a | 38.083 | 320 | 88 | 14 | 8.0 | 48.513 | 7600 | 305 | 475 | 46.5 | 2.24 |
| 32b | 43.107 | 320 | 90 | 14 | 10.0 | 54.913 | 8140 | 336 | 509 | 49.2 | 2.16 |
| 32c | 48.131 | 320 | 92 | 14 | 12.0 | 61.313 | 8690 | 374 | 543 | 52.6 | 2.09 |
| 36a | 47.814 | 360 | 96 | 16 | 9.0 | 60.910 | 11900 | 455 | 660 | 63.5 | 2.44 |
| 36b | 53.466 | 360 | 98 | 16 | 11.0 | 68.110 | 12700 | 497 | 703 | 66.9 | 2.37 |
| 36c | 59.118 | 360 | 100 | 16 | 13.0 | 75.310 | 13400 | 536 | 746 | 70.0 | 2.34 |
| 40a | 58.928 | 400 | 100 | 18 | 10.5 | 75.068 | 17600 | 592 | 879 | 78.8 | 2.49 |
| 40b | 65.208 | 400 | 102 | 18 | 12.5 | 93.068 | 18600 | 640 | 932 | 82.5 | 2.44 |
| 40c | 71.488 | 400 | 104 | 18 | 14.5 | 91.068 | 19700 | 688 | 986 | 86.2 | 2.42 |

## II.3 等边角钢

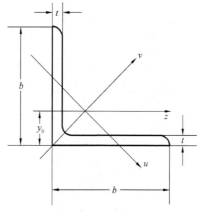

| 型号 | 单位长度质量 /(kg/m) | 宽度 $b$/mm | 厚度 $t$/mm | 面积 $A$/cm$^2$ | 惯性矩 $I_z$/cm$^4$ | 惯性矩 $I_v$/cm$^4$ | 惯性矩 $I_u$/cm$^4$ | $y_0$/cm |
|---|---|---|---|---|---|---|---|---|
| 2 | 0.889 | 20 | 3 | 1.132 | 0.40 | 0.63 | 0.17 | 0.60 |
|   | 1.145 |    | 4 | 1.459 | 0.50 | 0.78 | 0.22 | 0.64 |
| 2.5 | 1.124 | 25 | 3 | 1.432 | 0.82 | 1.29 | 0.34 | 0.73 |
|     | 1.459 |    | 4 | 1.859 | 1.03 | 1.62 | 0.43 | 0.76 |

续表

| 型号 | 单位长度质量 /(kg/m) | 宽度 b/mm | 厚度 t/mm | 面积 A/cm² | 惯性矩 $I_x$/cm⁴ | $I_y$/cm⁴ | $I_u$/cm⁴ | $y_0$/cm |
|---|---|---|---|---|---|---|---|---|
| 3.0 | 1.373 | 30 | 3 | 1.749 | 1.46 | 2.31 | 0.61 | 0.85 |
|  | 1.786 |  | 4 | 2.276 | 1.84 | 2.92 | 0.77 | 0.89 |
| 3.6 | 1.656 | 36 | 3 | 2.109 | 2.58 | 4.09 | 1.07 | 1.00 |
|  | 2.163 |  | 4 | 2.756 | 3.29 | 5.22 | 1.37 | 1.04 |
|  | 2.654 |  | 5 | 3.382 | 3.95 | 6.24 | 1.65 | 1.07 |
| 4.0 | 1.852 | 40 | 3 | 2.359 | 3.59 | 5.69 | 1.49 | 1.09 |
|  | 2.422 |  | 4 | 3.086 | 4.60 | 7.29 | 1.91 | 1.13 |
|  | 2.976 |  | 5 | 3.791 | 5.53 | 8.76 | 2.30 | 1.17 |
| 4.5 | 2.088 | 45 | 3 | 2.659 | 5.17 | 8.20 | 2.14 | 1.22 |
|  | 2.736 |  | 4 | 3.486 | 6.65 | 10.56 | 2.75 | 1.26 |
|  | 3.369 |  | 5 | 4.292 | 8.04 | 12.74 | 3.33 | 1.30 |
|  | 3.985 |  | 6 | 5.076 | 9.33 | 14.76 | 3.89 | 1.33 |
| 5 | 2.332 | 50 | 3 | 2.971 | 7.18 | 11.37 | 2.98 | 1.34 |
|  | 3.059 |  | 4 | 3.897 | 9.26 | 14.70 | 3.82 | 1.38 |
|  | 3.770 |  | 5 | 4.803 | 11.21 | 17.79 | 4.64 | 1.42 |
|  | 4.465 |  | 6 | 5.688 | 13.05 | 20.68 | 5.42 | 1.46 |
| 5.6 | 2.624 | 56 | 3 | 3.343 | 10.19 | 16.14 | 4.24 | 1.48 |
|  | 3.446 |  | 4 | 4.390 | 13.18 | 20.92 | 5.46 | 1.53 |
|  | 4.251 |  | 5 | 5.415 | 16.02 | 25.42 | 6.61 | 1.57 |
|  | 6.568 |  | 8 | 8.367 | 23.63 | 37.37 | 9.89 | 1.68 |
| 6.3 | 3.907 | 63 | 4 | 4.978 | 19.03 | 30.17 | 7.89 | 1.70 |
|  | 4.882 |  | 5 | 6.143 | 23.17 | 36.77 | 9.57 | 1.74 |
|  | 5.721 |  | 6 | 7.288 | 27.12 | 43.03 | 11.20 | 1.78 |
|  | 7.469 |  | 8 | 9.515 | 34.46 | 54.56 | 14.33 | 1.85 |
|  | 9.151 |  | 10 | 11.657 | 41.09 | 64.85 | 17.33 | 1.93 |
| 7 | 4.372 | 70 | 4 | 5.570 | 26.39 | 41.80 | 10.99 | 1.86 |
|  | 5.397 |  | 5 | 6.875 | 32.21 | 51.08 | 13.34 | 1.91 |
|  | 6.406 |  | 6 | 8.160 | 37.77 | 59.93 | 15.61 | 1.95 |
|  | 7.398 |  | 7 | 9.424 | 43.09 | 68.35 | 17.82 | 1.99 |
|  | 8.373 |  | 8 | 10.667 | 48.17 | 76.37 | 19.98 | 2.03 |
| 7.5 | 5.818 | 75 | 5 | 7.412 | 39.97 | 63.30 | 16.63 | 2.04 |
|  | 6.905 |  | 6 | 8.797 | 46.95 | 74.38 | 19.51 | 2.07 |
|  | 7.976 |  | 7 | 10.160 | 53.57 | 84.96 | 22.18 | 2.11 |
|  | 9.030 |  | 8 | 11.503 | 59.96 | 95.07 | 24.86 | 2.15 |
|  | 11.089 |  | 10 | 14.126 | 71.98 | 113.92 | 30.05 | 2.22 |
| 8 | 6.211 | 80 | 5 | 7.912 | 48.79 | 77.33 | 20.25 | 2.15 |
|  | 7.376 |  | 6 | 9.397 | 57.35 | 90.98 | 23.72 | 2.19 |
|  | 8.525 |  | 7 | 10.860 | 65.58 | 104.07 | 27.09 | 2.23 |
|  | 9.658 |  | 8 | 12.303 | 73.49 | 116.60 | 30.39 | 2.27 |
|  | 11.874 |  | 10 | 15.126 | 88.43 | 140.09 | 36.77 | 2.35 |
| 9 | 8.350 | 90 | 6 | 10.637 | 82.77 | 131.26 | 34.28 | 2.44 |
|  | 9.656 |  | 7 | 12.301 | 94.83 | 150.47 | 39.18 | 2.48 |
|  | 10.946 |  | 8 | 13.944 | 106.47 | 168.97 | 43.97 | 2.52 |
|  | 13.476 |  | 10 | 17.167 | 128.58 | 203.90 | 53.26 | 2.59 |
|  | 15.940 |  | 12 | 20.306 | 149.22 | 236.21 | 62.22 | 2.67 |

续表

| 型号 | 单位长度质量/(kg/m) | 宽度 b/mm | 厚度 t/mm | 面积 A/cm² | 惯性矩 $I_z$/cm⁴ | 惯性矩 $I_y$/cm⁴ | 惯性矩 $I_u$/cm⁴ | $y_0$/cm |
|---|---|---|---|---|---|---|---|---|
| 10 | 9.366 | 100 | 6 | 11.932 | 114.95 | 181.98 | 47.92 | 2.67 |
|    | 10.830 |    | 7 | 13.796 | 131.86 | 208.97 | 54.74 | 2.71 |
|    | 12.276 |    | 8 | 15.638 | 148.24 | 235.07 | 61.41 | 2.76 |
|    | 15.120 |    | 10 | 19.161 | 179.51 | 284.68 | 74.35 | 2.84 |
|    | 17.898 |    | 12 | 22.800 | 208.90 | 330.95 | 86.84 | 2.91 |
|    | 20.611 |    | 14 | 26.256 | 236.53 | 374.06 | 99.00 | 2.99 |
|    | 23.257 |    | 16 | 29.627 | 262.53 | 414.16 | 110.89 | 3.06 |
| 11 | 11.928 | 110 | 7 | 15.196 | 177.16 | 280.94 | 73.38 | 2.96 |
|    | 13.532 |    | 8 | 17.238 | 199.46 | 316.49 | 82.42 | 3.01 |
|    | 16.690 |    | 10 | 21.261 | 242.19 | 384.39 | 99.98 | 3.09 |
|    | 19.782 |    | 12 | 25.200 | 282.55 | 448.17 | 116.93 | 3.16 |
|    | 22.809 |    | 14 | 29.056 | 320.71 | 508.01 | 133.40 | 3.24 |
| 12.5 | 15.504 | 125 | 8 | 19.750 | 297.03 | 470.89 | 123.16 | 3.37 |
|    | 19.133 |    | 10 | 24.373 | 361.67 | 573.89 | 149.46 | 3.45 |
|    | 22.696 |    | 12 | 28.912 | 423.16 | 671.44 | 174.88 | 3.53 |
|    | 26.193 |    | 14 | 33.367 | 481.65 | 763.73 | 199.57 | 3.61 |
| 14 | 21.488 | 140 | 10 | 27.372 | 514.65 | 817.27 | 212.04 | 3.82 |
|    | 25.522 |    | 12 | 32.512 | 603.68 | 958.79 | 248.57 | 3.90 |
|    | 29.490 |    | 14 | 37.567 | 688.81 | 1093.56 | 284.06 | 3.98 |
|    | 33.393 |    | 16 | 42.539 | 770.24 | 1221.81 | 318.67 | 4.06 |
| 16 | 24.729 | 160 | 10 | 31.502 | 779.63 | 1237.30 | 321.76 | 4.31 |
|    | 29.391 |    | 12 | 37.441 | 916.58 | 1455.68 | 377.49 | 4.39 |
|    | 33.987 |    | 14 | 43.296 | 1048.36 | 1665.02 | 431.70 | 4.47 |
|    | 38.518 |    | 16 | 49.067 | 1175.08 | 1865.57 | 484.59 | 4.55 |
| 18 | 33.159 | 180 | 12 | 42.241 | 1321.35 | 2100.10 | 542.61 | 4.89 |
|    | 38.383 |    | 14 | 48.896 | 1514.48 | 2407.42 | 621.53 | 4.97 |
|    | 43.542 |    | 16 | 55.467 | 1700.99 | 2703.37 | 698.60 | 5.05 |
|    | 48.634 |    | 18 | 61.955 | 1875.12 | 2988.24 | 762.01 | 5.13 |
| 20 | 42.894 | 200 | 14 | 54.642 | 2103.55 | 3343.26 | 863.83 | 5.46 |
|    | 48.680 |    | 16 | 62.013 | 2366.15 | 3760.89 | 971.41 | 5.54 |
|    | 54.401 |    | 18 | 69.301 | 2620.64 | 4164.54 | 1076.74 | 5.62 |
|    | 60.056 |    | 20 | 76.505 | 2867.30 | 4554.55 | 1180.04 | 5.69 |
|    | 71.168 |    | 24 | 90.661 | 3338.25 | 5294.97 | 1381.53 | 5.87 |

## II.4 不等边角钢

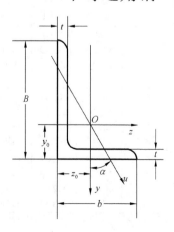

| 型号 | 单位长度质量 /(kg/m) | 宽度 B/mm | 宽度 b/mm | 厚度 t/mm | 面积 A/cm² | 惯性矩 $I_x$/cm⁴ | 惯性矩 $I_y$/cm⁴ | 惯性矩 $I_u$/cm⁴ | $z_0$/cm | $y_0$/cm | $\tan\alpha$ |
|---|---|---|---|---|---|---|---|---|---|---|---|
| 2.5/1.6 | 0.192 | 25 | 16 | 3 | 1.162 | 0.70 | 0.22 | 0.14 | 0.42 | 0.86 | 0.392 |
|  | 1.176 |  |  | 4 | 1.499 | 0.88 | 0.27 | 0.17 | 0.46 | 0.90 | 0.381 |
| 3.2/2 | 1.171 | 32 | 20 | 3 | 1.492 | 1.53 | 0.46 | 0.28 | 0.49 | 1.08 | 0.382 |
|  | 1.522 |  |  | 4 | 1.939 | 1.93 | 0.57 | 0.35 | 0.53 | 1.12 | 0.374 |
| 4/2.5 | 1.484 | 40 | 25 | 3 | 1.890 | 3.08 | 0.93 | 0.56 | 0.59 | 1.32 | 0.385 |
|  | 1.936 |  |  | 4 | 2.467 | 3.93 | 1.18 | 0.71 | 0.63 | 1.37 | 0.381 |
| 4.5/2.8 | 1.687 | 45 | 28 | 3 | 2.149 | 4.45 | 1.34 | 0.80 | 0.64 | 1.47 | 0.383 |
|  | 2.203 |  |  | 4 | 2.806 | 5.69 | 1.70 | 1.02 | 0.68 | 1.51 | 0.380 |
| 5/3.2 | 1.908 | 50 | 32 | 3 | 2.431 | 6.24 | 2.02 | 1.20 | 0.73 | 1.60 | 0.404 |
|  | 2.494 |  |  | 4 | 3.177 | 8.02 | 2.58 | 1.53 | 0.77 | 1.65 | 0.402 |
| 5.6/3.6 | 2.153 | 56 | 36 | 3 | 2.743 | 8.88 | 2.92 | 1.73 | 0.80 | 1.78 | 0.408 |
|  | 2.818 |  |  | 4 | 3.590 | 11.45 | 3.76 | 2.23 | 0.85 | 1.82 | 0.408 |
|  | 3.466 |  |  | 5 | 4.415 | 13.86 | 4.49 | 2.67 | 0.88 | 1.87 | 0.404 |
| 6.3/4 | 3.185 | 63 | 40 | 4 | 4.058 | 16.49 | 5.23 | 3.12 | 0.92 | 2.04 | 0.398 |
|  | 3.920 |  |  | 5 | 4.993 | 20.02 | 6.31 | 3.76 | 0.95 | 2.08 | 0.396 |
|  | 4.638 |  |  | 6 | 5.908 | 23.36 | 7.29 | 4.34 | 0.99 | 2.12 | 0.393 |
|  | 5.339 |  |  | 7 | 6.802 | 26.53 | 8.24 | 4.97 | 1.03 | 2.15 | 0.389 |
| 7/4.5 | 3.570 | 70 | 45 | 4 | 4.547 | 23.17 | 7.55 | 4.40 | 1.02 | 2.24 | 0.410 |
|  | 4.403 |  |  | 5 | 5.609 | 27.95 | 9.13 | 5.40 | 1.06 | 2.28 | 0.407 |
|  | 5.218 |  |  | 8 | 6.647 | 32.54 | 10.62 | 6.35 | 1.09 | 2.32 | 0.404 |
|  | 6.011 |  |  | 7 | 7.657 | 37.22 | 12.01 | 7.16 | 1.13 | 2.36 | 0.402 |
| (7.5/5) | 4.808 | 75 | 50 | 5 | 6.125 | 34.86 | 12.61 | 7.41 | 1.17.201 | 2.40 | 0.435 |
|  | 5.699 |  |  | 6 | 7.260 | 41.12 | 14.70 | 8.54 | 1 1.21 | 2.44 | 0.435 |
|  | 7.431 |  |  | 8 | 9.467 | 52.39 | 18.53 | 10.87 | 1.29 | 2.52 | 0.429 |
|  | 9.098 |  |  | 10 | 11.590 | 62.71 | 21.96 | 13.10 | 1.36 | 2.60 | 0.423 |
| 8/5 | 5.005 | 80 | 50 | 5 | 6.375 | 41.96 | 12.82 | 7.66 | 1.14 | 2.60 | 0.388 |
|  | 5.935 |  |  | 6 | 7.560 | 49.49 | 14.95 | 8.85 | 1.18 | 2.65 | 0.387 |
|  | 6.848 |  |  | 7 | 8.724 | 56.16 | 16.96 | 10.18 | 1.21 | 2.69 | 0.384 |
|  | 7.745 |  |  | 8 | 9.867 | 62.83 | 18.85 | 11.38 | 1.25 | 2.73 | 0.381 |
| 9/5.6 | 5.661 | 90 | 56 | 5 | 7.212 | 60.45 | 18.32 | 10.98 | 1.25 | 2.91 | 0.385 |
|  | 6.717 |  |  | 6 | 8.557 | 71.03 | 21.42 | 12.90 | 1.29 | 2.95 | 0.384 |
|  | 7.756 |  |  | 7 | 9.880 | 81.01 | 24.36 | 14.67 | 1.33 | 3.00 | 0.382 |
|  | 8.779 |  |  | 8 | 11.183 | 91.03 | 27.15 | 16.34 | 1.36 | 3.04 | 0.380 |
| 10/6.3 | 7.550 | 100 | 63 | 6 | 9.617 | 99.06 | 30.94 | 18.42 | 1.43 | 3.24 | 0.394 |
|  | 8.722 |  |  | 7 | 11.111 | 113.45 | 35.26 | 21.00 | 1.47 | 3.28 | 0.394 |
|  | 9.878 |  |  | 8 | 12.584 | 127.37 | 39.39 | 23.50 | 1.50 | 3.32 | 0.391 |
|  | 12.142 |  |  | 10 | 15.467 | 153.81 | 47.12 | 28.33 | 1.58 | 3.40 | 0.387 |
| 10/8 | 8.350 | 100 | 80 | 6 | 10.637 | 107.04 | 61.24 | 31.65 | 1.97 | 2.95 | 0.627 |
|  | 9.656 |  |  | 7 | 12.301 | 122.73 | 70.08 | 36.17 | 2.01 | 3.00 | 0.626 |
|  | 10.946 |  |  | 8 | 13.944 | 137.92 | 78.58 | 40.58 | 2.05 | 3.04 | 0.625 |
|  | 13.476 |  |  | 10 | 17.167 | 166.87 | 94.65 | 49.10 | 2.13 | 3.12 | 0.622 |
| 11/7 | 8.350 | 110 | 70 | 6 | 10.637 | 133.37 | 42.92 | 25.36 | 1.57 | 3.53 | 0.403 |
|  | 9.656 |  |  | 7 | 12.301 | 153.00 | 49.01 | 28.95 | 1.61 | 3.57 | 0.402 |
|  | 10.946 |  |  | 8 | 13.944 | 172.04 | 54.87 | 32.45 | 1.65 | 3.62 | 0.401 |
|  | 13.467 |  |  | 10 | 17.167 | 208.39 | 65.88 | 39.20 | 1.72 | 3.07 | 0.397 |

续表

| 型号 | 单位长度质量 /(kg/m) | 宽度 B/mm | 宽度 b/mm | 厚度 t/mm | 面积 A/cm² | 惯性矩 $I_x$/cm⁴ | 惯性矩 $I_y$/cm⁴ | 惯性矩 $I_u$/cm⁴ | $z_0$/cm | $y_0$/cm | $\tan\alpha$ |
|---|---|---|---|---|---|---|---|---|---|---|---|
| 12.5/8 | 11.066 | 125 | 80 | 7 | 14.096 | 227.98 | 74.42 | 43.81 | 1.80 | 4.01 | 0.408 |
|  | 12.551 |  |  | 8 | 15.989 | 256.77 | 83.49 | 49.15 | 1.84 | 4.06 | 0.407 |
|  | 15.474 |  |  | 10 | 19.712 | 312.04 | 100.67 | 59.45 | 1.92 | 4.14 | 0.404 |
|  | 18.330 |  |  | 12 | 23.351 | 364.41 | 116.67 | 69.35 | 2.00 | 4.22 | 0.400 |
| 14/9 | 14.160 | 140 | 90 | 8 | 18.038 | 365.64 | 120.69 | 70.83 | 2.04 | 4.50 | 0.411 |
|  | 17.475 |  |  | 10 | 22.261 | 445.50 | 140.03 | 85.82 | 2.12 | 4.58 | 0.409 |
|  | 20.724 |  |  | 12 | 26.400 | 521.59 | 169.79 | 100.21 | 2.19 | 4.66 | 0.406 |
|  | 23.908 |  |  | 14 | 30.456 | 594.10 | 192.10 | 114.13 | 2.27 | 4.74 | 0.403 |
| 16/10 | 19.872 | 160 | 100 | 10 | 25.315 | 668.69 | 205.03 | 121.74 | 2.28 | 5.24 | 0.390 |
|  | 23.592 |  |  | 12 | 30.054 | 784.91 | 239.06 | 142.33 | 2.36 | 5.32 | 0.388 |
|  | 27.247 |  |  | 14 | 34.709 | 896.30 | 271.20 | 162.23 | 2.43 | 5.40 | 0.385 |
|  | 30.835 |  |  | 16 | 39.281 | 1003.04 | 301.60 | 182.57 | 2.51 | 5.48 | 0.382 |
| 18/11 | 22.273 | 180 | 110 | 10 | 28.373 | 956.25 | 278.11 | 166.50 | 2.44 | 5.89 | 0.376 |
|  | 26.464 |  |  | 12 | 33.712 | 1124.72 | 325.03 | 194.87 | 2.52 | 5.98 | 0.374 |
|  | 30.589 |  |  | 14 | 38.967 | 1286.91 | 369.55 | 222.30 | 2.59 | 60.6 | 0.372 |
|  | 34.649 |  |  | 16 | 44.139 | 1443.06 | 411.85 | 248.94 | 2.67 | 6.14 | 0.369 |
| 20/12.5 | 29.761 | 200 | 125 | 12 | 37.912 | 1570.90 | 483.16 | 285.79 | 2.83 | 6.54 | 0.392 |
|  | 34.436 |  |  | 14 | 43.867 | 1800.97 | 550.83 | 326.58 | 2.91 | 6.62 | 0.390 |
|  | 39.045 |  |  | 16 | 49.739 | 2023.35 | 615.44 | 366.21 | 2.99 | 6.70 | 0.388 |
|  | 43.588 |  |  | 18 | 55.526 | 2238.30 | 677.19 | 404.83 | 3.06 | 6.78 | 0.385 |

# 附录Ⅲ 挠度曲线

| 梁和载荷 | 挠曲线方程 | 临界挠度 | 临界转角 |
|---|---|---|---|
| (悬臂梁，集中力F，距A为a处C点) | $w = \dfrac{Fx^2}{6EI}(3a - x)\ (0 \le x \le a)$ <br> $w = \dfrac{Fa^2}{6EI}(3x - a)\ (a \le x \le l)$ | $w_{max} = w_B = \dfrac{Fa^2}{6EI}(3l - a)$ <br> $w_C = \dfrac{Fa^3}{3EI}$ | $\theta_{max} = \theta_B = \theta_C = \dfrac{Fa^2}{2EI}$ |
| (悬臂梁，集中力偶$M_e$作用于C) | $w = \dfrac{M_e x^2}{2EI}\ (0 \le x \le a)$ <br> $w = \dfrac{M_e a}{2EI}(2x - a)\ (a \le x \le l)$ | $w_{max} = w_B = \dfrac{M_e a}{2EI}(2l - a)$ <br> $w_C = \dfrac{M_e a^2}{2EI}$ | $\theta_{max} = \theta_B = \theta_C = \dfrac{M_e a}{EI}$ |
| (悬臂梁，均布载荷q) | $w = \dfrac{qx^2}{24EI}(6l^2 - 4lx + x^2)$ | $w_{max} = w_B = \dfrac{ql^4}{8EI}$ | $\theta_{max} = \theta_B = \dfrac{ql^3}{6EI}$ |
| (简支梁，集中力F) | $w = \dfrac{Fbx}{6EIl}(l^2 - b^2 - x^2)$ <br> $(0 \le x \le a)$ <br> $w = \dfrac{F}{6EIl}[l(x-a)^3$ <br> $+ b(l^2 - b^2)x - bx^3]$ <br> $(a \le x \le l)$ | 设 $a \ge b$: <br> $w_{max} = w\left[\sqrt{\dfrac{1}{3}(l^2 - b^2)}\right]$ <br> $= \dfrac{Fb\sqrt{(l^2-b^2)^3}}{9\sqrt{3}EIl}$ <br> $w\left(\dfrac{1}{2}l\right) = \dfrac{Fb(3l^2 - 4b^2)}{48EI}$ | 设 $a \ge b$: <br> $\theta_A = \dfrac{Fab(l+b)}{6EIl}$ <br> $\theta_{max} = -\theta_B = \dfrac{Fab(l+a)}{6EIl}$ |
| (简支梁，力偶$M_e$) | $w = \dfrac{M_e x}{6EIl}(x^2 + 3b^2 - l^2)$ <br> $(0 \le x \le a)$ <br> $w = \dfrac{M_e}{6EIl}[x^3 - 3l(x-a)^2$ <br> $- (l^2 - 3b^2)x]$ <br> $(a \le x \le l)$ | 设 $a \ge b$: <br> $w_{max} = -w\left[\sqrt{\dfrac{1}{3}(l^2 - 3b^2)}\right]$ <br> $= \dfrac{M_e\sqrt{(l^2-3b^2)^3}}{9\sqrt{3}EIl}$ <br> $w\left(\dfrac{1}{2}l\right) = -\dfrac{M_e(l^2 - 4b^2)}{16EI}$ | $\theta_A = -\dfrac{M_e}{6EIl}(l^2 - 3b^2)$ <br> $\theta_B = -\dfrac{M_e}{6EIl}(l^2 - 3a^2)$ <br> $\theta_{max} = \theta_C = \dfrac{M_e}{6EIl}[3(a^2+b^2) - l^2]$ |
| (简支梁，均布载荷q) | $w = \dfrac{qx}{24EI}(l^3 - 2lx^2 + x^3)$ | $w_{max} = w\left(\dfrac{1}{2}l\right) = \dfrac{5ql^4}{384EI}$ | $\theta_{max} = \theta_A = -\theta_B = \dfrac{ql^3}{24EI}$ |

# 参 考 文 献

刘鸿文, 2011a. 材料力学 I [M]. 5 版. 北京: 高等教育出版社.
刘鸿文, 2011b. 材料力学 II [M]. 5 版. 北京: 高等教育出版社.
王开福, 2016. 工程力学(双语版)[M]. 2 版. 北京: 科学出版社.